ACTIVE FILTERS: LUMPED, DISTRIBUTED, INTEGRATED, DIGITAL, AND PARAMETRIC

INTER-UNIVERSITY ELECTRONICS SERIES

Books in Series

Vol. 1: *Jamieson et al.* Infrared Physics and Engineering, 1964
Vol. 2: *Bennett and Davey.* Data Transmission, 1965
Vol. 3: *Sutton.* Direct Energy Conversion, 1966
Vol. 4: *Schwartz, Bennett, and Stein.* Communications Systems and Techniques, 1966
Vol. 5: *Luxenberg and Kuehn.* Display Systems Engineering, 1968
Vol. 6: *Balakrishnan.* Communication Theory, 1968
Vol. 7: *Collin and Zucker.* Antenna Theory, Part 1, 1969
Collin and Zucker. Antenna Theory, Part 2, 1969
Vol. 8: *Zadeh and Polak.* System Theory, 1969
Vol. 9: *Schwan.* Biological Engineering, 1969
Vol. 10: *Clynes and Milsum.* Biomedical Engineering Systems, 1969
Vol. 11: *Huelsman.* Active Filters: Lumped, Distributed, Integrated, Digital, and Parametric, 1970

INTER-UNIVERSITY ELECTRONICS SERIES, VOL. 11

ACTIVE FILTERS: LUMPED, DISTRIBUTED, INTEGRATED, DIGITAL, AND PARAMETRIC

LAWRENCE P. HUELSMAN, Editor
Professor of Electrical Engineering
University of Arizona, Tucson

With Contributions by

WILLIAM J. KERWIN
Ames Research Center
Moffett Field, Calif.

B. J. LEON
Professor of Electrical Engineering
Purdue University, Lafayette, Ind.

JAN A. NARUD
Motorola Semiconductor Products Division
Phoenix, Ariz.

H. J. ORCHARD
Lenkurt Electric Company
San Carlos, Calif.

GRAHAM A. RIGBY
Assistant Professor of Electrical Engineering
University of California, Berkeley

JOHN V. WAIT
Professor of Electrical Engineering
University of Arizona, Tucson

McGraw-Hill Book Company

NEW YORK ST. LOUIS SAN FRANCISCO DÜSSELDORF
LONDON MEXICO PANAMA SYDNEY TORONTO

ACTIVE FILTERS: LUMPED, DISTRIBUTED, INTEGRATED, DIGITAL, AND PARAMETRIC

Library of Congress Catalog Card Number 78-95809

30847

1 2 3 4 5 6 7 8 9 0 MAMM 7 9 8 7 6 5 4 3 2 1 0

This book was set in Modern by The Maple Press Company, and printed on permanent paper and bound by The Maple Press Company. The designer was Edward Zytko; the drawings were done by B. Handelman Associates, Inc. The editors were B. G. Dandison, Jr. and M. E. Margolies. Paul B. Poss supervised the production.

INTER-UNIVERSITY ELECTRONICS SERIES

Series Purpose

The explosive rate at which knowledge in electronics has expanded in recent years has produced the need for unified state-of-the-art presentations that give authoritative pictures of individual fields of electronics.

The Inter-University Electronics Series is designed to meet this need by providing volumes that deal with particular areas of electronics where up-to-date reference material is either inadequate or is not conveniently organized. Each volume covers an individual area, or a series of related areas. Emphasis is upon providing timely and comprehensive coverage that stresses general principles, and integrates the newer developments into the overall picture. Each volume is edited by an authority in the field and is written by several coauthors, who are active participants in research or in educational programs dealing with the subject matter involved.

The volumes are written with a viewpoint and at a level that makes them suitable for reference use by research and development engineers and scientists in industry and by workers in governmental and university laboratories. They are also suitable for use as textbooks in specialized courses at graduate levels. The complete series of volumes will provide a reference library that should serve a wide spectrum of electronics engineers and scientists.

The organization and planning of the Series is being carried out with the aid of a Steering Committee, which operates with the counsel of an Advisory Committee. The Steering Committee concerns itself with the scope of the individual volumes and aids in the selection of editors for the different volumes. Each editor is in turn responsible for selecting his coauthors and deciding upon the detailed scope and content of his particular volume. Over-all management of the Series is in the hands of the Consulting Editor.

Frederick Emmons Terman

PREFACE

To most engineers a filter is a collection of resistors, capacitors, and inductors which has as its function the shaping or processing of some electric signal information. One of the purposes of this book is to extend such a viewpoint by showing how the techniques which have been developed in related areas of electrical engineering may also be used to effectively filter electric signals. Such techniques originate in many quite different basic electrical engineering subdisciplines, such as distributed elements, integrated circuits, sampling systems, and parametric devices. Because of their interdisciplinary nature these techniques present a challenging vista of new and different approaches to the basic problem of filtering electric signals.

In choosing the level of presentation for the technical material which is included in this book, considerable attention has been given to the fact that many readers may be encountering these topics for the first time. Thus introductory material has been included at the beginning of each chapter to enable the reader to understand the basic fundamentals, and also to develop some perspective on how the particular concepts developed in the chapter may be applied to the general problems of network filtering. Following such introductory material each chapter presents topics which, in general, include the results of the most modern research efforts. Many of these results have not previously appeared in hard cover form. Several of the authors have also included original research results, heretofore unpublished. Thus the treatment of each of the topics should be of interest to the researcher whose primary interest lies in these areas as well as to the engineer who may be encountering this material for the first time. A summary of the topics which are included in the various chapters may be found in Chap. 1, which serves as an introduction for the entire book.

The interdisciplinary nature of this book, together with the broad scope and variety of the topics treated, makes it most important that the individual subjects be presented by authors who are currently prominent and well versed in their specific disciplines. The brief descriptions of the backgrounds of the contributing authors given in Chap. 1 readily attest to their prominence in their chosen fields. A study of the individual chapters written by these authors will, I am sure, lead the reader to share my own enthusiasm for the excellent job that they have done.

Lawrence P. Huelsman

CONTENTS

CHAPTER 1

INTRODUCTION

Lawrence P. Huelsman
University of Arizona
Tucson

The general area of electrical engineering which has to do with selective processing of signal information, commonly referred to as filtering, is a field which has broadened tremendously in scope in the past decade. Some of the advances have been spurred by technological developments such as the tremendous increase in production and utilization of integrated circuits. Other advances have been brought about by the theoretical developments made by researchers in well-established disciplines. Taken as a whole, these advances have tremendously broadened the scope of filter theory, which, not too many years ago, was primarily concerned only with methods for interconnecting lumped resistors, capacitors, and inductors. Thus, a modern view of filter theory must necessarily include current results concerning the effects of active, distributed, and nonlinear phenomena as well as the more familiar passive, lumped, and linear situations. In addition, modern filter theory must not only consider the processing of signals by continuous filters but also by digital or sampling techniques. Finally, the limitations of integrated-circuit technology place restrictions on the filter designer which must be considered if practical realizations are to result.

Considering all the above, it is easy to agree that today the subject of filtering is indeed a complex and challenging one. These challenges, however, are not without their rewards. The current broad scope of filter theory presents a wide range of intriguing concepts. A few examples include the use of a single distributed *RC* network element to replace a twin-T of lumped-network elements; active *RC* realizations specified by conventional *RLC* design charts but with the inductors replaced by capacitors and gyrators; state-variable bandpass realizations with *Q*'s in the hundreds and with low sensitivities, etc. In this volume, a treatment of the wide and diverse methods which currently constitute modern filter theory is presented. A discussion of the content of the various chapters follows.

Chapter 2 discusses the synthesis of a broad class of active *RC* networks. Specifically, it treats passive networks whose elements are restricted to being lumped resistors, lumped capacitors, or distributed *RC* networks, and whose active elements are voltage-controlled voltage sources. This class of networks provides a model for the general category of circuits which may be referred to as integrated circuits. Thus, it is an extremely practical class of networks. The presentation of the theory recognizes the well-known fact that in the frequency range below 50–100 MHz, inductors are impracticable to integrate. Not only is their quality poor, but they are unrealistic from space and weight considerations. Thus, this chapter represents a very practical approach to the theory of linear integrated networks. The active element which is treated in this chapter is realized by an operational amplifier which is connected as a positive- or negative-gain voltage-controlled voltage source. Due to the low output impedance of such a connection it is relatively simple to realize high-degree network functions through the use of cascade connections of circuits. The chapter treats the optimization of circuit realizations from the viewpoint of the minimization of the number of network elements. In addition, considerable attention is paid to realizable ranges of Q and to sensitivity. Much of the material presented in the chapter has not previously appeared in book form. For example, methods for using distributed elements to minimize the number of circuit components are described. The realizations treated in this chapter include techniques which have the capability of producing Q's of up to 1,000 with theoretical gain sensitivities of less than unity.

A specific area of active *RC* circuits which has received considerable attention in recent years is the use of gyrators and capacitors to replace inductors in the general frequency range below 50–100 MHz. In Chap. 3, the specific techniques which lead to such inductorless realizations are presented and discussed. These techniques in general consist of applying well-known passive *RLC* synthesis methods directly to the realizations of network functions, and then utilizing gyrator-capacitance networks to replace the inductors. One of the major advantages of such techniques are the extremely low sensitivities which are attained in the resulting network realizations. The chapter gives an extensive treatment of various methods for realizing and utilizing gyrators, and a comparison of the advantages and disadvantages of these methods. Descriptions of some integrated gyrator realizations are also included.

In the preceding two chapters, the theoretical aspects of linear active networks and filters are presented. In Chap. 4, the viewpoint changes to encompass the problems and techniques presently encountered in the integration of linear active circuit configurations. Although the techniques emphasize integrated realizations they are also directly applicable to the practical implementation of theoretical realizations in lumped component form. Considerable emphasis is given to the limitations placed on the network realization by practical considerations. The chapter also gives a comprehensive treatment of the properties of basic circuit elements which are realized by integrated-circuit techniques. The use of appropriate models for these elements is discussed. The chapter also points out some of the methods by means of which digital computers may be used to analyze circuits comprised of these elements. Considerable treatment is made of standard integrated-circuit building blocks. Although the discussion is primarily oriented to linear circuits, many of the conclusions are also applicable to digital integrated circuits.

To this point in the book the primary concern has been with linear circuits. In Chap. 5, however, the attention of the reader is focused on the techniques of digital filtering. The advantages of such filters are well known. For example they provide excellent characteristics of accuracy, stability, and reliability. In addition, they may be superior to their linear counterparts when judged by such criteria as size, cost, and weight, especially in the low-frequency range. The chapter presents an integrated treatment of the theory and application of digital filtering techniques. A discussion is given of the approximation of continuous-time filters in a sampled environment. The relations between the s or complex frequency domain and the z or sampled domain are developed. The chapter is especially thorough in its description of the practical details of data-reconstruction filters and the errors associated with such filters. An analysis of the reconstruction errors for various types of digital filters is given. The chapter also includes design methods for both nonrecursive (finite memory) and recursive filters. Both time- and frequency-domain topics are included.

In Chap. 6 a still different aspect of signal filtering is presented. This is the use of sinusoidal power sources to amplify low-power signals. Such techniques are of great interest in the high-frequency spectrum. A comprehensive treatment is given of the nonlinear effects which arise in related circuits. The Manley-Rowe formulas are presented and related to the energy-conversion processes. The chapter includes a discussion of design considerations pertinent to linear two-frequency parametric amplifiers. Treatments of gain-bandwidth and noise are also included.

In the final chapter of the book a discussion is given of present and future trends in integrated circuits. This chapter presents a discussion of integrated-circuit-fabrication techniques and also a discussion of the types of logic circuits which are currently in use in digital integrated realizations. In addition to predicting future trends in component sizes and circuit complexity, a discussion is also given of the implications of LSI (large-scale integration). The role and usage of the techniques of computer-aided design are also mentioned.

Each of the chapters described above was prepared by a well-known authority currently active in research in his subject area. Chapter 2 on active *RC* synthesis was written by W. J. Kerwin, Chief of the Electronics Research Branch, Ames Research Center, National Aeronautics and Space Administration, Moffett Field, Calif. Chapter 3 on gyrator circuits was written by H. J. Orchard of the Lenkurt Electric Company, San Carlos, Calif. Chapter 4 on the electronic-circuit aspects of active filters was written by Graham Rigby, Assistant Professor of Electrical Engineering, University of California, Berkeley. Chapter 5 on digital filters was written by J. V. Wait, Professor of Electrical Engineering, University of Arizona, Tucson. Chapter 6 was written by B. J. Leon, Editor of the IEEE Transactions on Circuit Theory and Professor of Electrical Engineering, Purdue University, Lafayette, Ind. Chapter 7 was written by J. A. Narud, Director of Computer-aided Design, Motorola Semiconductor Products Division, Phoenix, Ariz.

The editor wishes to express his appreciation to these authors for their enthusiasm for this project and for their concerted efforts to make this volume an up-to-date coverage of a broad range of active circuit techniques. In addition, the editor gratefully acknowledges the support and encouragement given to this project by Dr. R. H. Mattson, Head of the Department of Electrical Engineering, University of Arizona, Tucson.

CHAPTER 2

ACTIVE *RC* NETWORK SYNTHESIS USING VOLTAGE AMPLIFIERS

William J. Kerwin
Ames Research Center
Moffett Field, Calif.

2.1 Introduction

In this chapter we will be concerned primarily with active *RC* networks in which the active element is an operational amplifier connected as a voltage-controlled voltage source (VCVS). The VCVS will be assumed to have ideal characteristics, that is, infinite input impedance, zero output impedance, zero reverse transmission, and ideal phase shift (either zero or 180°). The voltage gain required will generally be low and will be determined by the synthesis method used in the various cases, as will the gain polarity. The Burr-Brown handbook (1963) and Pande and Shukla (1965) are good references for high-gain realizations (usually assumed to be infinite), as are Morse (1964), Hakim (1965a), and Chap. 6 of Huelsman (1968). The equivalence between active *RC* synthesis methods using amplifiers and those using negative impedance converters is discussed by Gorski-Popiel (1965). A good discussion of the realization of various controlled sources is given by Brugler (1966). A more theoretical treatment of various methods of active *RC* network synthesis can be found in Su (1965). The passive elements will be resistors and capacitors only, both lumped and distributed, but linear and constant, for these are the prime passive elements of linear integrated circuits. Emphasis will be given to those structures which minimize the number of passive elements (particularly capacitors) for a given-order transfer function, and to practical design equations for the various methods of synthesis, as well as their range of usefulness (Q and frequency). The inductor is conspicuously absent since at frequencies below about 50 MHz, it is impractical to integrate, and even above that frequency occupies too much space. A discussion of the state of the art in fabricating thin-film inductors is given in Holland (1965, p. 36) and Warner and Fordemwalt (1965, p. 267). Useful design methods must be predicted on the unavailability of inductors.

In nearly all cases considered, synthesis will be confined to the realization of the various second-order transfer functions needed for low-pass, high-pass, or bandpass functions in normalized form. This is to alleviate two fundamental problems of active *RC* synthesis: (1) attempts to obtain higher than second-order functions in a single structure with a single active element lead to a higher sensitivity of the pole positions to changes in the elements than necessary; and (2) the pole positions become highly sensitive functions of the polynomial coefficients in high-order polynomials, thereby making it extremely difficult to realize a desired set of poles by achieving a given set of polynomial coefficients. The latter is a particularly serious problem in the use of state-variable synthesis methods. Both of these problems are minimized by restricting the synthesis to either first- or second-order functions only. Higher-order systems are then synthesized by cascade connection of the individual sections. Of course, any first-order functions needed can be conveniently realized by standard methods of passive *RC* network synthesis and then cascaded with the active sections to obtain the complete function desired. The network configurations are chosen so that the output is taken from the VCVS. Any required cascade connections can then be made without the addition of isolation amplifiers, since the VCVS prevents interaction between the individual networks. This is the primary advantage of the VCVS over other controlled sources and is one of the reasons for choosing the VCVS approach. The classic structures of this type were given by Sallen and Key (1955) for the case of single-loop feedback. We will first consider these cases, and then turn our attention to additional structures which produce $j\omega$-axis zeros and which have independently adjustable zeros and poles. In addition, we will consider multiple-loop feedback structures using both positive- and negative-gain VCVSs.

The use of phantom zeros (transmission zeros of the feedback network) to reduce the sensitivity of the poles to changes in the active or passive elements will be considered next. This will include the use of negative-sigma-axis phantom zeros (Bialko 1967), left-half-plane phantom zeros (Hakim, December 1965; Ghausi and Pederson 1961; Ghausi and Pederson 1962), and the use of right-half-plane phantom zeros which allow a trade off between sensitivity and VCVS gain so as to allow higher-frequency performance for a given gain-bandwidth product (Kerwin and Shaffer 1968).

The last lumped network we will consider involves synthesis using integrators—the state-variable approach (Kerwin, Huelsman, and Newcomb 1967). These are particularly interesting because of their exceptionally low sensitivity and the use of a minimum number of capacitors. In addition, they provide simple and independent resistive control of frequency and Q over a wide range. Finally, we will discuss the use of distributed *RC* elements in place of some of the lumped *RC* elements.

This is particularly advantageous when $j\omega$-axis zeros are needed, and in realizing appropriate phantom zeros (Kerwin 1967; and Kerwin and Shaffer 1968). The relation between sensitivity to changes in amplifier gain and the amplifier gain required will be given in a set of design charts.

Sensitivity. In each of the various synthesis methods we will be primarily concerned with the sensitivity of the resulting transfer function to changes in the active and passive elements. Bode's classical definition of sensitivity as modified by Mason (Truxal 1955) is

$$S_K{}^T = \frac{\partial T/T}{\partial K/K} = \frac{\partial \ln T}{\partial \ln K} \tag{2.1}$$

where T is the transfer function and K is the element in question. This is not the best for our purpose as it is a function of frequency. However, McVey tabulates the maximum value of the sensitivity $S_K{}^T$ for various active RC networks (McVey 1965). *Root sensitivity* to an element W has been defined as

$$S_W{}^{p_j} = \frac{\partial p_j}{\partial W/W} = W\frac{\partial p_j}{\partial W}; \quad p_j = \sigma_j + j\omega_j \tag{2.2}$$

(Truxal and Horowitz 1956), where p_j is a particular root, and this definition has been used to compare a number of active RC configurations (Blecher 1960). In addition, a percentage change in pole position has been used (Uzunoglu 1964; and Newell 1961). This is given as

$$S_W{}^{p_j} = \frac{\partial p_j/p_j}{\partial W/W} = \frac{W}{p_j}\frac{\partial p_j}{\partial W} \tag{2.3}$$

A more suitable definition for our purpose is (Kerwin, Huelsman, and Newcomb 1967)

$$S_X{}^{p_j} = \frac{\partial\sigma_j/\sigma_j}{\partial X/X} + j\frac{\partial\omega_j/\omega_j}{\partial X/X} = \frac{X}{\sigma_j}\frac{\partial\sigma_j}{\partial X} + j\frac{X}{\omega_j}\frac{\partial\omega_j}{\partial X} \tag{2.4}$$

where p_j is the root in question and X is any active or passive element. This is the one we will use for lumped networks in which the transfer function is rational. We will call it the *pole-position sensitivity*. The real part of $S_X{}^{p_j}$ gives the component of pole motion parallel to the sigma axis, and the imaginary part of $S_X{}^{p_j}$ gives the component of pole motion parallel to the $j\omega$ axis. The resultant vector combination of the two components (when multiplied by σ_j and ω_j respectively) shows the magnitude and direction of pole motion due to a given change in a particular component. This sensitivity is not a function of frequency, and it has the additional advantage that the effect of changes in a number of parameters can be conveniently shown on a complex-plane plot by vectors emanating from the pole in question and showing the magnitude and direction of pole motion, for, say, a 1 percent change in each of the parameters.

In the case of networks containing distributed elements, the transfer function is not rational, so that the calculation of $S_X^{p_i}$ is impractical. In this case, we will use the Q sensitivity (since the Q can be determined from an amplitude-response plot), which is defined as (Bown 1967; and Geffe 1967)

$$S_X{}^Q = \frac{\partial Q/Q}{\partial X/X} = \frac{X}{Q}\frac{\partial Q}{\partial X} \tag{2.5}$$

For those cases which approximate a second-order system and for values of $Q > 2$ in which ω (the imaginary part of the pole position) is not a function of X, we find that

$$S_X{}^Q = \frac{X}{Q}\frac{\partial Q}{\partial X} \approx \frac{2\sigma X}{\omega}\frac{\partial(\omega/2\sigma)}{\partial X} = -\frac{X}{\sigma}\frac{\partial \sigma}{\partial X} \tag{2.6}$$

where σ is the real part of the pole position. Comparison of (2.6) and (2.4) shows that, for the above conditions,

$$S_X{}^Q \approx -\mathrm{Re}\, S_X{}^{p_i} \tag{2.7}$$

These two sensitivity definitions [(2.4) and (2.5)] are thus sufficient for an evaluation of both lumped and distributed active RC systems. Additional definitions of sensitivity functions are given in Antoniou (1967).

2.2 Synthesis Using a VCVS and Lumped RC Elements

Positive-gain VCVS, Low-pass. The positive-gain VCVS, low-pass network is the first of the structures originated by Sallen and Key (1955). Figure 2.1 shows the basic configuration. The transfer function is

$$\frac{V_{\text{out}}}{V_{\text{in}}} = T(p) = \frac{K/RC}{p^2 + [(1 + RC + C - K)/RC]p + 1/RC} \tag{2.8}$$

This is not a completely general configuration, but it is a practical one. For a more complete analysis, see McVey (1965). If we now equate (2.8) to

$$T(p) = \frac{K\gamma}{p^2 + \beta p + \gamma} \tag{2.9}$$

we obtain the design equations

$$R = \frac{1}{\gamma C} \tag{2.10}$$

and

$$K = 1 + C - \frac{\beta}{\gamma} + \frac{1}{\gamma} \tag{2.11}$$

in which C can be chosen arbitrarily (possibly so as to minimize K, or to make the capacitors equal in value). Of course, (2.11) shows that

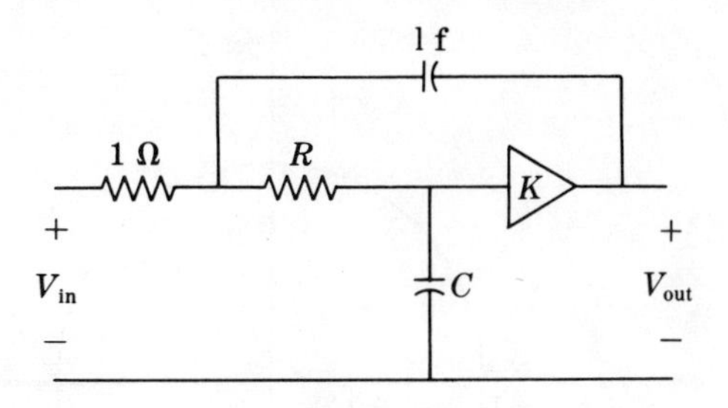

FIG. 2.1 Positive-gain VCVS low-pass filter section.

K is minimized as C approaches zero, but from a practical standpoint, little is gained by going below $C = 0.1$ farad, and the element spread is thereby held to 10 to 1. Since γ will normally be near unity for normalized transfer functions, the gain K required will be near 2 for most cases of interest. It is sometimes advantageous to use equal-value resistors to determine the feedback ratio of an operational amplifier (Kerwin, April 1966), and in this case, $K \approx 2$. This value of K can then be used to determine C from (2.11) for specified values of β and γ. A minimum number of capacitors is used (two) and the gain is low so that good stabilization and wideband operation are possible. However, the sensitivity is not good. If we look at the effect of changes in K in a typical case ($R = 1$ ohm, $C = 1$ farad), for which the transfer function is

$$T(p) = \frac{K}{p^2 + (3 - K)p + 1} \tag{2.12}$$

we obtain the root-locus plot shown in Fig. 2.2. Note that the poles become complex at $K = 1$ and follow a circular locus of unit radius reaching the $j\omega$ axis (oscillation at $\omega = 1$) when $K = 3$. The perpendicular approach to the $j\omega$ axis at high Q gives a high sensitivity of Q to changes in K while the frequency of maximum response is relatively unaffected by changes in K. The sensitivity to changes in K can be determined as follows (and is easily modified to include other values of R and C). From (2.12) the pole positions are

$$p_j = \frac{-(3 - K) \pm j\sqrt{4 - (3 - K)^2}}{2} \tag{2.13}$$

and therefore

$$\sigma_j = \frac{K - 3}{2} \tag{2.14}$$

and

$$\omega_j = \sqrt{1 - \frac{(3 - K)^2}{4}} \tag{2.15}$$

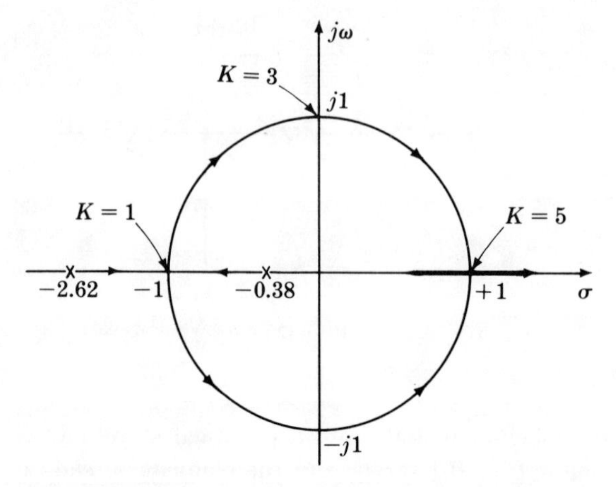

FIG. 2.2 Root-locus plot for the network of Fig. 2.1.

Now, rewriting (2.4) as

$$S_K^{p_j} = \frac{K}{\sigma_j}\frac{\partial \sigma_j}{\partial K} + j\,\frac{K}{\omega_j}\frac{\partial \omega_j}{\partial K} \tag{2.16}$$

we obtain by differentiation and substitution

$$S_K^{p_j} = \frac{K}{K-3} + j\,\frac{K(3-K)}{4-(3-K)^2} \tag{2.17}$$

The fact that Re $S_K^{p_j}$ is negative (for $K < 3$) indicates that $|\sigma_j|$ decreases for increasing K as expected. At $Q = 10$ [that is, $K = 2.9$ from (2.12)], Re $S_K^{p_j} = -29$ and Im $S_K^{p_j} = 0.073$ or $S_K^{p_j} = -29 + j0.073$. The change in σ_j (a Q change primarily) is obviously the important one and the pole-position-sensitivity vector is essentially horizontal.

A choice of $C = 0.1$ farad, $R = 10$ ohms (thereby reducing gain, and sensitivity) would change (2.12) to

$$T(p) = \frac{K}{p^2 + (2.1 - K)p + 1} \tag{2.18}$$

and for this case $Q = 10$, $K = 2.0$ results in

$$S_K^{p_j} = -20 + j0.05 \tag{2.19}$$

Since this is an element spread of 10 to 1, and since $(2.0 - K)$ is the minimum possible value of the p coefficient for the network of Fig. 2.1 (when $\gamma = 1/RC = 1$), the minimum practical sensitivity is Re $S_K^{p_j} \approx -2Q$. Therefore, values of Q higher than 5 or 10 become impractical for most applications.

A similar investigation of the pole-position sensitivity to changes in C for the transfer function of (2.8), for $C = 0.1$ farad, $R = 10$ ohms, $K = 2.0$, $Q = 10$, results in

$$\sigma_j = -\frac{-1 + 11C}{20C} \tag{2.20}$$

and

$$\omega_j = \sqrt{\frac{1}{10C} - \frac{1}{4}\left(\frac{11C - 1}{20C}\right)^2} \tag{2.21}$$

and therefore

$$S_C^{p_1} = \frac{C}{\sigma}\frac{\partial\sigma}{\partial C} + j\frac{C}{\omega}\frac{\partial\omega}{\partial C} = -5.0 + j0.013 \tag{2.22}$$

Similar calculations show that the sensitivities to the remaining passive elements are of like magnitude, and, although a Q sensitivity of essentially 5 at a Q of 10 is not at all negligible, it is considerably less of a problem than the sensitivity to gain change of 20 given in (2.19). If the R's and C's change in opposite directions, with temperature for example, so that all RC products remain constant, the entire quadratic is invariant with temperature except for the effect of K.

As an example of the use of this method, we will design a third-order Butterworth (maximally-flat magnitude) low-pass filter. Of course, any function could be chosen; equal-ripple, maximally-flat delay, etc. A convenient source of these functions is Ghausi (1965), or, for a complete tabulation of elliptic functions in factored form, Skwirzynski (1965). The transfer function to be synthesized is

$$T(p) = \frac{1}{p^3 + 2p^2 + 2p + 1} = \frac{1}{p + 1}\,\frac{1}{p^2 + p + 1} \tag{2.23}$$

In this case, the quadratic factor has $\beta = 1$, $\gamma = 1$ [as defined in (2.9)], and, choosing $C = 0.1$ farad to minimize gain, we obtain from (2.10) and (2.11), $R = 10$ ohms, $K = 1.1$. The quadratic factor is therefore realized by the network of Fig. 2.3. The realization of the first-order

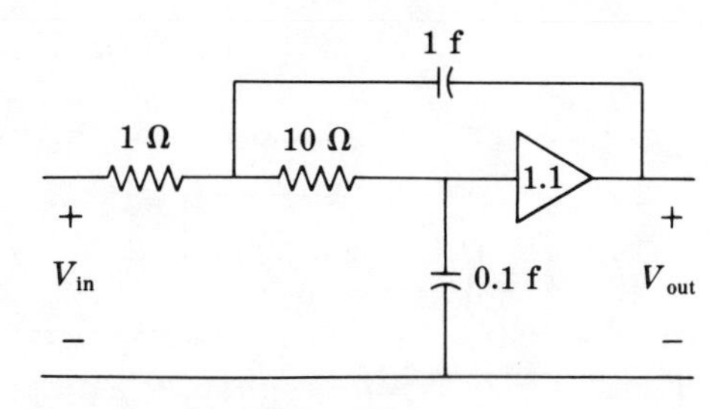

FIG. 2.3 Second-order section of the Butterworth filter.

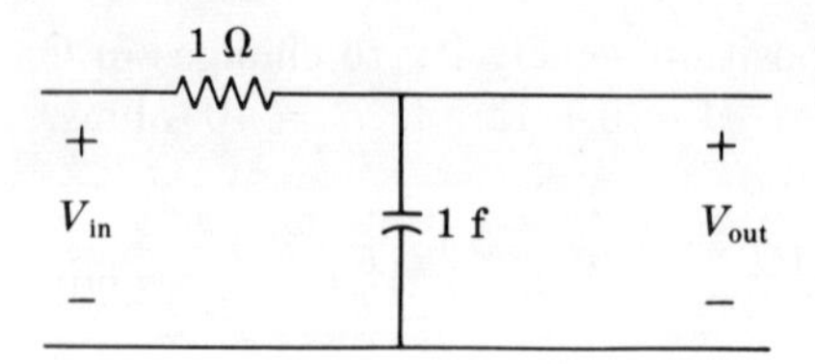

Fig. 2.4 First-order section of the Butterworth filter.

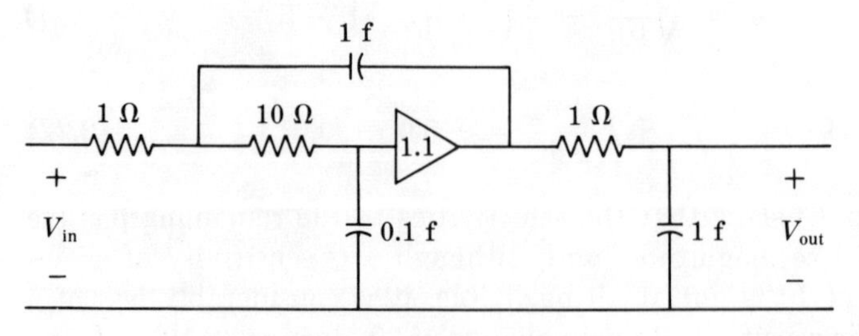

Fig. 2.5 Third-order Butterworth low-pass filter.

factor is achieved by the network of Fig. 2.4, and the complete filter is obtained by a cascade connection of the two, and is shown in Fig. 2.5. Note that the voltage amplifier prevents interaction between the two sections when they are cascaded. The overall multiplier can be obtained by observing that the dc gain is 1.1 and therefore that the original function has been realized within this constant multiplier. The 3-db-down cutoff frequency is $\omega_{3db} = 1$ rad/sec. The realization of more complex functions consists of realizing each quadratic factor separately and cascading the individual networks, with a final first-order section if the transfer function is of odd degree.

Positive-gain VCVS, High-pass. In this case, the structure required is obtained by a simple transformation of the low-pass configuration of Fig. 2.1, resulting in the network shown in Fig. 2.6 (Sallen and Key 1955). The transfer function of the network is

$$T(p) = \frac{Kp^2}{p^2 + (1 + 1/R + 1/RC - K)p + 1/RC} \tag{2.24}$$

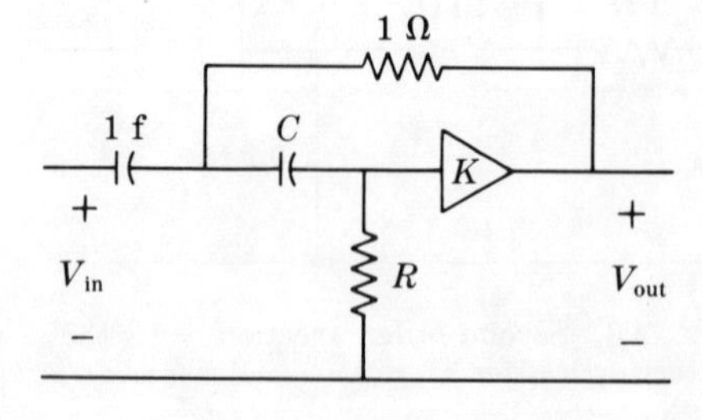

Fig. 2.6 Positive-gain VCVS high-pass filter section.

Equating this function to

$$T(p) = \frac{Hp^2}{p^2 + \beta p + \gamma} \tag{2.25}$$

we obtain the following design equations for any specified β and γ in which R can be chosen arbitrarily

$$C = \frac{1}{\gamma R} \tag{2.26}$$

$$K = 1 + \gamma + \frac{1}{R} - \beta \tag{2.27}$$

The value of H cannot be specified independently, but of course is equal to K in (2.27). In this case, K is minimized as R approaches infinity, although $R = 10$ ohms is about as far as is reasonable. The pole-position sensitivity to changes in K is

$$S_K^{p_i} = -\frac{K}{\beta} + j\frac{\beta}{2\sqrt{4\gamma - \beta^2}} \tag{2.28}$$

This structure is in all respects similar to the low-pass case, and has the same root locus.

Using $R = 10$ ohms in (2.24), (2.26), and (2.27), and cascading the appropriate first-order factor, the network of Fig. 2.7 is obtained for the third-order Butterworth high-pass filter function, given in (2.29).

$$T(p) = \frac{p^3}{p^3 + 2p^2 + 2p + 1} \tag{2.29}$$

Positive-gain VCVS, Bandpass. The structure for this case (Sallen and Key 1955) is shown in Fig. 2.8. Other networks are possible, but this one is illustrative and practical, and uses only two variable elements and the gain K. The transfer function of this network is

$$T(p) = \frac{Kp}{p^2 + (2 + 1/RC + 1/R - K)p + 2/RC} \tag{2.30}$$

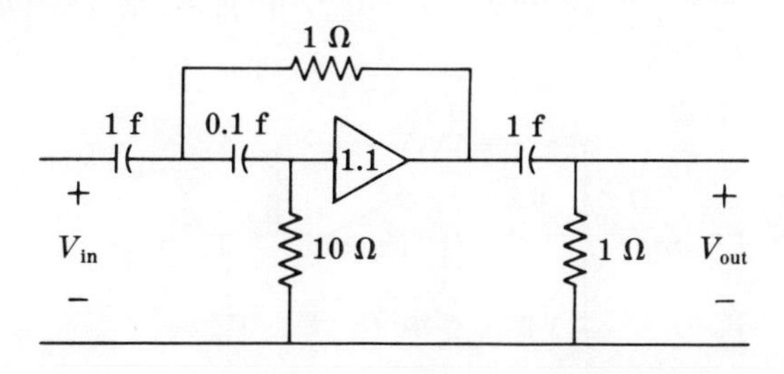

FIG. 2.7 Third-order Butterworth high-pass filter.

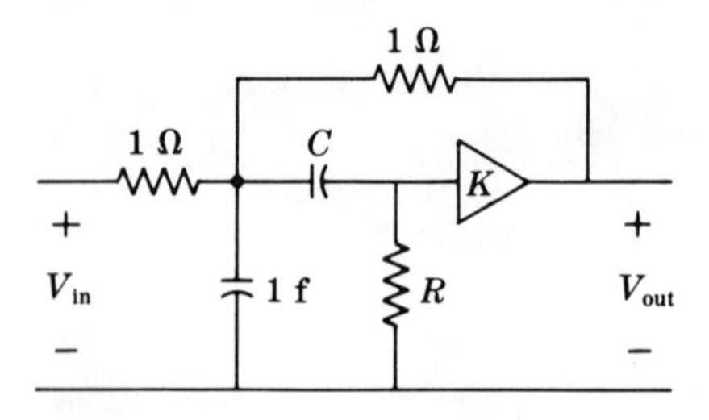

FIG. 2.8 Positive-gain VCVS bandpass filter section.

In this case, the poles become complex (for $RC = 2$ sec, $R = 10$ ohms) when $K = 0.6$, at which point a double pole occurs at $\sigma = -1$. Oscillation occurs at $\omega_0 = 1$ rad/sec when $K = 2.6$. The root locus is again circular and crosses the $j\omega$ axis at right angles. The design equations for

$$T(p) = \frac{Hp}{p^2 + \beta p + \gamma} \tag{2.31}$$

are

$$C = \frac{2}{R\gamma} \tag{2.32}$$

and

$$K = 2 + \frac{\gamma}{2} + \frac{1}{R} - \beta \tag{2.33}$$

The value of H is again given by K in (2.33). Also, a choice of $R = 10$ ohms is about as high as is reasonable, since further increases in R do not appreciably reduce K.

As an example, consider the design of a bandpass filter centered at $\omega_0 \approx 1$ rad/sec and having a $Q = 10$. From (2.31), $p^2 + \gamma = 0$ at $\omega_0 = 1$ rad/sec, and therefore $\gamma = 1.0$; and, since $Q = \sqrt{\gamma}/\beta$, $\beta = 0.1$. Therefore, from (2.32) and (2.33) (using $R = 10$ ohms to minimize K), we find $C = 0.2$ farad and $K = 2.5$. The resulting network is shown in Fig. 2.9. If we look at the real part of the pole-position sensitivity for this example, we find that Re $S_K^{p_i} = -25$; that is, Re $S_K^{p_i} = -2.5Q$. This is essentially the negative of the Q sensitivity and again we see that a Q of 10 requires about as much amplifier stability as can be reasonably achieved (± 0.04 percent change in K produces a ± 1 percent change

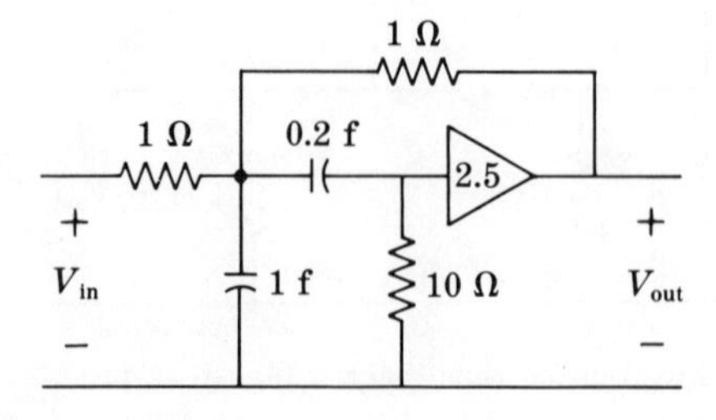

FIG. 2.9 Bandpass filter example.

in Q). The change in frequency of peak response with change in gain is negligible at high Q due to the circular root locus.

Positive-gain VCVS, Finite $j\omega$-axis Zeros. The preceding networks have had their zeros restricted to the origin or infinity, whereas a large class of filter functions require finite $j\omega$-axis zeros, and these must not be dependent on the pole positions. The networks developed by Sallen and Key did not include this possibility. Later work (Jagoda 1962; McVey 1962; Balabanian and Patel 1963; and Holt and Sewell 1965) provided networks of this type; however, Jagoda's network required two inputs of different amplitude and was therefore slightly more complex than necessary. If an adjustable zero position is needed, however, it has the decided convenience of a single potentiometric adjustment of zero position. Later work (Kerwin and Huelsman 1966) has provided additional networks as shown in Fig. 2.10*a*, with two special cases shown in Figs. 2.10*b* and 2.10*c* (Kerwin, April 1966). These latter two cases cover the entire $j\omega$ axis, and, since fewer elements are required, are more practical. They produce transfer functions of the type

$$T(p) = \frac{H(p^2 + \alpha)}{p^2 + \beta p + \gamma} \tag{2.34}$$

For the network of Fig. 2.10*b*, we obtain the following transfer function

$$T(p) = \frac{k(p^2 + 1/a^2)}{p^2 + [(k+1)/a][1/R + (2-K)/k]p + [1 + (k+1)/R]/a^2} \tag{2.35}$$

which is capable of achieving only those transfer functions in which $\gamma > \alpha$ in (2.34); that is, the poles are beyond the zeros. Equating the transfer function of (2.35) to (2.34), we obtain the following design equations for the network of Fig. 2.10*b* in which k can be chosen arbitrarily, possibly to obtain a particular value of K or to minimize K.

$$a = \sqrt{\frac{1}{\alpha}} \tag{2.36}$$

$$R = \frac{k+1}{\gamma/\alpha - 1} \tag{2.37}$$

$$K = 2 + \frac{k}{k+1}\left(\frac{\gamma}{\alpha} - 1 - \frac{\beta}{\sqrt{\alpha}}\right) \tag{2.38}$$

$$H = K \tag{2.39}$$

These equations are applicable only to the case in which $\gamma > \alpha$ in (2.34). Of course, element-spread considerations may dictate that k be reasonably close to unity. Equations (2.38) and (2.39) determine H; only α, β, and γ are specifiable in (2.34).

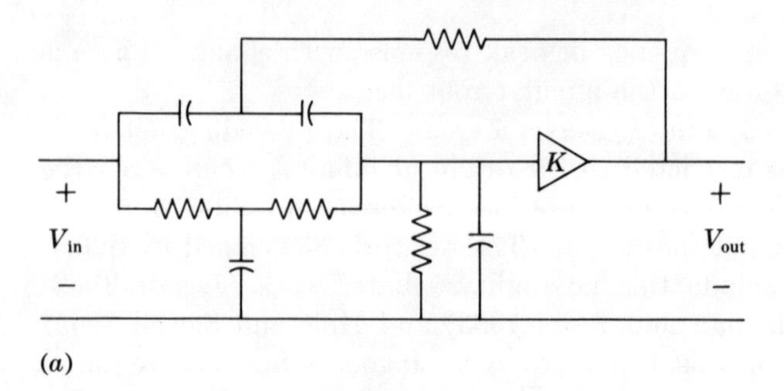

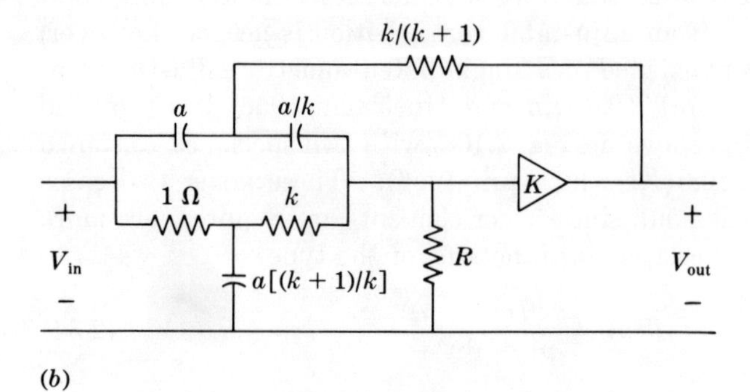

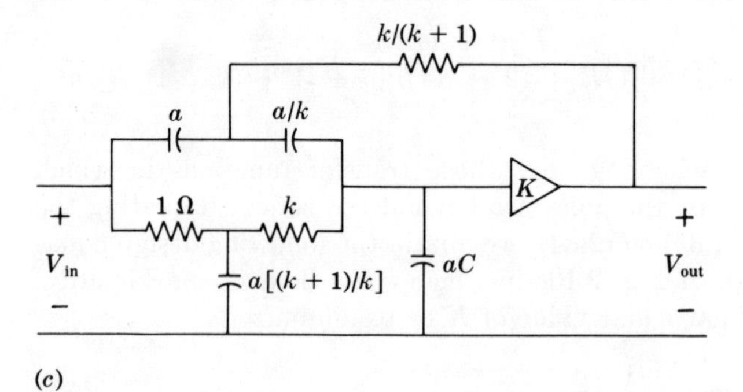

FIG. 2.10 Two-pole, two-$j\omega$-axis-zero networks using a positive-gain VCVS.

The transfer function of the network of Fig. 2.10c is

$$T(p) = \frac{\{K/[(k+1)C+1]\}(p^2 + 1/a^2)}{p^2 + \{(k+1)[C + (2-K)/k]/a[(k+1)C+1]\}p + 1/a^2[(k+1)C+1]} \quad (2.40)$$

In this case it is evident that $\alpha > \gamma$ in (2.34), and therefore that the zeros must lie beyond the poles. Any two-pole, two-$j\omega$-axis-zero function can thus be obtained by one or the other of the two networks shown in Fig. 2.10*b* and Fig. 2.10*c*.

Equating the coefficients of (2.40) to those of (2.34), we obtain the following design equations for the network of Fig. 2.10*c* ($\alpha > \gamma$):

$$a = \sqrt{\frac{1}{\alpha}} \tag{2.41}$$

$$C = \frac{\alpha/\gamma - 1}{k + 1} \tag{2.42}$$

$$K = 2 + \frac{k}{k+1}\left(\frac{\alpha}{\gamma} - 1 - \frac{\beta\sqrt{\alpha}}{\gamma}\right) \tag{2.43}$$

$$H = \frac{\gamma}{\alpha} K \tag{2.44}$$

The factor k can be chosen arbitrarily as before. If we take, for example, the transfer function (in which $\alpha > \gamma$)

$$T(p) = H \frac{p^2 + 1.2}{p^2 + 0.1p + 1} \tag{2.45}$$

we obtain the network shown in Fig. 2.11 for $k = 1$. The multiplier obtained is $H = 1.71$. For the case of the network of Fig. 2.11 (for which $Q = 10$), we find that the pole-position sensitivity is

$$S_K{}^{p_i} = -37.4 + j0.093 \tag{2.46}$$

As an example of a complete design using the positive-gain VCVS, we will take a six-pole, four-zero elliptic-function low-pass filter having 0.18 ripple in the passband and 39.3-db ripple in the stop band (Kerwin, April 1966). The transfer function is [normalized to cutoff (−0.18 db)

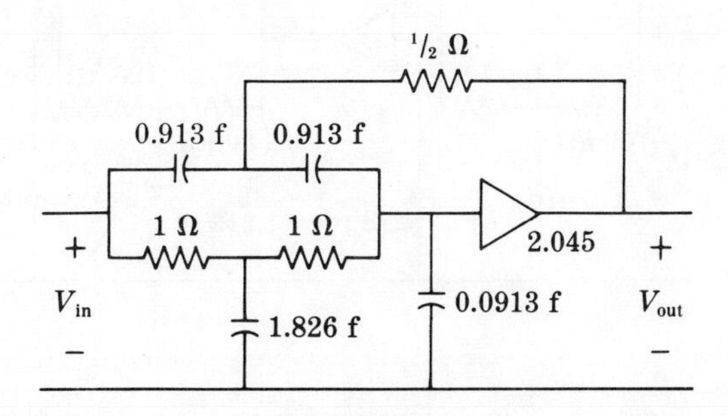

FIG. 2.11 Network realizing equation (2.45).

at $\omega = 1.0$ rad/sec]:

$$T(p) = \frac{0.268(p^2 + 2.449)(p^2 + 1.522)}{(p^2 + 0.113p + 1.058)(p^2 + 0.492p + 0.822)(p^2 + 1.155p + 0.407)} \tag{2.47}$$

A low-pass second-order section is used to realize the last biquadratic factor in the denominator using a network of the type shown in Fig. 2.1. The other two-zero, two-pole factors are realized individually using the network shown in Fig. 2.10*c*. Cascade connection of the individual network sections is shown in Fig. 2.12*a*. After impedance scaling to 50 $k\Omega$

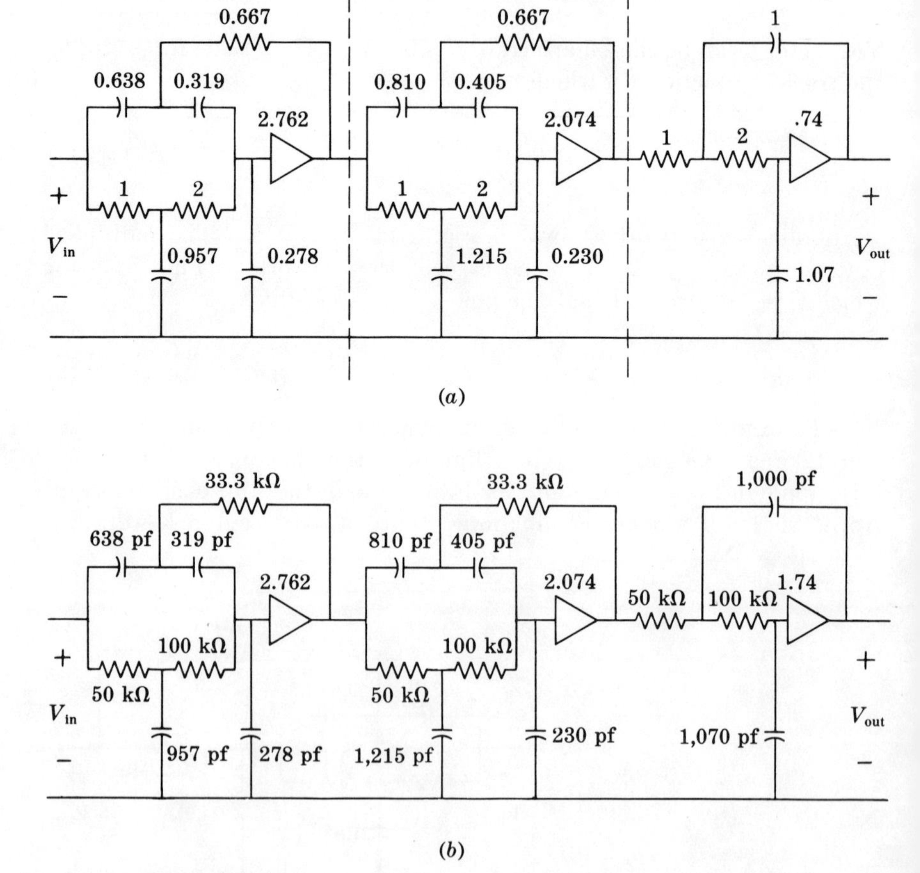

FIG. 2.12 Elliptic-function low-pass filter. (*a*) Normalized (ohms, farads); (*b*) scaled to 50 kΩ and 3,180 Hz.

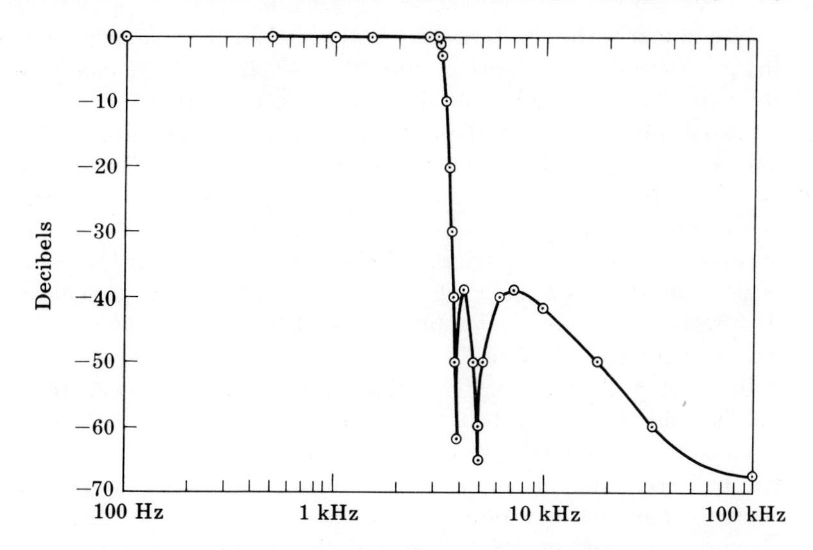

FIG. 2.13 Measured performance of the filter of Fig. 2.12*b*.

and frequency scaling to a cutoff frequency of 3,180 Hz, we obtain the network shown in Fig. 2.12*b*. Figure 2.13 shows the measured frequency response of the network of Fig. 2.12*b*. No measurable deviation from the theoretical curve exists except for the finite attenuation at the zeros. The amplifiers used were designed specifically for use in active *RC* VCVS realizations (Kerwin and Huelsman 1966), and are shown in Fig. 2.14 with equal-feedback resistors R_a and R_b for a gain of 2.00.

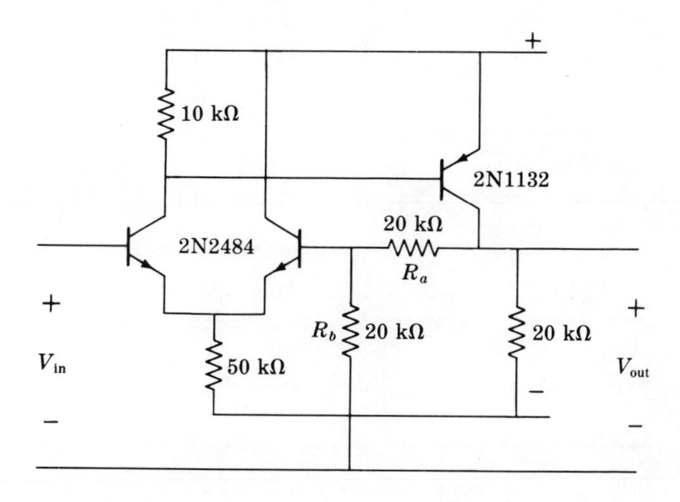

FIG. 2.14 Positive-gain VCVS.

In summary, the positive-gain VCVS can be used to achieve most of the pole-zero combinations of interest (in fact, by slight modification of the twin-T design used to produce $j\omega$-axis zeros, either left-half-plane or right-half-plane zeros can be obtained), and relatively simple structures are used which can be cascaded without interaction. The total capacitance required for a given cutoff frequency is reasonably low; that is, it is no greater than that required in a passive *RC* network. The primary disadvantage of the positive-gain VCVS networks is their high sensitivity of pole position to amplifier-gain change and to passive-element change. This restricts the practical application of these circuits to a Q of 10 or less and in many cases to 5 or less.

Unity-gain Structures. For those cases where a low-Q function is needed and therefore sensitivity is not a serious problem, it is possible to substitute increased sensitivity to gain change for reduced gain. This can be particularly useful if the gain can be reduced to unity or below, since the amplifier is simplified to an emitter follower or voltage follower (Hogin 1966; McVey 1965; Foss and Green 1966). The low-pass function previously described, Eq. (2.18), is practically limited to $\beta > 1.1$ for $K = 1$. This can be modified if we add an additional degree of freedom as in the network shown in Fig. 2.15. An analysis of this configuration results in the following transfer function:

$$T(p) = \frac{K/RC_1C_2}{p^2 + [C_2(1 + R) + C_1(1 - K)]p + 1/RC_1C_2} \tag{2.48}$$

If we now choose $K = 1$, we obtain the following design equations [capacitance in farads, β, γ as defined in (2.9)]

$$C_1 = \frac{1 + R}{R\beta\gamma} = \frac{2}{\beta\gamma}\Big]_{R=1\text{ ohm}} \tag{2.49}$$

$$C_2 = \frac{\beta}{1 + R} = \frac{\beta}{2}\Big]_{R=1\text{ ohm}} \tag{2.50}$$

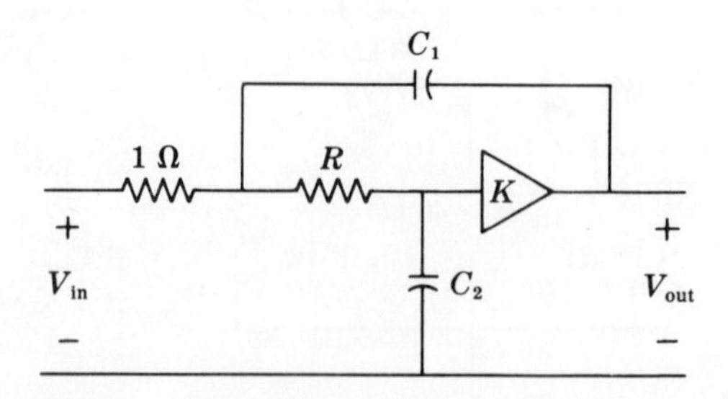

Fig. 2.15 Modified positive-gain VCVS low-pass filter section.

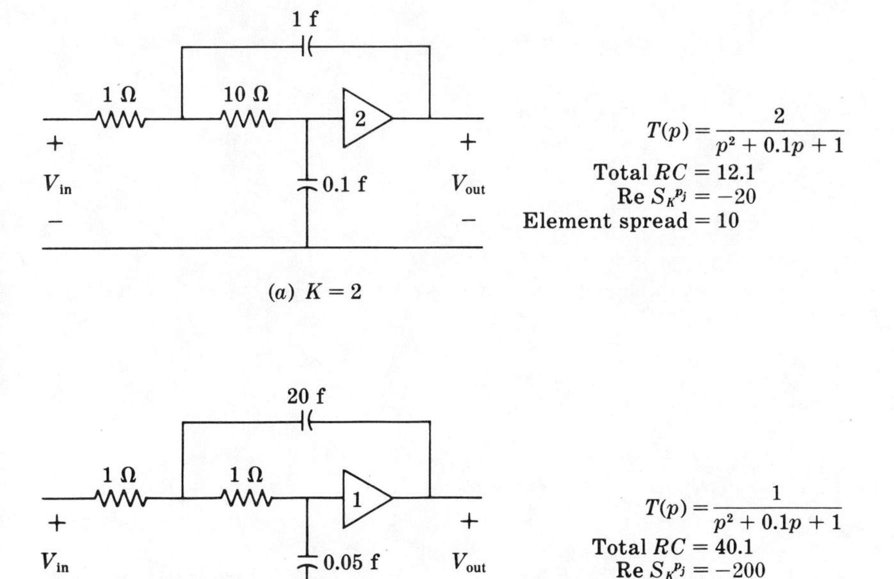

FIG. 2.16 Comparison of positive-gain VCVS realizations for two different gains.

and we are now free to choose R arbitrarily, but the capacitors can no longer be made equal. Although this appears to be an attractive result, the price paid is considerable if the Q is at all high. A comparison of unity and nonunity gains at $Q = 10$ is shown in Fig. 2.16. As can be seen, the total capacitance required, the element spread, and the sensitivity are all seriously worsened. The practical range of usefulness is approximately $\frac{1}{2} < Q < 3$ from the standpoint of sensitivity, and even then the element spread is excessive at the upper Q limit. Calculation of the pole-position sensitivity for $R = 1$ ohm shows that Re $S_K{}^{p_i} = -2Q^2$. The Im $S_K{}^{p_i}$ is negligible. Additional details are given in Bown (1967). There are, however, many other networks using unity-gain amplifiers which are capable of high-Q performance. In most cases, these still show very high sensitivity, but since the unity-gain amplifier can be very stable, they may be quite practical (Hogin 1966; Zai 1967).

Negative-gain VCVS, Low-pass. The basic negative-gain VCVS realizations were given by Sallen and Key (1955), and, with more generality, by Taylor (1963). The low-pass structure is shown in Fig. 2.17. The

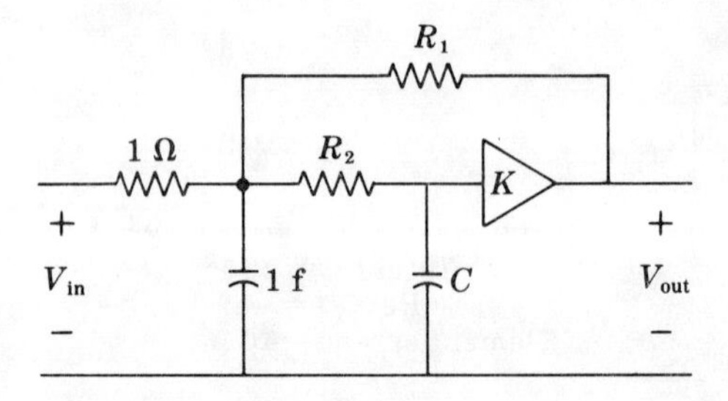

Fig. 2.17 Negative-gain VCVS low-pass filter section.

transfer function is

$$T(p) = \frac{KR_1}{R_1R_2Cp^2 + [R_2C(1 + R_1) + R_1(1 + C)]\,p + 1 + R_1 - K} \tag{2.51}$$

A typical root locus (for $R_1 = R_2 = 1$ ohm, $C = 1$ farad) is shown in Fig. 2.18. Rearranging (2.51) and introducing a constant γ, we obtain

$$T(p) = \frac{KR_1\,\gamma/(1 + R_1 - K)}{\dfrac{R_1R_2C\gamma}{1 + R_1 - K}\,p^2 + \left(\dfrac{R_2C(1 + R_1) + R_1(1 + C)}{1 + R_1 - K}\right)\gamma p + \gamma} \tag{2.52}$$

and if we make the substitution

$$p = s\sqrt{\frac{1 + R_1 - K}{R_1R_2C\gamma}} \tag{2.53}$$

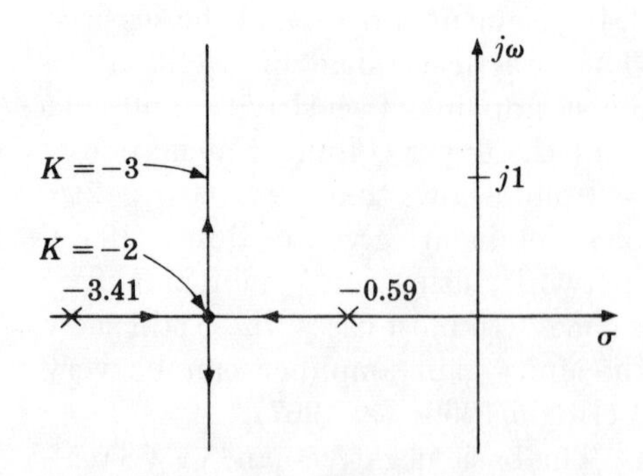

Fig. 2.18 Root-locus plot for the negative-gain VCVS filter of Fig. 2.17.

we obtain

$$T_1(s) = \frac{KR_1\gamma/(1 + R_1 - K)}{s^2 + \left(\dfrac{R_2C(1 + R_1) + R_1(1 + C)}{\sqrt{R_1R_2C(1 + R_1 - K)}}\right)\sqrt{\gamma}\, s + \gamma} \tag{2.54}$$

An investigation of the effect of various values of R_1, R_2, and C shows that to minimize the gain required, reduce the element spread, and reduce the total capacity required, a good choice is $R_1 = \frac{1}{2}$ ohm, $R_2 = 2$ ohms, and $C = \frac{1}{4}$ farad; thus, from (2.51) we then have

$$T(p) = \frac{K/2}{0.25p^2 + 1.375p + 1.5 - K} \tag{2.55}$$

For these element values, the relation between p and s, (2.53), is

$$p = 2\sqrt{\frac{1.5 - K}{\gamma}}\, s \tag{2.56}$$

and (2.54) becomes

$$T_1(s) = \frac{0.5K\gamma/(1.5 - K)}{s^2 + [2.75\sqrt{\gamma}/(\sqrt{1.5 - K})]s + \gamma} \tag{2.57}$$

We are now in a position to equate the transfer function of (2.57) to

$$T_1(s) = \frac{H}{s^2 + \beta s + \gamma} \tag{2.58}$$

to obtain a set of design equations; however, in doing this we must also scale the resulting network back to the p plane by the p-to-s ratio given in (2.56). This results in the network shown in Fig. 2.19 for which the following design equations apply for any given transfer function of the

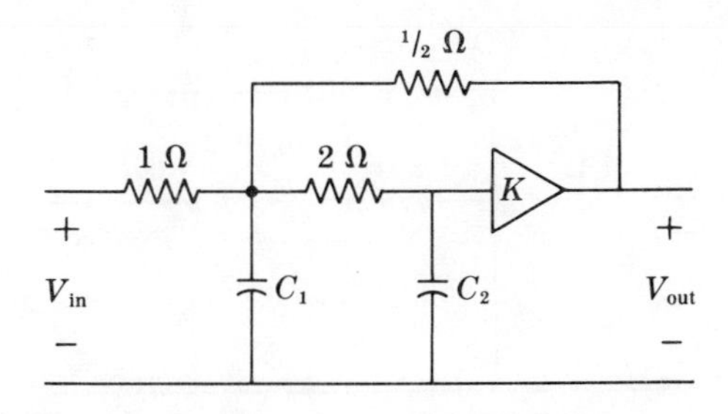

FIG. 2.19 Negative-gain VCVS low-pass filter section.

form of (2.59) (capacitances in farads):

$$T(p) = \frac{H}{p^2 + \beta p + \gamma} \tag{2.59}$$

$$C_1 = \frac{5.5}{\beta} \tag{2.60}$$

$$C_2 = \frac{1.375}{\beta} \tag{2.61}$$

$$K = 1.5 - \frac{7.56\gamma}{\beta^2} \tag{2.62}$$

$$H = 0.099\beta^2 - 0.5\gamma \tag{2.63}$$

In this way, we obtain a very practical and simple set of design equations, which clearly show the problems inherent in this method of realization. The capacitors are greatly increased in value as Q increases (small β) as is the value of gain K required, but the gain increases as Q^2. In using these design equations, the values of β and γ are taken directly from the specified $T(p)$; and C_1, C_2, and K are calculated using (2.60) to (2.62), but the value of H cannot be specified. It is determined by the values of β and γ as given by (2.63), and there is no choice here. The constant multiplier H approaches -0.5γ for high Q and indicates the severity of the system-stabilization problem at high Q (and therefore high-gain K) due to the very large amount of negative feedback which reduces the amplifier gain below unity at dc. The pole-position sensitivity to amplifier-gain change for the poles of (2.55) is

$$S_K^{p_i} = 0 + j\frac{K}{0.8 + 2K} \tag{2.64}$$

which approaches $S_K^{p_i} = 0 + j0.5$ for large K.

As an example, we will design a second-order low-pass filter having a $Q = 5$ and a frequency of maximum response $\omega_0 \approx 1$ rad/sec;

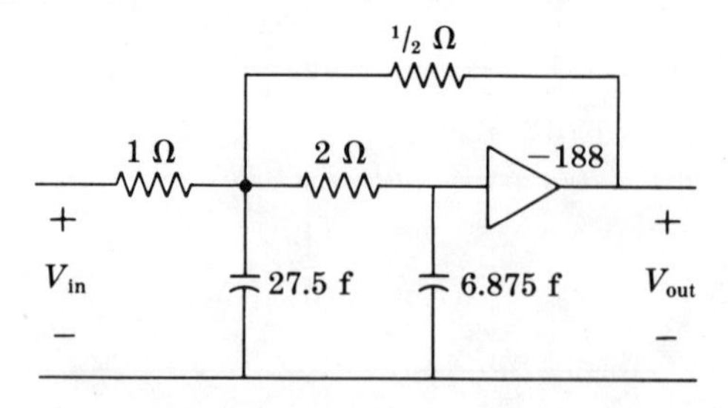

FIG. 2.20 Negative-gain VCVS low-pass filter.

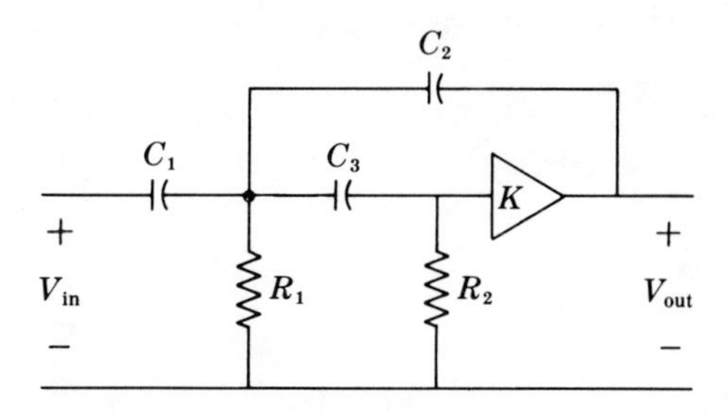

FIG. 2.21 Negative-gain VCVS high-pass filter section.

($\beta = 1/Q = 0.2$, $\gamma = 1.0$) whose transfer function is

$$T(p) = \frac{H}{p^2 + 0.2p + 1} \tag{2.65}$$

From (2.60) to (2.63) we find that $C_1 = 27.5$ farads, $C_2 = 6.875$ farads, $K = -188$, and $H = -0.496$ as shown in Fig. 2.20. The total RC product required is 100 sec (neglecting the feedback resistor R_1) compared to $RC = 12.1$ sec for the equivalent positive-gain realization.

Negative-gain VCVS, High-pass. The basic network for this case is shown in Fig. 2.21. A new problem exists here, namely, the need to use three capacitors for a second-order function. In this case analysis, transformation, and minimization of gain and element spread for the transfer function

$$T(p) = \frac{Hp^2}{p^2 + \beta p + \gamma} \tag{2.66}$$

results in the following design equations for the network shown in Fig.

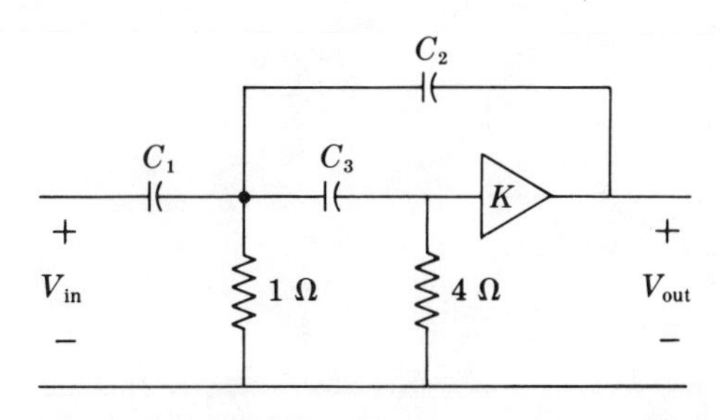

FIG. 2.22 Optimized negative-gain VCVS high-pass filter section.

2.22 (capacitance values in farads):

$$C_1 = 0.4\frac{\beta}{\gamma} \tag{2.67}$$

$$C_3 = 0.4\frac{\beta}{\gamma} \tag{2.68}$$

$$C_2 = 0.1\frac{\beta}{\gamma} \tag{2.69}$$

$$K = 1.06 - \frac{6.25\gamma}{\beta^2} \tag{2.70}$$

$$H = \frac{K}{5 - 4K} \tag{2.71}$$

We now find a happy result: the capacitor sizes decrease as β decreases (Q increases). However, the gain required still increases as Q^2, and it is difficult to achieve much more than $Q = 10$ because of the high gain required.

Negative-gain VCVS, Bandpass, Case 1. The basic network configuration (Sallen and Key 1955) for this case is shown in Fig. 2.23. The transfer function is

$$T(p) = \frac{Kp}{RCp^2 + (1 + C + RC)p + 1 - K} \tag{2.72}$$

The root-locus plot for this case is similar to the previous negative-gain cases, i.e., parallel to the $j\omega$ axis. Rearranging (2.72) and making the substitution

$$p = \sqrt{1 - K}\, s \tag{2.73}$$

results in

$$T_1(s) = \frac{(K/RC\sqrt{1 - K})s}{s^2 + [(1 + C + RC)/RC\sqrt{1 - K}]s + 1/RC} \tag{2.74}$$

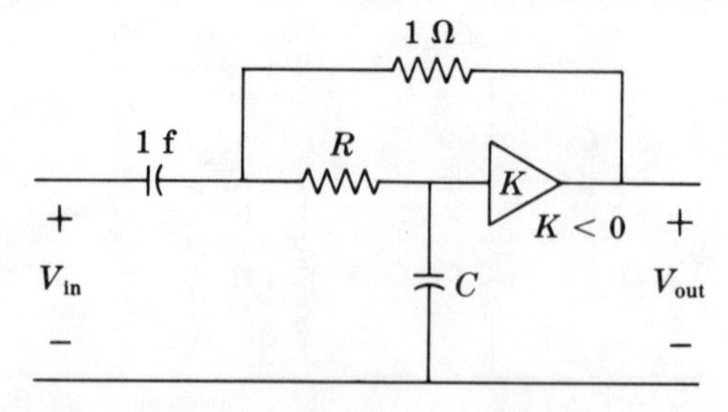

FIG. 2.23 Negative-gain VCVS bandpass filter section, case 1.

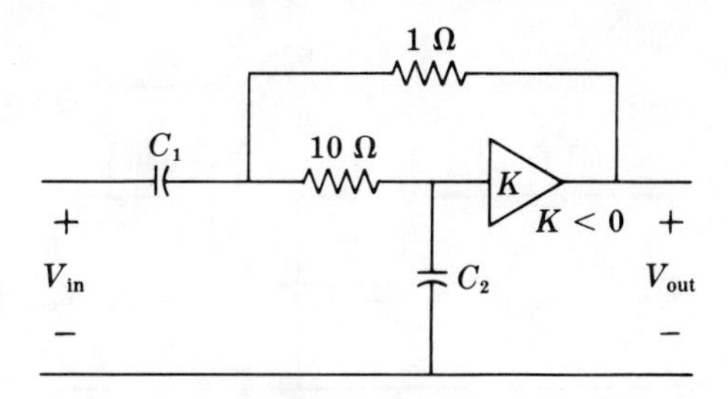

FIG. 2.24 Optimized negative-gain VCVS bandpass filter section, case 1.

which clearly shows the mechanism by which the Q is increased, and that K must be negative. If we now choose a value for R based on minimizing K and element spread, for example, $R = 10$ ohms, and scale the network by multiplying the capacitors by $\sqrt{1-K}$ (to return to the p plane), we obtain the following design equations (capacitance in farads) for the network shown in Fig. 2.24, where the transfer function to be realized is

$$T(p) = \frac{Hp}{p^2 + \beta p + \gamma} \tag{2.75}$$

$$C_1 = \sqrt{1-K} \tag{2.76}$$

$$C_2 = \frac{\sqrt{1-K}}{10\gamma} \tag{2.77}$$

$$K = 1 - \left(\frac{1.1+\gamma}{\beta}\right)^2 \tag{2.78}$$

$$H = \frac{K\beta\gamma}{\gamma + 1.1} \tag{2.79}$$

As an example, let $Q = 10$, $\beta = 0.1$, $\gamma = 1.0$, for which $\omega_0 = 1$ rad/sec. Then we find $K = -440$, $C_1 = 21$ farads, $C_2 = 2.1$ farads, $H = -21$, and that the center-frequency gain H/β is -210; so that a good share of the amplifier gain is available at band center, in marked contrast to the previous two cases. The final network is shown in Fig. 2.25. The pole-position sensitivity to amplifier-gain change is

$$S_K{}^{p_i} = 0 - j\frac{K}{1.4 - 2K} \tag{2.80}$$

which, as expected for a negative-gain system of this type, is very low and independent of Q. At high gain $S_K{}^{p_i} = 0 + j0.5$.

Negative-gain VCVS, Bandpass, Case 2. A second alternative exists for the negative-gain VCVS bandpass realization and is shown in Fig.

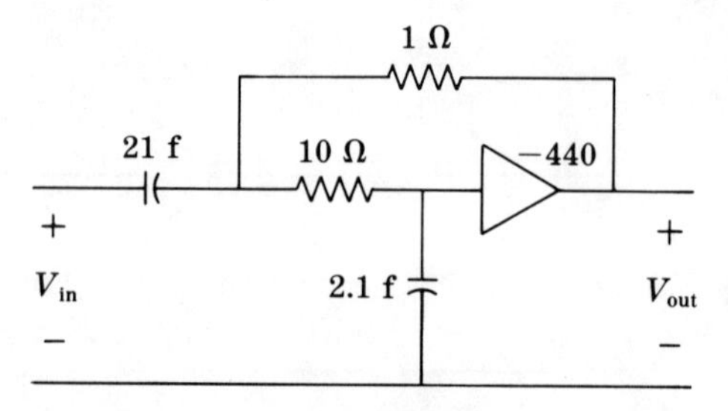

FIG. 2.25 Negative-gain VCVS bandpass filter, case 1.

2.26. Analysis results in the following transfer function:

$$T(p) = \frac{Kp}{(1 - K)p^2 + (1 + 1/R + 1/RC)p + 1/RC} \tag{2.81}$$

In this case, the root locus is not of the preceding kind, but is a circle terminating, as $K \to \infty$, at the origin as shown in Fig. 2.27, for $RC = 1$ sec, $R = 10$ ohms. Frequency-scaling effects and the gain required are similar to the preceding negative-gain realizations, except that the capacitor sizes are reduced as the gain is increased, rather than increased as in the preceding network. This can be decidedly advantageous in integrated-circuit synthesis. Substituting

$$p = \frac{s}{\sqrt{1 - K}} \tag{2.82}$$

into (2.81), we obtain

$$T_1(s) = \frac{(K/\sqrt{1 - K})s}{s^2 + [(1 + 1/R + 1/RC)/\sqrt{1 - K}]s + 1/RC} \tag{2.83}$$

If we now frequency-scale the network back to the p plane by dividing the capacitors by $\sqrt{1 - K}$, we obtain the following design equations

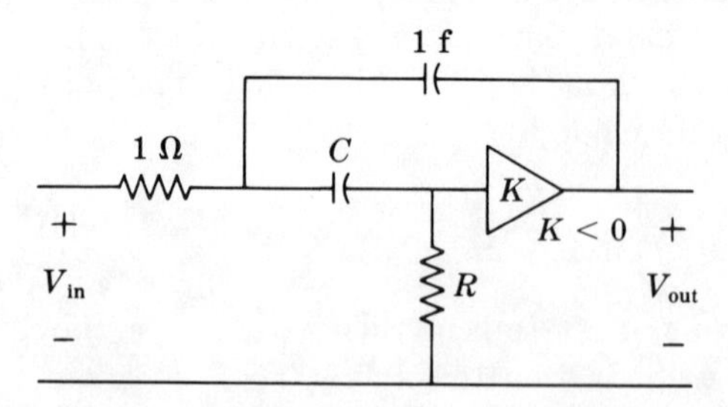

FIG. 2.26 Negative-gain VCVS bandpass filter section, case 2.

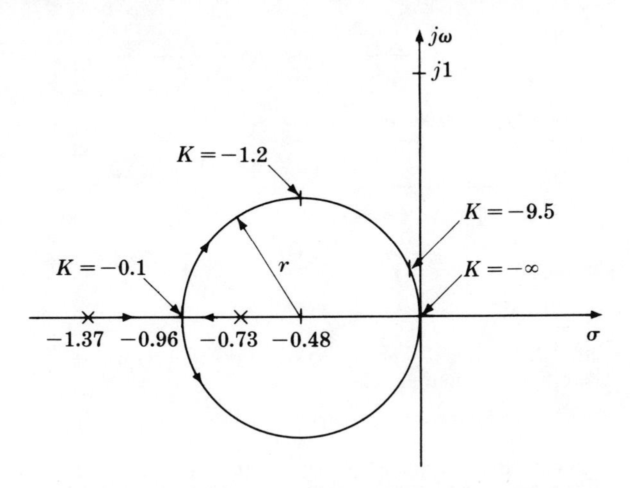

FIG. 2.27 Root-locus plot for the network of Fig. 2.26, R = 10 ohms, C = 0.1 farad.

for the network of Fig. 2.28 (in which we have chosen R = 10 ohms) for any given transfer-function specification:

$$T(p) = \frac{Hp}{p^2 + \beta p + \gamma} \tag{2.84}$$

$$C_1 = \frac{1}{10\gamma\sqrt{1 - K}} \tag{2.85}$$

$$C_2 = \frac{1}{\sqrt{1 - K}} \tag{2.86}$$

$$K = 1 - \left(\frac{1.1 + \gamma}{\beta}\right)^2 \tag{2.87}$$

$$H = \frac{K\beta}{1.1 + \gamma} \tag{2.88}$$

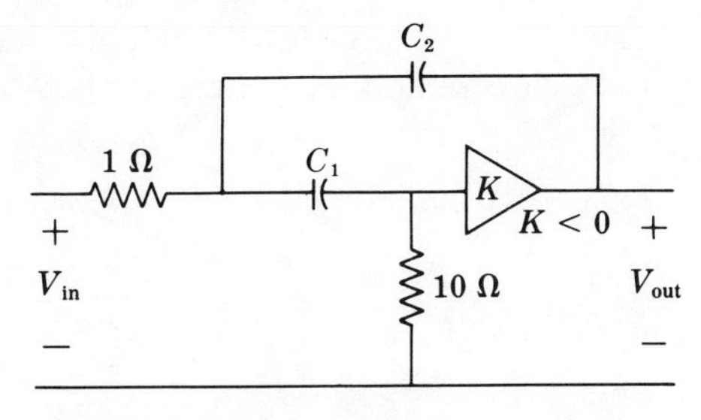

FIG. 2.28 Optimized negative-gain VCVS bandpass filter section, case 2.

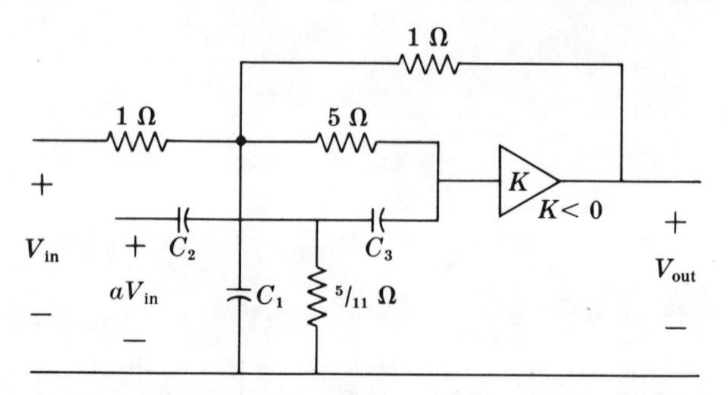

FIG. 2.29 Negative-gain VCVS two-pole, two-$j\omega$-axis-zero filter section.

If we now choose as a design example the same conditions as were used in case 1: $Q = 10$, $\beta = 0.1$, and $\gamma = 1.0$, we obtain $K = -440$, $C_1 = 1/210$ farad, $C_2 = 1/21$ farad, and $H = -21$ for the network shown in Fig. 2.28. The resulting center-frequency (1 rad/sec) gain is -210. This is a reduction of total capacitance required from $C_T = 23.1$ farad to $C_T = 0.052$ farad for the same resonant frequency! The pole-position sensitivity to changes in amplifier gain is (for $R = 10$ ohms, $C = 0.1$ farad)

$$S_K{}^{p_i} = \frac{K}{1-K} + j\left(\frac{K}{1-K} + \frac{K}{2K+0.20}\right) \approx -1 - j0.5 \qquad |K| \gg 1 \tag{2.89}$$

Negative-gain VCVS, Finite $j\omega$-axis Zeros. In this case the frequency-scaling effect produced by changes in amplifier gain makes it difficult to keep the poles and zeros conveniently close to each other using the usual parallel ladder networks as in Fig. 2.10*a*, *b*, and *c*, but modified to operate with negative-gain amplifiers. A modification to the twin-T network solves this problem (Jagoda 1962). It does, however, require two inputs. This is not too serious a problem in this case, since the second input required is at a very low level and can easily be obtained from the first by resistive attenuation. This configuration is shown in Fig. 2.29. The design equations for the general function

$$T(p) = \frac{H(p^2+\alpha)}{p^2+\beta p+\gamma} \tag{2.90}$$

are

$$C_1 = \frac{4.83}{\beta} \tag{2.91}$$

$$C_2 = \frac{4.39}{\beta} \tag{2.92}$$

$$C_3 = \frac{0.439}{\beta} \tag{2.93}$$

$$a = 0.103\frac{\beta^2}{\alpha} \tag{2.94}$$

$$K = 2 - \frac{9.68\gamma}{\beta^2} \tag{2.95}$$

$$H = \frac{0.206\beta^2 - \gamma}{\alpha} \tag{2.96}$$

An interesting result is that the network elements are determined only by β and that for $\beta < 1$ (and $\alpha \approx 1$), the value of a is less than 0.1. The value of a required is conveniently obtained by a resistive divider from V_{in}. The capacitor size increases directly with Q and the gain required increases as Q^2, so that this network is again limited to low-Q poles.

An example of the use of this network to achieve a three-pole, two-zero elliptic-function filter in which the passband ripple is 0 db (maximally-flat magnitude) and the ratio of rejection frequency to cut off frequency (−3 db) is 2 is shown in Fig. 2.30*a*. This network has been impedance-scaled to 20 kΩ and frequency-scaled to give a cutoff (−3 db) frequency of 5 kHz. The *normalized* transfer function of this network as realized is

$$T(p) = -0.23\frac{p^2 + 4}{p^2 + 0.89p + 1.09}\frac{1.21}{p + 1.21} \tag{2.97}$$

The single negative-sigma-axis pole has been obtained by cascading a simple *RC* section following the VCVS as shown in Fig. 2.30*a*, thereby realizing the last factor of (2.97). As can be seen in Fig. 2.30*b*, the experimental measurements confirm the predicted performance of this filter.

In summary, we see that for all the negative-gain VCVS realizations described, the improvement in sensitivity is obtained at too high a price, since it is in exactly the region where we need reduced sensitivity to gain change (that is, $Q > 5$) that these realizations become impractical due not to sensitivity problems but to the total capacitance and the high gain required, as well as to practical difficulties due to the large amount of negative feedback. For relatively low-Q applications, this structure is practical, since an amplifier of low negative gain is simpler than a positive-gain amplifier whose gain is greater than one, and the sensitivity to gain change is smaller for the negative-gain case. The unity-gain structure should also be considered in choosing the most appropriate network for a specific low-Q requirement.

Multiloop-feedback Low-pass Filter. The excessive sensitivity of the low positive-gain VCVS networks and the excessive gain of the low-sensitivity negative-gain VCVS networks can both be alleviated by a combination of the two networks so as to obtain a more reasonable sensitivity and gain, and allow the practical attainment of much higher Q systems. A direct cascade combination of the positive-gain low-pass structure of Fig. 2.1 and the negative-gain low-pass structure of Fig. 2.17 results in the network of Fig. 2.31 if we cascade the positive- and nega-

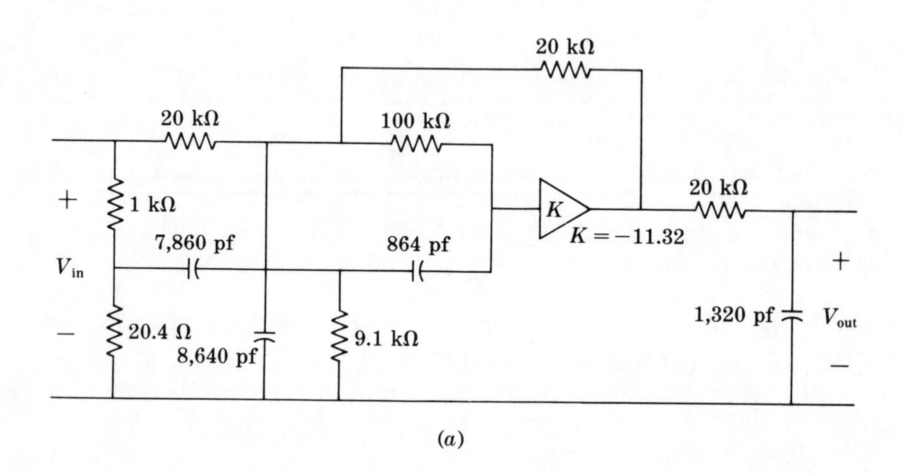

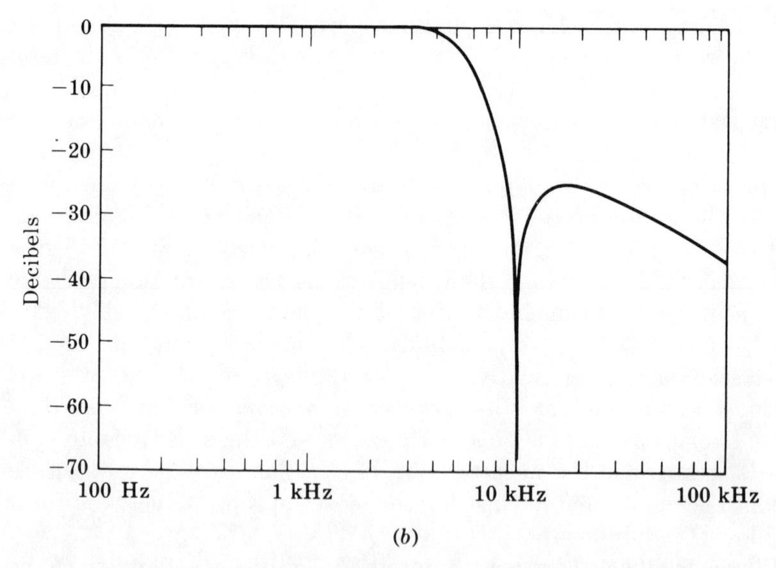

FIG. 2.30 Third-order elliptic-function filter using a negative-gain VCVS. (a) Schematic; (b) measured amplitude response.

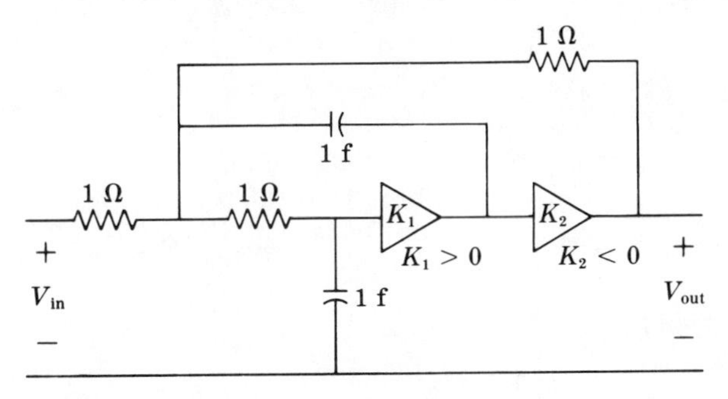

FIG. 2.31 Multiloop low-pass filter section.

tive-gain amplifiers so that the combination gain is negative (Kerwin, June 1966). Although this leads to a rather complex amplifier structure, the reduction in gain required greatly simplifies the negative-gain VCVS and reduces the loop-stabilization problem, since the amount of overall gain reduction as a result of the negative feedback is less. Due to the reduction in gain of the negative-gain VCVS for a given Q, the scaling effect on capacitor size is also reduced, and the high-frequency capability is improved. In addition, the negative-gain VCVS need not have a high input impedance; that is, it need not be ideal since it is driven directly by the positive-gain VCVS. The number of network elements is also increased, but not the number of capacitors. Higher-order multiloop structures are given in Hazony and Lagerlof (1966).

Analysis of the network of Fig. 2.31 results in the following transfer function

$$T(p) = \frac{K_1 K_2}{p^2 + (4 - K_1)p + 2 - K_1 K_2} \tag{2.98}$$

A root-locus plot with K_2 only as a variable shows a straight-line locus parallel to the $j\omega$ axis, and therefore low sensitivity to changes in K_2 as shown in Fig. 2.32*a* (for $K_1 = 1$) and Fig. 2.32*b* (for $K_1 = 3$). However, if we use K_1 as the variable, the locus leaves the sigma axis with a much larger radius of curvature than that of the single positive-gain VCVS system, because of the much larger than normal constant term in (2.98) (the radius of curvature is the $\sqrt{\gamma}$ in a quadratic of the form $p^2 + \beta p + \gamma$). Since the radius is approximately proportional to $\sqrt{K_1}$, the curvature increases as K_1 increases as shown in Fig. 2.33*a* and *b* for two different values of K_2. This, of course, decreases the rate of approach to the $j\omega$ axis and reduces the real part of the pole-position sensitivity to changes in K_1. Note that the locus crosses the $j\omega$ axis at $\omega = 6.46$ when $K_1 = 4.0$ for the case shown in Fig. 2.33*a*. If we use the positive-gain amplifier to produce a Q of 10, then we start with a value of the first-degree coefficient

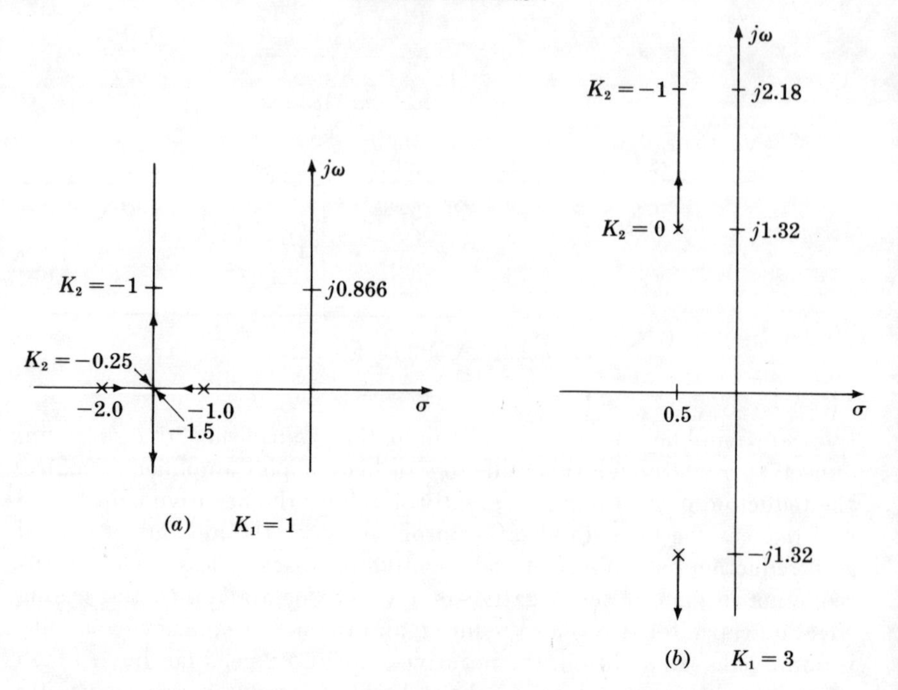

FIG. 2.32 Root locus vs. K_2 of the multiloop filter section of Fig. 2.31.

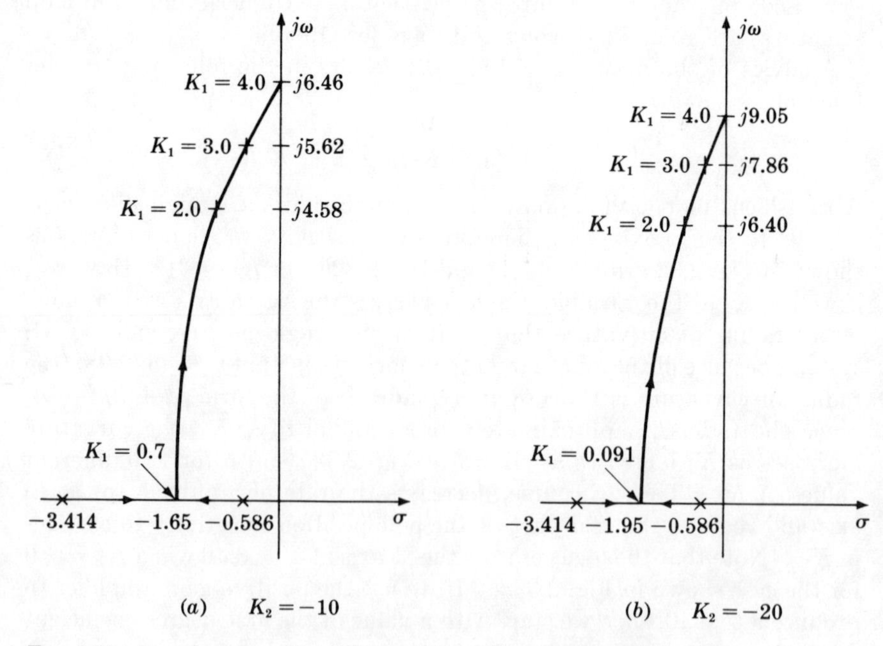

FIG. 2.33 Root locus vs. K_1 of the multiloop filter section of Fig. 2.31.

which is more than an order of magnitude smaller than the usual (passive) value of the first-degree coefficient which a negative-gain amplifier must reduce. A further factor-of-10 increase in Q can then be achieved with a negative gain of 100 to give a final Q of 100. The pole sensitivity to amplifier-gain change, however, is essentially that of a $Q = 10$ positive-gain VCVS system. The multiloop system is thus a reasonable compromise between the high-gain, low-sensitivity negative-gain VCVS network and the low-gain, high-sensitivity positive-gain VCVS network.

If we now multiply (2.98) by γ and substitute

$$p = \sqrt{\frac{2 - K_1K_2}{\gamma}}\, s \tag{2.99}$$

into (2.98), we obtain

$$T_1(s) = \frac{K_1K_2\gamma/(2 - K_1K_2)}{s^2 + [(4 - K_1)\sqrt{\gamma}/\sqrt{2 - K_1K_2}]\, s + \gamma} \tag{2.100}$$

Equating this to (2.58) and then transforming back to the p plane we obtain the following design equations for the network shown in Fig. 2.34 for the transfer function of (2.101):

$$T(p) = \frac{H}{p^2 + \beta p + \gamma} \tag{2.101}$$

$$C_1 = C_2 = \sqrt{\frac{2 - K_1K_2}{\gamma}} \tag{2.102}$$

$$K_1K_2 = 2 - \gamma\left(\frac{4 - K_1}{\beta}\right)^2 \tag{2.103}$$

$$H = \frac{K_1K_2}{2 - K_1K_2} \tag{2.104}$$

where $K_1 > 0$, $K_2 < 0$.

We now have a choice between K_1 and K_2 and since large K_1 will minimize the K_1K_2 product as shown by (2.103) and this will reduce the

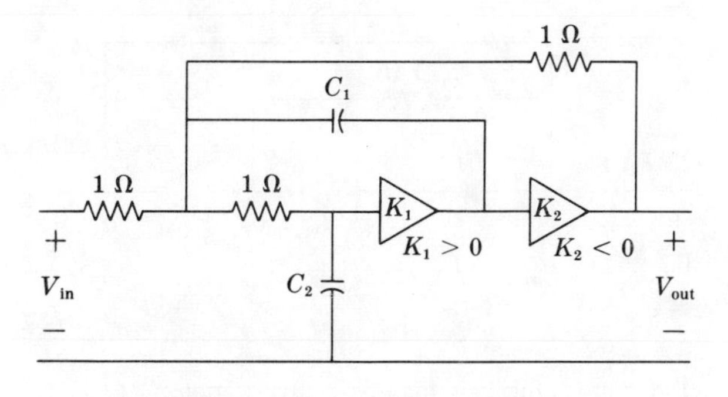

FIG. 2.34 Multiloop low-pass filter section.

capacitor sizes, it is advantageous to make K_1 as large as possible subject only to the pole-position-sensitivity requirements. The pole-position sensitivities to K_1 and K_2 for (2.98) are

$$S_{K_1}{}^{p_j} = -\frac{K_1}{4 - K_1} + j\frac{4 - K_1 - 2K_2}{4(2 - K_1K_2) - (4 - K_1)^2} \tag{2.105}$$

$$S_{K_2}{}^{p_j} = 0 - j\frac{2K_1K_2}{4(2 - K_1K_2) - (4 - K_1)^2} \tag{2.106}$$

The allowable pole-position sensitivity will therefore determine how large K_1 can be; then β and γ will determine K_1K_2, and therefore K_2 from (2.103).

Due to the low gains required, this network performs well in practice and is capable of high-frequency operation at reasonably high Q. Additional optimization is possible by varying the resistive and capacitive values, but the network of Fig. 2.34 is illustrative of the principles involved.

Multiloop-feedback High-pass Filter. In this case we combine the networks of Figs. 2.6 and 2.21 to obtain the normalized multiloop circuit of Fig. 2.35. The transfer function is

$$T(p) = \frac{K_1K_2p^2}{(2 - K_1K_2)p^2 + (4 - K_1)p + 1} \tag{2.107}$$

Derivation of a set of design equations in a similar manner to that used in the low-pass case previously described results in the following set of design equations for the network of Fig. 2.36 for the generalized high-pass transfer function of (2.108):

$$T(p) = \frac{Hp^2}{p^2 + \beta p + \gamma} \tag{2.108}$$

$$C_1 = C_2 = C_3 = \frac{1}{\sqrt{(2 - K_1K_2)\gamma}} \tag{2.109}$$

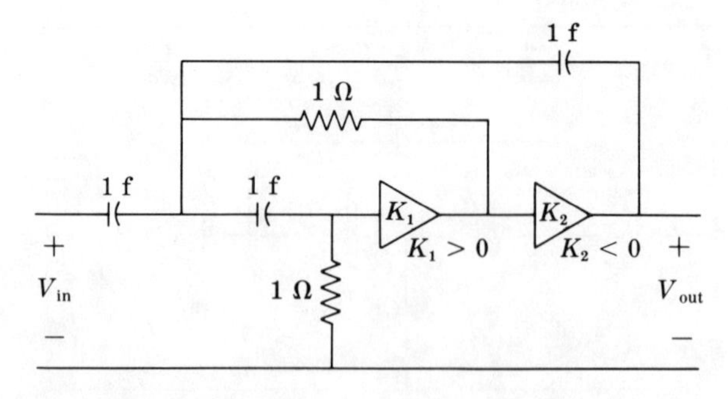

FIG. 2.35 Multiloop high-pass filter section.

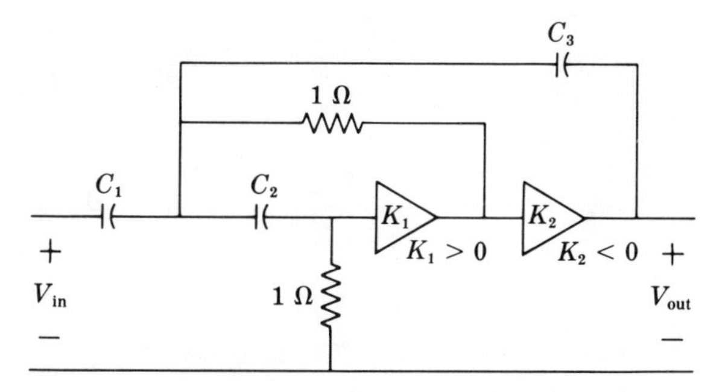

FIG. 2.36 Multiloop high-pass filter section.

$$K_1K_2 = 2 - \gamma\left(\frac{4 - K_1}{\beta}\right)^2 \tag{2.110}$$

$$H = \frac{K_1K_2}{2 - K_1K_2} \tag{2.111}$$

where $K_1 > 0$; $K_2 < 0$.

As before, we have a choice between K_1 and K_2, and if we choose K_1 as large as possible within the pole-position-sensitivity limits, K_2 is minimized for a given Q. The pole-position sensitivities are

$$S_{K_1}{}^{p_i} = \frac{-K_1(2 - 4K_2)}{(4 - K_1)(2 - K_1K_2)} + j\left[\frac{K_1K_2}{2 - K_1K_2} + \frac{K_1(4 - K_1 - 2K_2)}{4(2 - K_1K_2) - (4 - K_1)^2}\right] \tag{2.112}$$

$$S_{K_2}{}^{p_i} = \frac{K_1K_2}{2 - K_1K_2} + j\left[\frac{K_1K_2}{2 - K_1K_2} - \frac{2K_1K_2}{4(2 - K_1K_2) - (4 - K_1)^2}\right] \tag{2.113}$$

These design equations and sensitivities can be modified by choosing other element values for the fixed elements of the network of Fig. 2.36; however, these results are representative of this method of synthesis.

Multiloop-feedback Bandpass Filter. A bandpass multiloop structure can similarly be obtained by combining the positive-gain structure shown in Fig. 2.8 with the negative-gain structure of Fig. 2.26 to obtain the network shown in Fig. 2.37. Analysis of this network gives the transfer function

$$T(p) = \frac{K_1K_2p}{(1 - K_1K_2)p^2 + (4 - K_1)p + 2} \tag{2.114}$$

The effects of changes in K_1 and K_2 are similar to the low-pass case

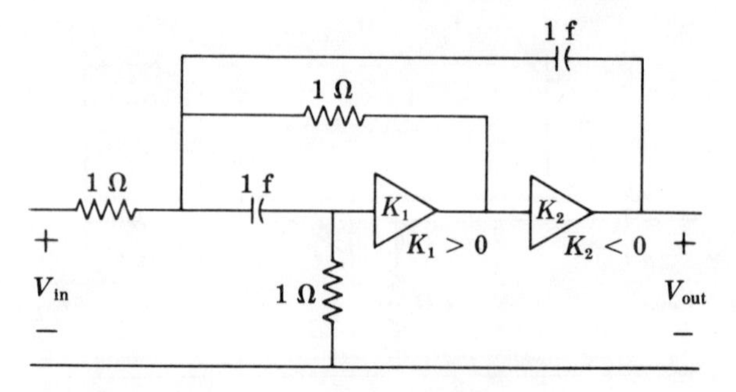

FIG. 2.37 Multiloop bandpass filter section.

and the center frequency and center-frequency gain are

$$\omega_0 = \sqrt{\frac{2}{1 - K_1K_2}} \tag{2.115}$$

$$|T(p)|_{\omega_0} = \frac{K_1K_2}{4 - K_1} \tag{2.116}$$

An alternate configuration is shown in Fig. 2.38. The transfer function in this case is

$$T(p) = \frac{K_1K_2p}{2p^2 + (4 - K_1)p + 1 - K_1K_2} \tag{2.117}$$

The pole-position sensitivities to K_1 and K_2 for the multiloop bandpass filter of Fig. 2.37 are

$$S_{K_1}{}^{p_j} = -\frac{K_1(1 - 4K_2)}{(4 - K_1)(1 - K_1K_2)} + j\left[\frac{K_1K_2}{1 - K_1K_2} + \frac{K_1(4 - K_1 - 4K_2)}{8(1 - K_1K_2) - (4 - K_1)^2}\right] \tag{2.118}$$

$$S_{K_2}{}^{p_j} = \frac{K_1K_2}{1 - K_1K_2} + j\left[\frac{K_1K_2}{1 - K_1K_2} - \frac{4K_1K_2}{8(1 - K_1K_2) - (4 - K_1)^2}\right] \tag{2.119}$$

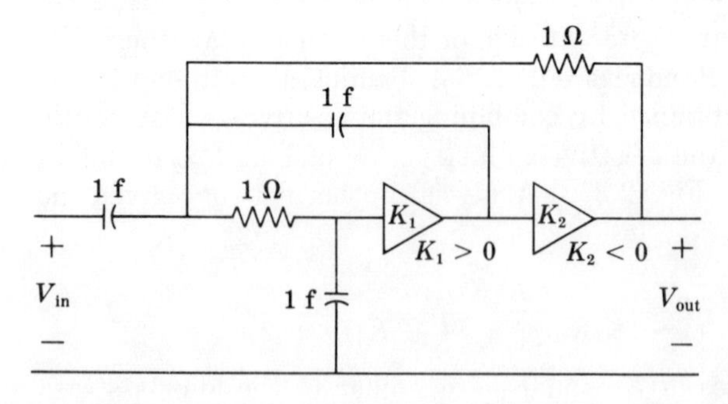

FIG. 2.38 Alternate multiloop bandpass filter section.

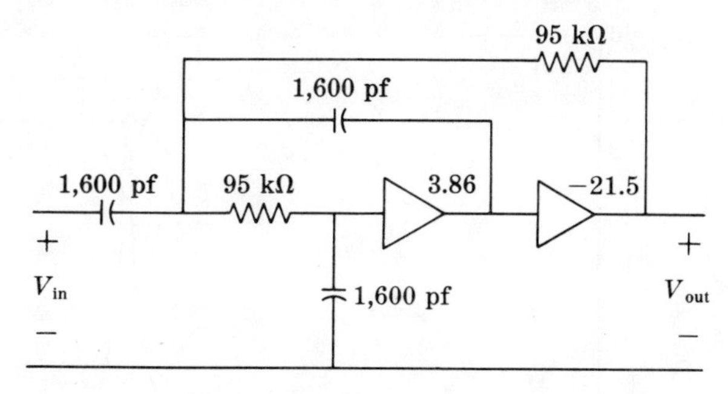

FIG. 2.39 Multiloop bandpass filter example.

The schematic of a tested multiloop bandpass filter is shown in Fig. 2.39. The gains used for this example were $K_1 = 3.86$ and $K_2 = -21.5$. The amplifier used to provide K_1 and K_2 is shown in Fig. 2.40. For this amplifier, the gains are $K_1 \approx 1 + R_3/R_4$ and $K_2 \approx R_6/R_5$ for the usual range of values of K_1 and K_2. The use of the negative-gain amplifier directly in cascade with the positive-gain amplifier eliminates the necessity of providing a high input impedance in the negative-gain amplifier and therefore simplifies the overall design of this type of multiloop system. The measured response of the network of Fig. 2.39 is shown in Fig. 2.41.

The use of combined positive and negative feedback provides a considerable improvement in Q possible for a given pole-position sensitivity, typically a tenfold increase in Q. The gains required are reasonably low and therefore this method is practical for high Q and high frequency.

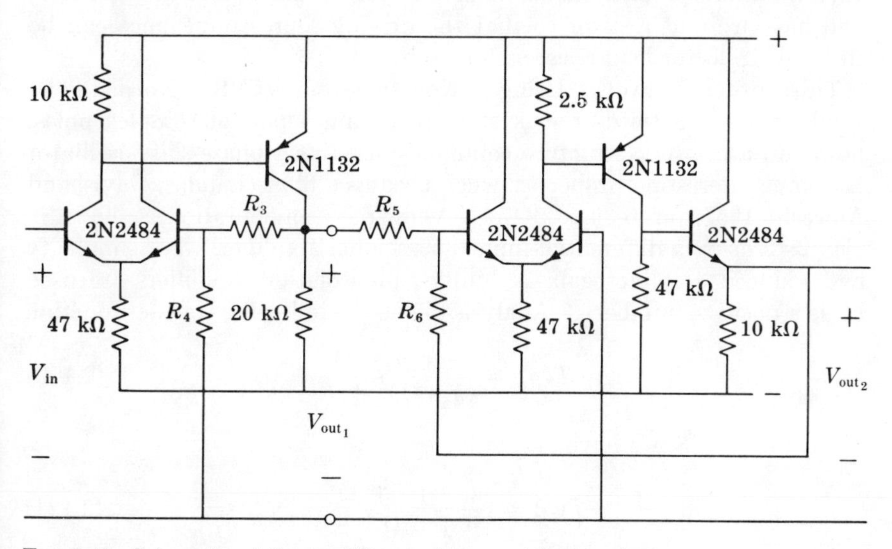

FIG. 2.40 Schematic of the amplifier used to provide K_1 and K_2.

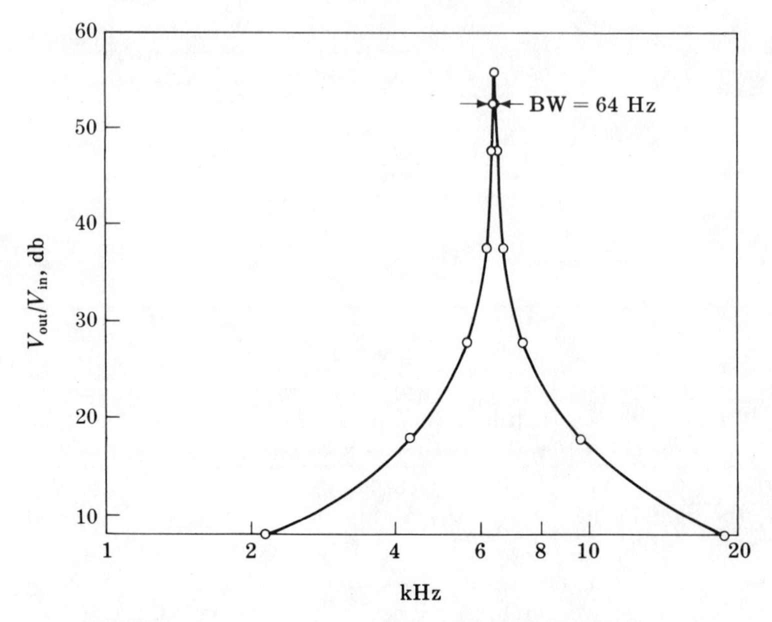

FIG. 2.41 Measured performance of the network of Fig. 2.39.

The disadvantages are primarily the increased complexity of the amplifier and the increased number of passive elements. However, the low-pass and bandpass second-order functions can be obtained with only two capacitors. Of the two bandpass filters, the circuit of Fig. 2.37 is probably the more practical, since it uses a minimum number of capacitors and has an input resistor so that the driving source impedance can be utilized (collector load resistor, for example).

Third-order Networks Using a Negative-gain VCVS. Normally we would not use a third-order system to obtain a pair of complex poles; however, a network structure commonly used as a phase-shift oscillator has some interesting properties when used as a filter (Landee, Davis, and Albrecht 1957, pp. 6-39 to 6-47). A low-pass configuration is shown in Fig. 2.42 using a differential-input operational amplifier. For simplicity and reduced amplifier gain K, unity-gain isolation amplifiers are used (K is a positive number). Analysis yields the following transfer function

$$T(p) = \frac{-K(p+1)^3}{(p+1)^3 + K} \tag{2.120}$$

Now, for $K = 8$, we have

$$T(p) = \frac{-8(p+1)^3}{(p+3)(p^2+3)} \tag{2.121}$$

and oscillation occurs at $\omega = \sqrt{3}$. In general, we can factor the denominator of (2.120) and obtain

$$T(p) = \frac{K(p+1)^3}{(p+\alpha)[p^2 + (3-\alpha)p + 3 - \alpha(3-\alpha)]} \tag{2.122}$$

where

$$K = \alpha^3 - 3\alpha^2 + 3\alpha - 1 \tag{2.123}$$

and

$$Q \approx \frac{\sqrt{3 - \alpha(3-\alpha)}}{3-\alpha} \tag{2.124}$$

or

$$\alpha^2 - 3\,\frac{2Q^2-1}{Q^2-1}\,\alpha + 3\,\frac{3Q^2-1}{Q^2-1} \approx 0 \tag{2.125}$$

If we now determine the value of α required for any specified Q and substitute this value into (2.123) to find K, we have a simple design for a high-Q pair of poles, and of course one real pole at $p = -\alpha$. We must choose the value of α such that $\alpha < 3$, if we use (2.124) to find Q, as can be seen from (2.122). This network is only of value if it offers reduced sensitivity, considering that three capacitors are required and that higher gain is necessary as compared to the positive-gain structure. Since we are dealing with a cubic here, the computation of the sensitivity is not quite as simple as in the previous cases. The result of this calculation, however, shows that

$$\operatorname{Re} S_K{}^{p_i} \approx -S_K{}^Q \approx -\tfrac{1}{2}Q \tag{2.126}$$

This is an improvement by about a factor of 6 compared to the positive-gain VCVS networks (for equal capacitors) discussed earlier. The poles reach the $j\omega$ axis at $\omega = \sqrt{3}$, from (2.121). The total C is 3 farads and the total RC product is 9 sec. This compares to a total RC product of 4 sec for the simple Sallen and Key structure whose poles cross the $j\omega$ axis at $\omega = 1$. We therefore pay the price of $\frac{9}{4}\sqrt{3} = 3.89$ times more capacity for a given high-Q pole position in order to achieve this reduction in sensitivity. The gain required is never more than 8 and therefore

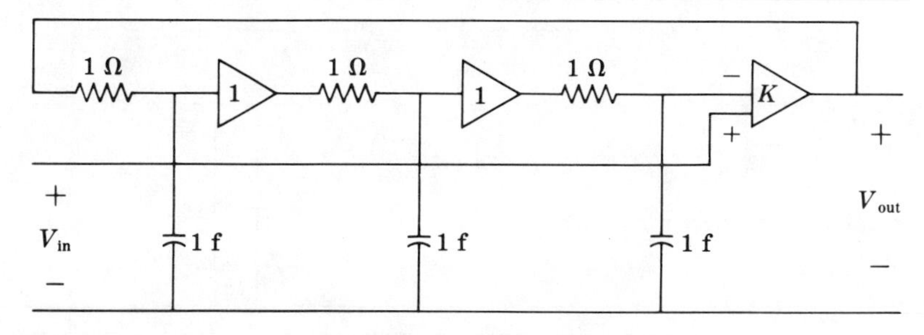

FIG. 2.42 Third-order phase-shift low-pass filter section.

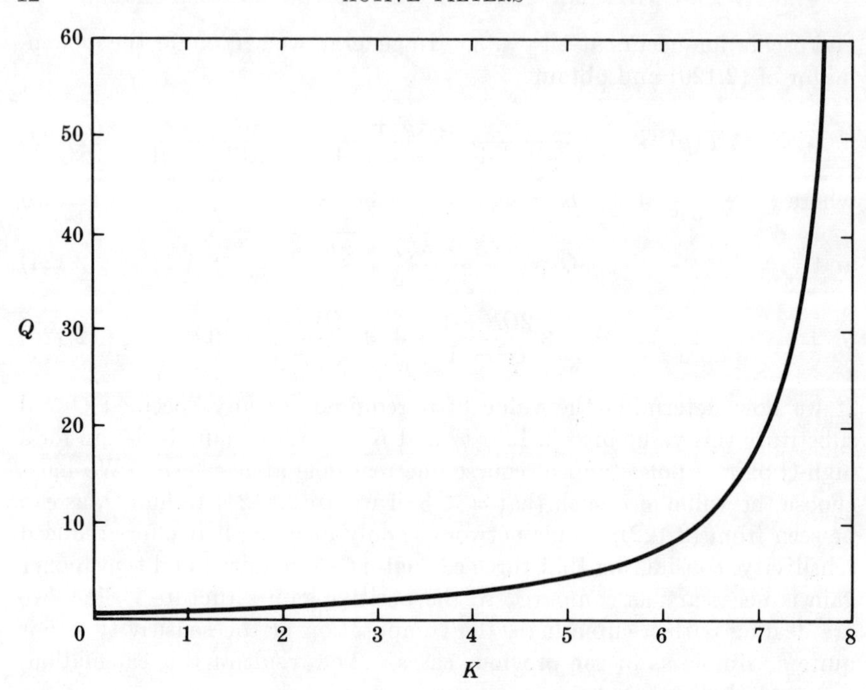

FIG. 2.43 Q vs. gain for the network of Fig. 2.42.

the high-frequency capability of this network is excellent. The primary limitation is that imposed by the sensitivity of the system to gain change. Figure 2.43 shows a plot of Q vs. amplifier gain for this network. A more advantageous connection (which does not change the Q, the gain required, or the pole positions) is to modify the input so that a zero is produced at the origin and a zero at infinity. Figure 2.44 shows this modification. It is always useful to consider a given topology, whose poles are fixed, from the standpoint of the various zeros that can be

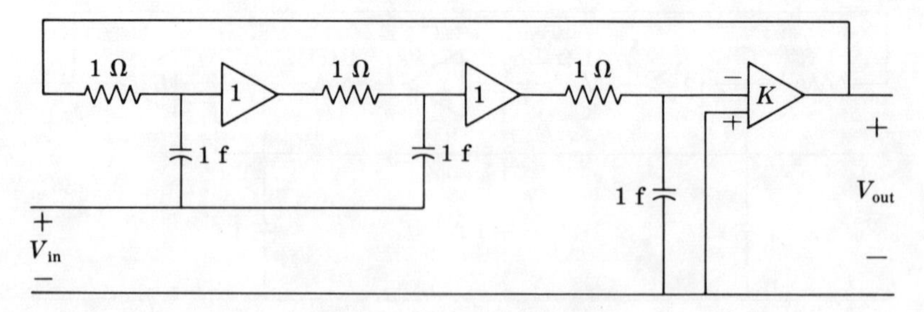

FIG. 2.44 Alternate input for the network of Fig. 2.42.

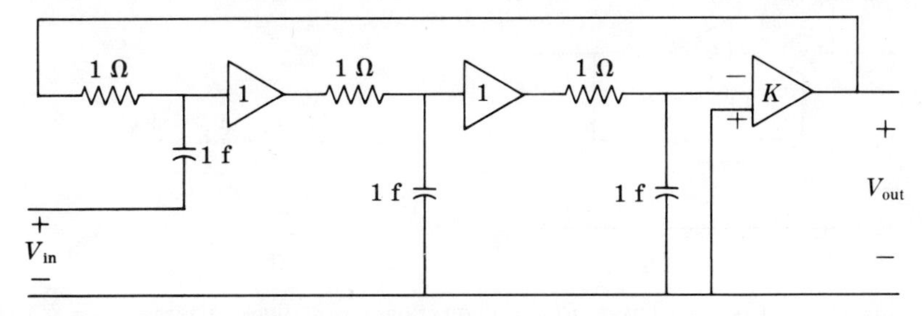

Fig. 2.45 Alternate input for the network of Fig. 2.42.

obtained. A further advantage of the network of Fig. 2.44 is that a single-input amplifier is all that is now required. Of course, two zeros at infinity can easily be obtained as shown in Fig. 2.45.

Third-order Networks Using a Positive-gain VCVS. Another useful third-order system uses a positive-gain VCVS in which the VCVS is located at the output. This network is shown in Fig. 2.46. The transfer function when the VCVS gain $K = 1$ is

$$T(p) = \frac{1}{R_1C_1R_2C_2R_3C_3p^3 + [R_1C_1R_3C_3 + (R_1 + R_2)C_2R_3C_3 + R_1C_1R_2C_3]p^2 + [R_1C_1 + (R_1 + R_2 + R_3)C_3]p + 1} \tag{2.127}$$

The low-pass third-order Butterworth (maximally-flat magnitude) function is obtained ($\omega_{3db} = 1$ rad/sec) when $R_1 = 1$ ohm, $R_2 = R_3 = 2$ ohms, $C_1 = 1.21$ farads, $C_2 = 1.30$ farads, $C_3 = 0.158$ farad. Another combination giving third-order maximally-flat magnitude response and using a unity-gain VCVS is $R_1 = R_2 = R_3 = 1$ ohm, $C_1 = 1.39$ farads, $C_2 = 3.54$ farads, $C_3 = 0.203$ farad. The optimum-transient-response third-order function

$$T(p) = \frac{1}{p^3 + 1.75p^2 + 2.15p + 1} \tag{2.128}$$

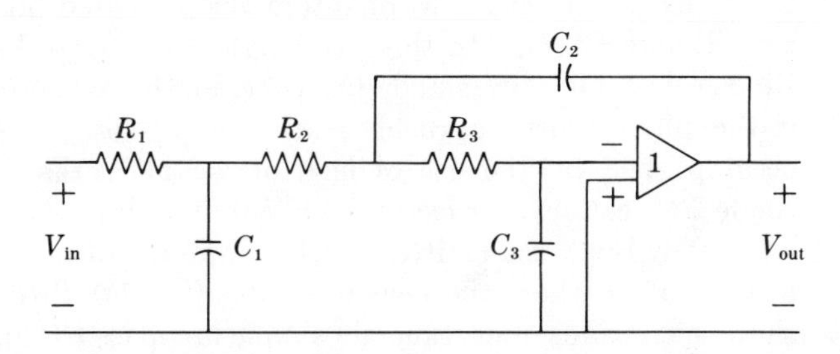

Fig. 2.46 Positive-gain VCVS third-order low-pass filter section.

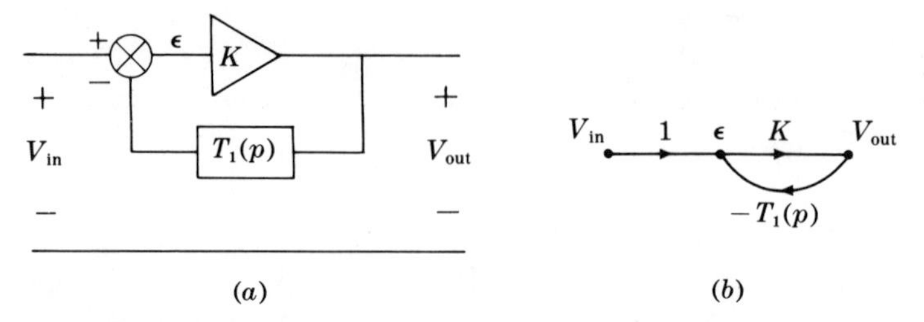

Fig. 2.47 General active *RC* configuration. (*a*) Block diagram; (*b*) flowgraph.

is obtained for $R_1 = 1$ ohm, $R_2 = R_3 = 2$ ohms, when $C_1 = 1.52$ farads, $C_2 = 1.30$ farads, $C_3 = 0.126$ farad, for $K = 1$ as before. This function is down 3 db at $\omega = 1.03$ rad/sec.

Since these networks use a positive-gain VCVS, the real part of the pole-position sensitivity is proportional to Q and thus they are limited practically to the low-Q region. In addition, if a unity-gain VCVS is used as in the examples above, the sensitivity is proportional to Q^2 so that the Q should be reasonably low, which is the case for both of the above functions.

VCVS Synthesis Using Negative-sigma-axis Phantom Zeros. A network consisting of a forward transmission path K and a feedback path $T_1(p)$ as shown in Fig. 2.47*a* and *b* has the overall transfer function

$$T(p) = \frac{K}{1 + KT_1(p)} \tag{2.129}$$

The zeros of transmission of $T_1(p)$, which do not appear in $T(p)$, are termed the phantom zeros of the system. As the gain K approaches infinity, these phantom zeros become the system poles and serve to terminate the root loci. This termination allows us to tailor the system sensitivity and stability as desired. Bialko has used the general configuration of a phase-shift oscillator in which the pole loci do not cross the $j\omega$ axis, but return to phantom zeros located on the negative-sigma axis (Bialko 1967). In this case, high-Q poles can be obtained in which Re $S_K{}^{p_i}$ is arbitrarily small (however, at the expense of higher Im $S_K{}^{p_i}$) if the phase shift approaches arbitrarily close to 180°. This can be accomplished by the use of lead or lag networks in combination with single *RC* sections, for example, in combination with an amplifier having a partially bypassed emitter resistor or an operational amplifier operating in the region where the response falls off at 20 db/decade. Figure 2.48 shows a cascade connection of a simple low-pass *RC* and two lag (integral) networks with unity-gain amplifiers used for isolation. This isolation is quite practical since the required gain and the total capacity are both

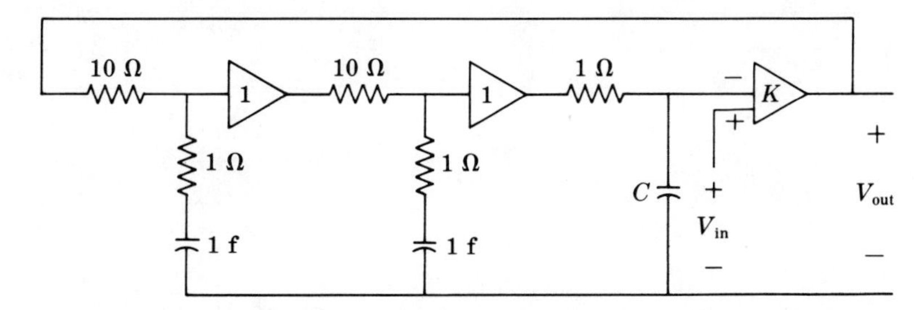

FIG. 2.48 Negative-sigma-axis phantom-zero network.

reduced. For convenience in analysis we will assume that the isolation is ideal. The transfer function of this network is

$$T(p) = \frac{K}{1 + K(p+1)^2/(11p+1)^2(Cp+1)} \tag{2.130}$$

A root-locus plot of this function is shown in Fig. 2.49, in which C is varied to change the root loci. As can be seen, the complex root loci approach the $j\omega$ axis and then reverse direction (ultimately returning to the negative-sigma axis). Note that a constant-Q line is tangent to

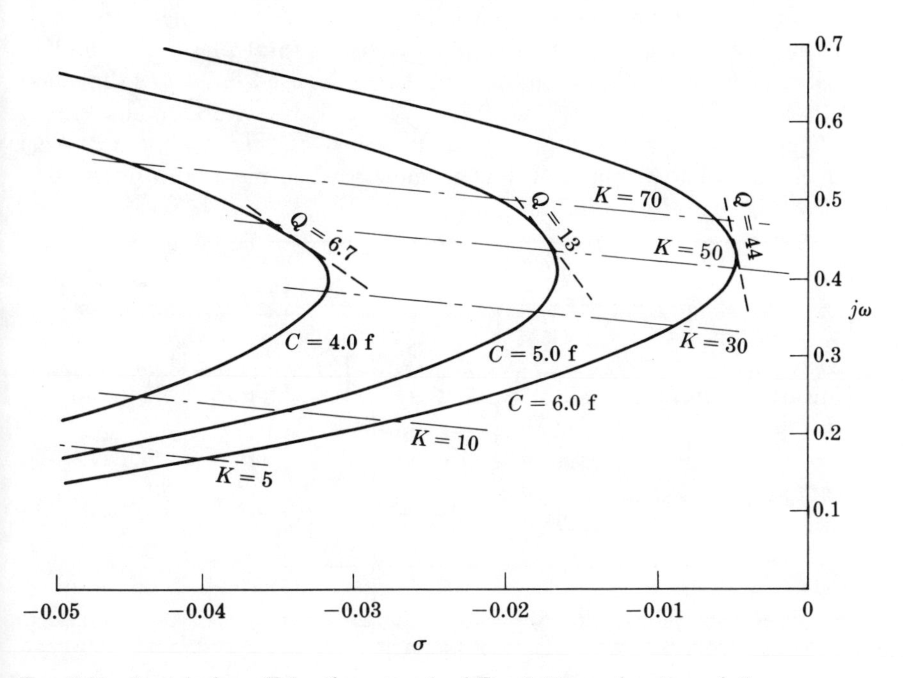

FIG. 2.49 Root loci vs. K for the network of Fig. 2.48 as a function of C.

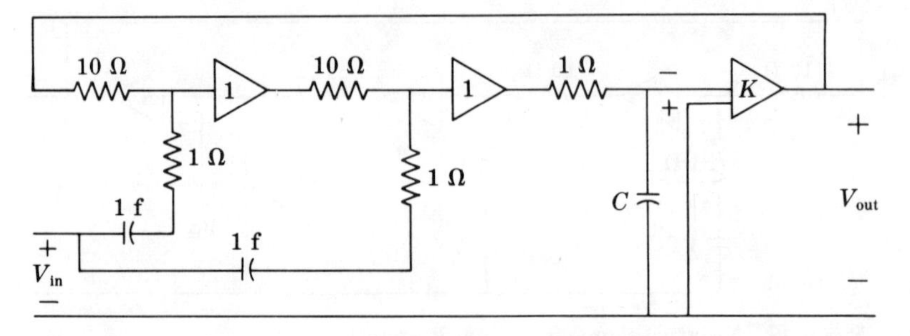

FIG. 2.50 Alternate input for the network of Fig. 2.48.

the root locus at the highest Q point and the Q is independent of amplifier gain at this point! The sensitivity to the passive components is increased to a considerable extent as can be seen from Fig. 2.49 by noting the Q change as a function of C. This is the price paid for low-Q sensitivity to amplifier gain. A simple modification to this network involves changing the zeros so as to obtain a zero at the origin and a zero at infinity as shown in Fig. 2.50. Of course, only a single-input amplifier is now needed.

A network of the type shown in Fig. 2.50 is shown in Fig. 2.51 with the measured performance given in Fig. 2.52. A convenient approach to the design of these networks is to consider total phase shift and to realize that any combination of elements can be used so long as the phase approaches but does not reach 180°. Two lead networks and a simple RC high-pass is another possibility as shown in Fig. 2.53. A typical set of normalized element values is shown. The maximum phase as a

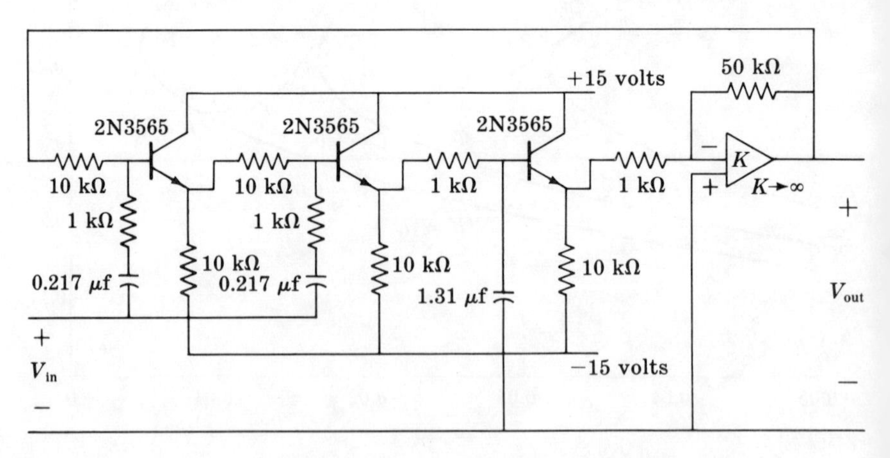

FIG. 2.51 Test circuit using phantom zeros on the negative-sigma axis.

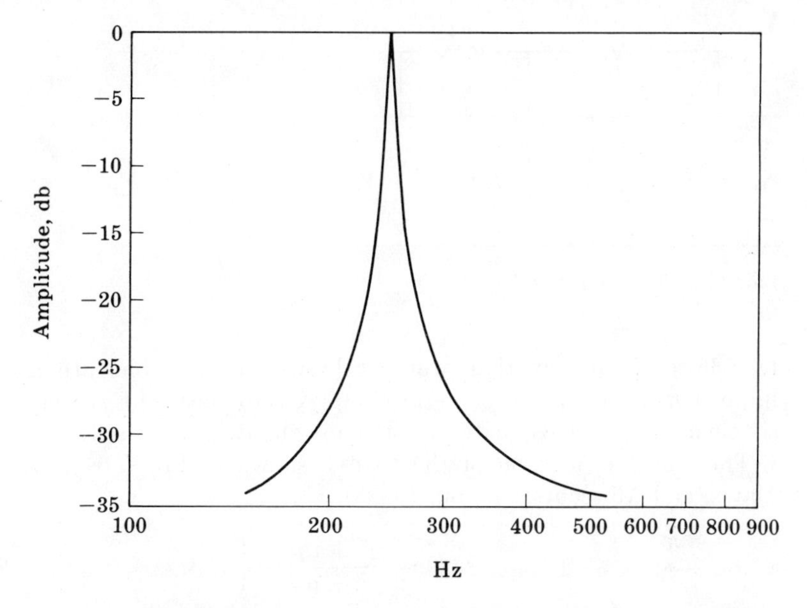

FIG. 2.52 Measured performance of the circuit shown in Fig. 2.51.

function of the ratio of the break frequency of a lead network is given by Joyce and Clarke (1961, p. 126). A plot of the phase vs. frequency as a function of the ratio of the break frequencies is also given (p. 122).

VCVS Synthesis Using Complex Phantom Zeros. Hakim has discussed active *RC* synthesis using left-half-plane phantom zeros (including the $j\omega$ axis) (Hakim, December 1965), as also have Bachmann (1959) and Horowitz (1960), and complex phantom zeros have been used in feedback-amplifier design (Ghausi and Pederson 1961). In this case, high Q can only be obtained with high gain and the method is thereby restricted

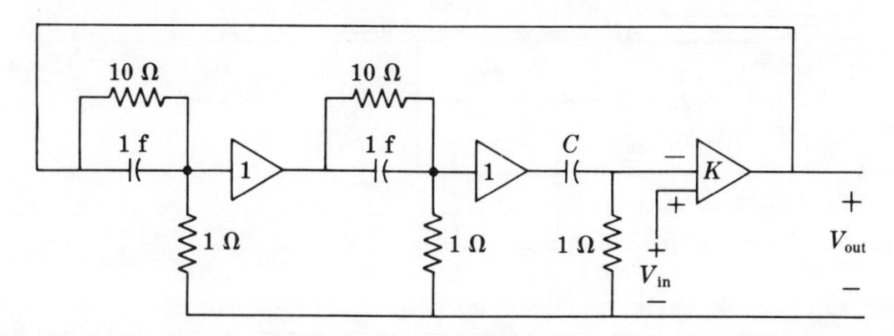

FIG. 2.53 Lead networks used to produce phantom zeros on the negative-sigma axis.

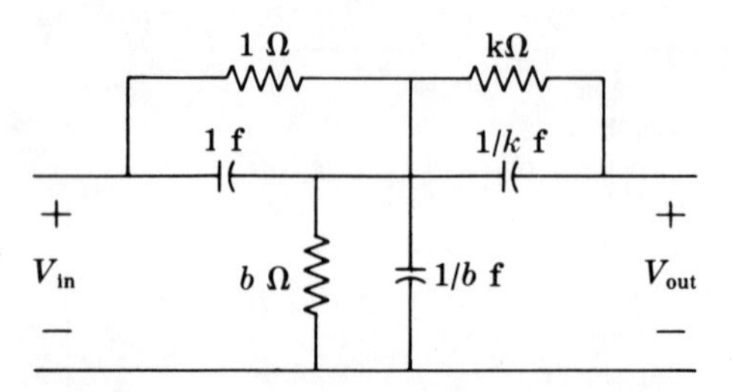

FIG. 2.54 Twin-T network.

to a lower frequency than would otherwise be possible. In addition, lumped RC networks capable of complex zeros near the $j\omega$ axis require six elements (three capacitors). Usually the twin-T is used.

The transfer function of the twin-T shown in Fig. 2.54 is (Landee, Davis, and Albrecht 1957, pp. 16–23)

$$T(p) = \frac{p^2 + \alpha p + 1}{p^2 + \beta p + 1} \tag{2.131}$$

where

$$\alpha = b + \frac{b}{k} - 1 \tag{2.132}$$

$$\beta = b + \frac{b}{k} + \frac{1}{k} + \frac{1}{b} \tag{2.133}$$

Figure 2.55 shows a typical network structure using the twin-T, in which the transmission zeros are located on the $j\omega$ axis. The transfer function for this network is ($K > 0$)

$$T(p) = \frac{K(p^2 + \beta p + 1)}{(1 + K)\{p^2 + [(\beta + \alpha K)/(1 + K)]p + 1\}} \tag{2.134}$$

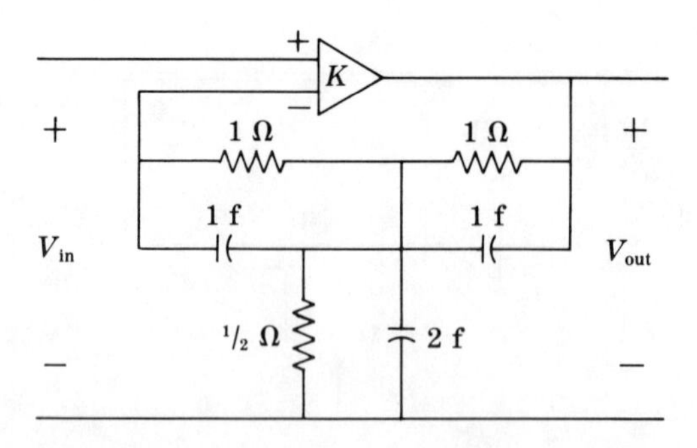

FIG. 2.55 Network using a twin-T to produce $j\omega$-axis phantom zeros.

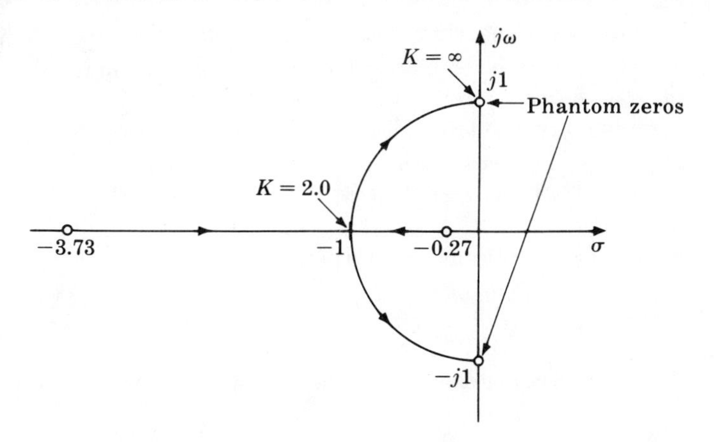

FIG. 2.56 Root locus plot vs. K of the network of Fig. 2.53.

Figure 2.56 shows the root-locus plot vs. amplifier gain for the case in which the phantom zeros are located on the $j\omega$ axis (that is, $\alpha = 0$), for $k = 1$, $b = \frac{1}{2}$. Because of the inherent frequency restrictions in the use of this method (due to the need for high amplifier gain), and because the pole-position sensitivity to amplifier gain is generally *very* low, it is frequently possible to reduce the amplifier gain required by increasing the pole-position sensitivity. This can be achieved by placing the phantom zeros in the right-half plane (Kerwin and Shaffer 1968).

If we choose, for example, $b = \frac{1}{2}$, $k = 2$, for the element values of the twin-T, Fig. 2.54 [for which, from (2.132) and (2.133), $\alpha = -\frac{1}{4}$, $\beta = 3\frac{1}{4}$], so as to place the zeros in the right-half plane, we obtain for the network of Fig. 2.57 the following transfer function

$$T(p) = \frac{[K/(1+K)](p^2 + 3.25p + 1)}{p^2 + [(3.25 - 0.25K)/(1+K)]p + 1} \tag{2.135}$$

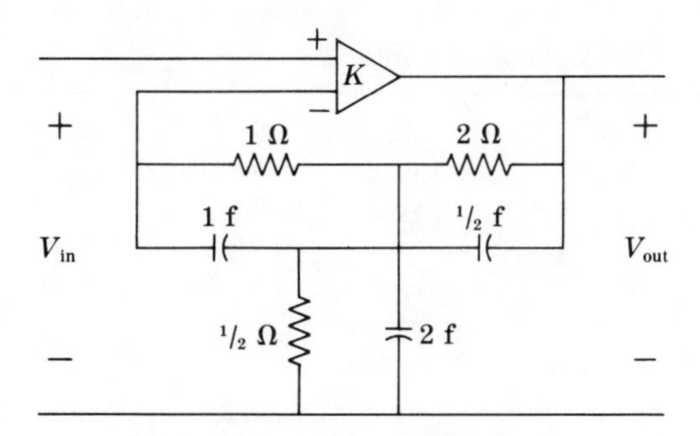

FIG. 2.57 Network using a twin-T to produce right-half-plane complex phantom zeros.

We now have the very desirable property of a direct reduction in the p coefficient as K increases due both to the denominator term $(1 + K)$ and to the subtraction which occurs in the numerator. In addition, there is no frequency-scaling effect as K changes (since the root locus is circular). The root-locus plot for this system is shown in Fig. 2.58. Note that oscillation occurs at $3.25 - 0.25K = 0$, or $K = 13$. For the transfer function (2.134), the pole-position sensitivity to changes in K [α, β as in (2.132) and (2.133)] is

$$S_K{}^{p_i} = \frac{K(\alpha - \beta)}{(\beta + \alpha K)(1 + K)} + j\frac{K(\alpha - \beta)(\beta + \alpha K)}{(1 + K)(\beta + \alpha K)^2 + 4(1 + K)^3} \tag{2.136}$$

For the preceding example, Fig. 2.57, we thereby obtain for $K = 12$

$$S_K{}^{p_i} = -12.95 + j0.92 \times 10^{-4} \tag{2.137}$$

and

$$Q \approx \frac{1 + K}{\beta + \alpha K} = 52 \tag{2.138}$$

Note the almost complete insensitivity of center frequency to changes in K. If we compare this result to a positive-gain VCVS system (Fig. 2.1), we find that for $R = 10$ ohms, $C = 0.1$ farad, $K = 2.08$, for which $Q = 52$

$$\text{Re } S_K{}^{p_i} = -104 \tag{2.139}$$

The phantom-zero synthesis method has therefore reduced the pole-position sensitivity by a factor of 8 in this case, while still maintaining a reasonably low VCVS gain. The pole-position sensitivity to changes in the passive elements is increased when complex phantom zeros are used and should be carefully checked in any specific case to select the best compromise between active- and passive-element sensitivities.

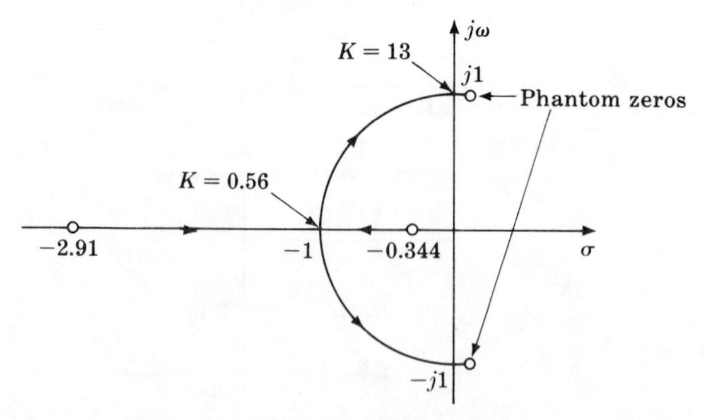

FIG. 2.58 Root-locus plot vs. K of the network of Fig. 2.57.

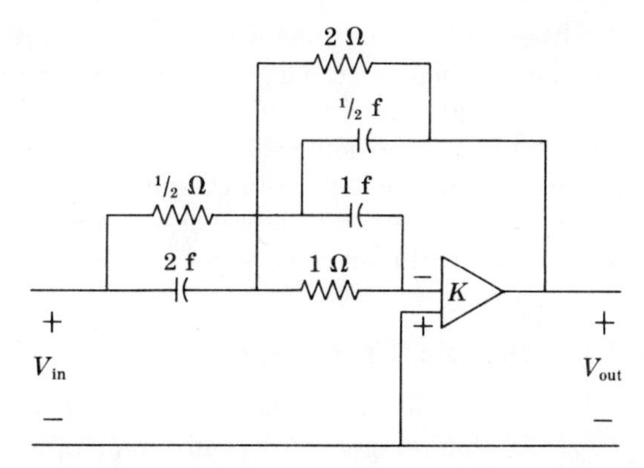

Fig. 2.59 Modified right-half-plane phantom-zero network.

It is a simple matter to select α (by appropriate choice of b, k) of (2.132) to place the phantom zeros so as to increase or decrease the required gain for any given Q. Of course, the sensitivity is inversely affected. The gain required for any given Q as a function of α, β is

$$K \approx \frac{\beta Q - 1}{1 - \alpha Q} \tag{2.140}$$

The resulting pole-position sensitivity as a function of α, β, K is then given directly by (2.136). If the result is not satisfactory, a new value of α can be selected and a second iteration made.

This network can also be modified so as to use a single-ended amplifier and so as to produce a zero at the origin and a zero at infinity as shown in Fig. 2.59 for the network of Fig. 2.57 without changing the pole positions. The result is then a true second-order bandpass function.

Summary. The use of phantom zeros is a powerful tool in modifying the sensitivity to variation in both the active and passive elements. The use of negative-sigma-axis zeros provides absolutely stable systems requiring moderate gain, but having nearly zero pole-position sensitivity to gain change with increased sensitivity to passive-element variation. In addition, the normal operational amplifier roll-off (20 db/decade) can be incorporated directly into the synthesis procedure. In this way, an amplifier having a gain of only 50 at 10 MHz, for example, and falling at 6 db/octave at this point, is capable of high-Q performance and ± 20 percent change in gain is relatively insignificant. The system is much more sensitive to the passive elements in this case, but in general these are more controllable. An initial trim adjustment should be sufficient with stable RC components.

The use of *complex* phantom zeros also provides a network configuration having high Q stability to changes in amplifier gain, but in addition provides high stability of frequency to changes in gain. The gain required is high for left-half or $j\omega$-axis complex phantom zeros, but if the phantom zeros are located in the right-half plane, this gain can be greatly reduced (at the expense, however, of increased pole-position sensitivity to VCVS gain and passive-element variation).

2.3 Synthesis Using Integrators

The phantom-zero method of active *RC* synthesis has made it possible to realize high-Q poles with only moderate sensitivity to amplifier-gain change, but at the expense of higher-gain requirements or increased sensitivity to the passive elements. The number of capacitors is not a minimum, however, and the sensitivities are still not as small as might be desired. Good (1957) used integrators to design an oscillator, as did Sutcliffe (1964). The first use of integrators specifically in the design of active *RC* networks was apparently that of Somerville and Tomlinson (1962), and later Prabhaker (1963). Certain results from state-variable theory (Kalman 1963) give a general transfer-function synthesis configuration which for linear systems uses a minimum number of capacitors and has extremely low sensitivity to both the passive and the active elements (Kerwin, Huelsman, and Newcomb 1966; Kerwin, Huelsman, and Newcomb 1967; Geffe 1967). Figure 2.60 shows a flowgraph of a second-order system realizing

$$T(p) = \frac{a_3p^2 + a_2p + a_1}{p^2 + b_2p + b_1} \tag{2.141}$$

Although extension to higher order is possible (Kalman 1963; Kerwin, Huelsman, and Newcomb 1967), this is not a practical approach for the synthesis of high-order systems since the pole positions of a high-order

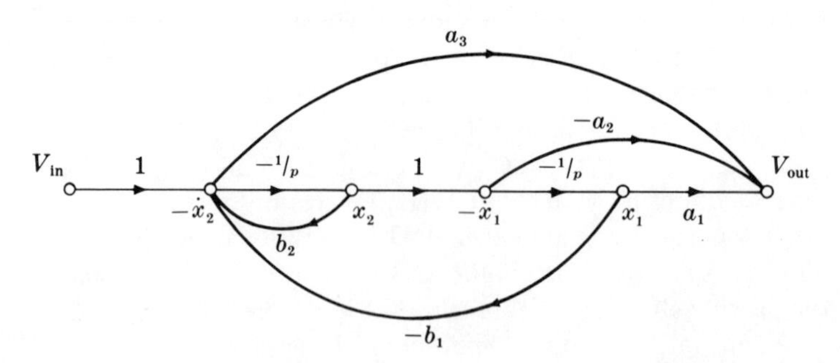

FIG. 2.60 State-variable flowgraph of a general second-order system.

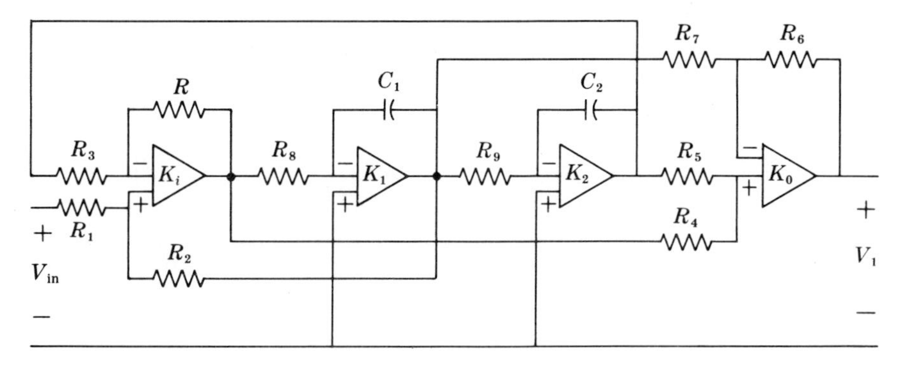

FIG. 2.61 State-variable synthesis of a general second-order function.

polynomial are very sensitive to the polynomial coefficients. Any method, then, in which specific element values determine a particular coefficient, as would be true here, is in difficulty. This problem can be circumvented if we restrict our consideration to the second-order system, as we have done in previous realizations; however, in the previous cases, the reason was primarily minimization of pole-position sensitivity to active-element variation, whereas here it is to minimize the pole-position sensitivity to passive-element variation as well. Figure 2.61 shows the complete schematic for the general second-order system using conventional negative-gain integrators. This method uses a minimum number of capacitors; in fact, it is the only lumped RC element network capable of realizing the completely general second-order system with only two capacitors. Analysis of this network (all gains are assumed infinite) results in the following transfer function

$$\frac{V_1}{V_{in}} = T_1(p) = \frac{R_2(R + R_3)R_5(R_6 + R_7)}{(R_1 + R_2)R_3(R_4 + R_5)R_7} \frac{R_8C_1R_9C_2p^2 + [(R_4 + R_5)R_6/R_5(R_6 + R_7)]R_9C_2p + R_4/R_5}{R_8C_1R_9C_2p^2 + [R_1(R + R_3)/(R_1 + R_2)R_3]R_9C_2p + R/R_3} \quad (2.142)$$

Upon normalization ($R_1 = R_3 = R_5 = R_6 = 1$ ohm, $R_8C_1 = R_9C_2 =$ 1 sec), (2.142) reduces to

$$T_1(p) = \frac{R_2(1 + R)(1 + R_7)}{(1 + R_2)(1 + R_4)R_7} \frac{p^2 + [(1 + R_4)/(1 + R_7)]p + R_4}{p^2 + [(1 + R)/(1 + R_2)]p + R} \quad (2.143)$$

The radial distance to the zeros is controlled by a single resistor R_4, and after setting R_4, the horizontal position of the zeros is controlled directly by R_7. A sign reversal at the output summer to make the numerator p coefficient negative is all that is necessary to move the zeros into the right-half plane. The frequency of maximum response is determined by

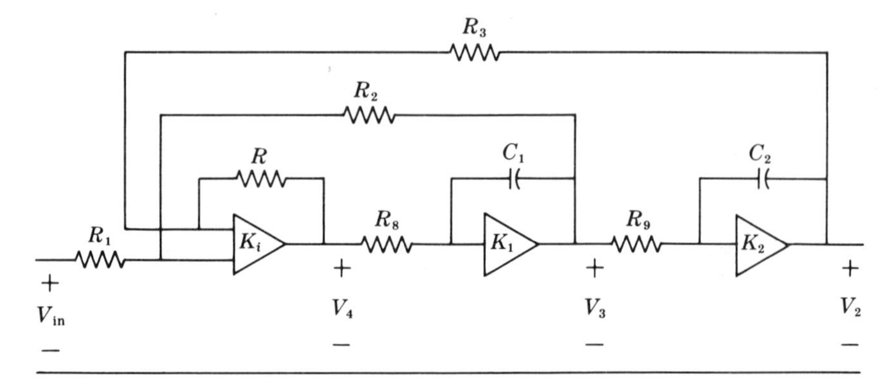

FIG. 2.62 State-variable synthesis of a low-pass, bandpass, and high-pass second-order function.

R, and after setting R, the Q is determined by R_2. R_2 does not affect the frequency of maximum response or the position of the zeros. The most important feature of this method of synthesis, however, is the absence of any subtractions in either the numerator or denominator p coefficient. The pole- and zero-position sensitivity to passive-element change is therefore minimized.

The network of Fig. 2.61 can be simplified as shown in Fig. 2.62 for low-pass, bandpass, or high-pass second-order functions, all of which are available simultaneously. The transfer functions are

$$\frac{V_2}{V_{in}} = T_2(p) = \frac{R_2(R + R_3)}{(R_1 + R_2)R_3} \frac{R_9C_2}{R_8C_1R_9C_2p^2 + [R_1(R + R_3)/(R_1 + R_2)R_3]R_9C_2p + R/R_3} \tag{2.144}$$

$$\frac{V_3}{V_{in}} = T_3(p) = -\frac{R_2(R + R_3)}{(R_1 + R_2)R_3} \frac{R_9C_2p}{R_8C_1R_9C_2p^2 + [R_1(R + R_3)/(R_1 + R_2)R_3]R_9C_2p + R/R_3} \tag{2.145}$$

$$\frac{V_4}{V_{in}} = T_4(p) = \frac{R_2(R + R_3)}{(R_1 + R_2)R_3} \frac{R_8C_1R_9C_2p^2}{R_8C_1R_9C_2p^2 + [R_1(R + R_3)/(R_1 + R_2)R_3]R_9C_2p + R/R_3} \tag{2.146}$$

The bandpass center frequency (2.145) is

$$\omega_0 = \sqrt{\frac{R}{R_8C_1R_9C_2R_3}} \tag{2.147}$$

and the amplitude $|T(p)|$ at center frequency is

$$|T(p)|_{j\omega_0} = \frac{R_2}{R_1} \tag{2.148}$$

This is particularly advantageous in integrated-circuit synthesis since only the ratio of the resistors is important. The normalized low-pass, bandpass, and high-pass transfer functions ($R_1 = R_3 = R_5 = R_6 = 1$ ohm, $R_8C_1 = R_9C_2 = 1$ sec) are

$$\frac{V_2}{V_{\text{in}}} = T_2(p) = \frac{R_2(1+R)}{(1+R_2)} \frac{1}{p^2 + [(1+R)/(1+R_2)]p + R} \tag{2.149}$$

$$\frac{V_3}{V_{\text{in}}} = T_3(p) = -\frac{R_2(1+R)}{(1+R_2)} \frac{p}{p^2 + [(1+R)/(1+R_2)]p + R} \tag{2.150}$$

$$\frac{V_4}{V_{\text{in}}} = T_4(p) = \frac{R_2(1+R)}{(1+R_2)} \frac{p^2}{p^2 + [(1+R)/(1+R_2)]p + R} \tag{2.151}$$

with $\omega_0 = \sqrt{R}$ in all cases and $|T_3(j\omega_0)| = R_2$. These expressions for T_1, T_2, and T_3 clearly show which parameters are important for obtaining a desired pole-zero pattern. The most important property of this network, however, is the low sensitivity obtained. Analysis shows that the pole-position sensitivities are (for $R = 1$ ohm, that is, $\omega_0 = 1$ rad/sec; and $R_2 \gg 1$ ohm, that is, high Q)

$$S_R{}^{p_i} = \tfrac{1}{2}(1+j) \tag{2.152}$$

$$S_{R_1}{}^{p_i} = 1 \tag{2.153}$$

$$S_{R_2}{}^{p_i} = -1 \tag{2.154}$$

$$S_{R_3}{}^{p_i} = -\tfrac{1}{2}(1+j) \tag{2.155}$$

$$S_{R_4}{}^{p_i} = 0 \tag{2.156}$$

$$S_{R_5}{}^{p_i} = 0 \tag{2.157}$$

$$S_{R_6}{}^{p_i} = 0 \tag{2.158}$$

$$S_{R_7}{}^{p_i} = 0 \tag{2.159}$$

$$S_{R_8}{}^{p_i} = -\tfrac{1}{2}(2+j) \tag{2.160}$$

$$S_{R_9}{}^{p_i} = 0 - j\tfrac{1}{2} \tag{2.161}$$

$$S_{C_1}{}^{p_i} = -\tfrac{1}{2}(2+j) \tag{2.162}$$

$$S_{C_2}{}^{p_i} = 0 - j\tfrac{1}{2} \tag{2.163}$$

For the functions obtainable with the network of Fig. 2.61, R_4, R_5, R_6, R_7 are not used, and with the exception of the integrating resistors R_8, R_9, we find that the sum of the sensitivities of the remaining resistors R, R_1, R_2, and R_3 under high-Q conditions is *zero!*

All of the amplifier gains have been assumed infinite and therefore the transfer functions are independent of these gains; however, if we use an integrator of finite gain K (as of course must be done in practice), as shown in Fig. 2.63*a*, we obtain the following transfer function:

$$T(p) = \frac{-K}{1 + (1 + K)RCp} \tag{2.164}$$

If we now include the open-loop gains K_1, K_2 of the integrator amplifiers, as shown in Fig. 2.62, we find that the denominator of the transfer function for all outputs is

$$D(p) = R_8C_1R_9C_2p^2 + \left[\frac{1}{K_1} + \frac{1}{K_2}\frac{R_8C_1}{R_9C_2} + \frac{R_1(R + R_3)}{R_3(R_1 + R_2)}\right]R_9C_2p + \frac{R}{R_3} \tag{2.165}$$

or in normalized form ($R_8C_1 = R_9C_2 = 1$ sec, $R_1 = R_3 = 1$ ohm)

$$D_n(p) = p^2 + \left(\frac{1}{K_1} + \frac{1}{K_2} + \frac{1 + R}{1 + R_2}\right)p + R \tag{2.166}$$

The effect of finite operational amplifier gains is then to limit the maximum Q attainable. In particular, if we let R_2 approach infinity in (2.166), the p coefficient is $(1/K_1 + 1/K_2)$ and therefore the maximum Q is (for $R = 1$ ohm)

$$Q_{\max} = \frac{K_1K_2}{K_1 + K_2} \tag{2.167}$$

Another characteristic of this method of synthesis is the absolute stability

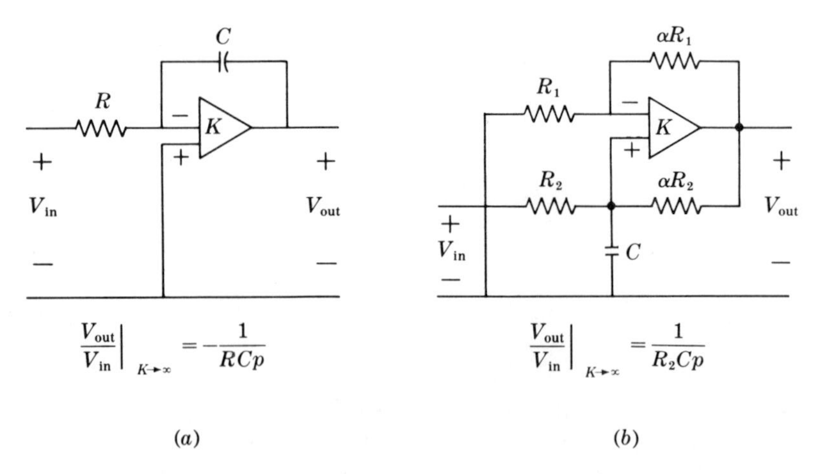

FIG. 2.63 Negative- and positive-gain integrators. (*a*) Negative gain; (*b*) positive gain.

of the system (assuming ideal integrators). We are now in a position to determine the pole-position sensitivity to changes in amplifier gains, K_1 and K_2 for (2.166). These sensitivities apply for any of the three outputs of the network of Fig. 2.62 and are

$$S_{K_1}{}^{p_i} = -\frac{Q}{K_1} + j\frac{1}{4QK_1(1 - 1/4Q^2)} \tag{2.168}$$

$$S_{K_2}{}^{p_i} = -\frac{Q}{K_2} + j\frac{1}{4QK_2(1 - 1/4Q^2)} \tag{2.169}$$

Even for very high Q values, then, we find quite practical pole-position sensitivities. For an open-loop amplifier gain of 2,000, which is not at all difficult to obtain, at least in the region below 100 kHz, we find for the maximum Q of 1,000 a sensitivity to gain change of only $-\frac{1}{2}$! The imaginary part of the sensitivity is completely negligible at high Q and high gain.

Even though a number of operational amplifiers are required when this method of synthesis is used, the present rapid reduction in the cost of operational amplifiers makes this method very attractive for most applications within a frequency range limited only by the gain-bandwidth product attainable and the Q required. Another integrator is shown in Fig. 2.63*b*. This integrator was described by Deboo (1967) and has the advantage of positive gain and a grounded capacitor; however, the resistors must be accurately balanced. The transfer function for an ideal operational amplifier having an infinite gain is

$$T(p) = \frac{1 + \alpha}{R_2Cp} \tag{2.170}$$

This integrator will provide the correct feedback signal polarity so that a dual polarity summer is no longer needed and the single polarity summation needed can be accomplished with a summing integrator. This configuration for a normalized second-order low-pass transfer function is shown in Fig. 2.64. Analysis of this network results in the flowgraph of Fig. 2.65 from which we obtain the following transfer function

$$\frac{V_{\text{out}_2}}{V_{\text{in}}} = \frac{-(1 + R_2)/R_1}{p^2 + (1/R_1)p + (1 + R_2)/2} \tag{2.171}$$

Of course, if the output is taken from the first integrator V_{out_1}, we obtain the bandpass function

$$\frac{V_{\text{out}_1}}{V_{\text{in}}} = \frac{[(1 + R_2)/R_1]p}{p^2 + (1/R_1)p + (1 + R_2)/2} \tag{2.172}$$

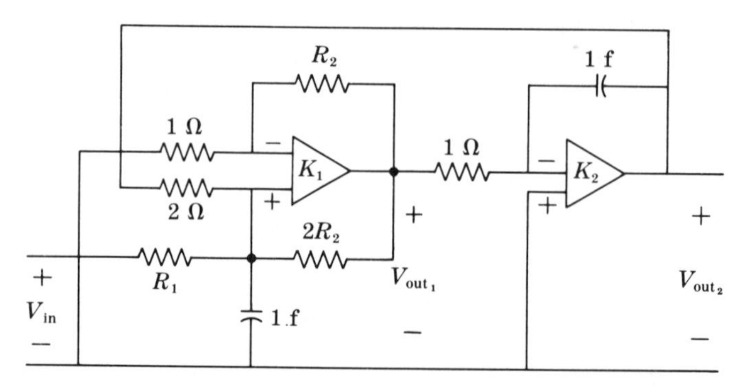

Fig. 2.64 Synthesis of a low-pass second-order function using positive- and negative-gain integrators.

The most important advantage of this network configuration (Fig. 2.64) for many applications may well be the reduction in the power required, although, of course there would also be an economic advantage.

Two additional normalized configurations are shown in Figs. 2.66 and 2.67. The transfer function for Fig. 2.66 is

$$\frac{V_{out}}{V_{in}} = \frac{-2}{p^2 + (1/R_1)p + 2/R_2} \tag{2.173}$$

and for Fig. 2.67 is

$$\frac{V_{out}}{V_{in}} = \frac{2(1 + R_2)/R_1R_2}{p^2 + (1/R_1)p + 2[2 + (1/R_1)]/R_2} \tag{2.174}$$

The use of two positive integrators is particularly advantageous since both capacitors are grounded.

For $j\omega$-axis zeros, a potentiometer in place of R_4 and R_5 (Fig. 2.61) with the output taken from the potentiometer tap allows continuous adjustment of the $j\omega$-axis zeros. Since no connection is made to the

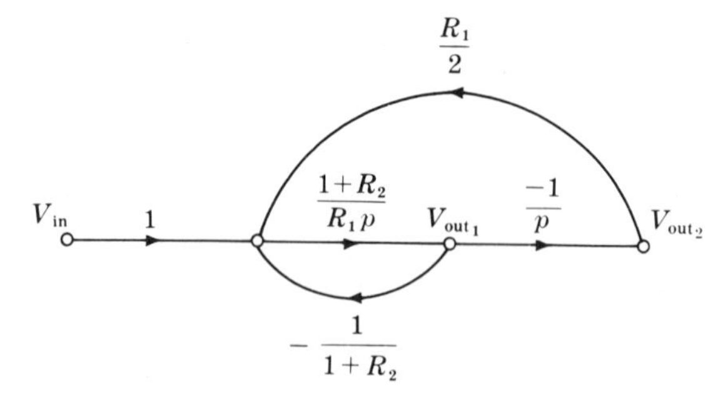

Fig. 2.65 Flowgraph for the network of Fig. 2.64.

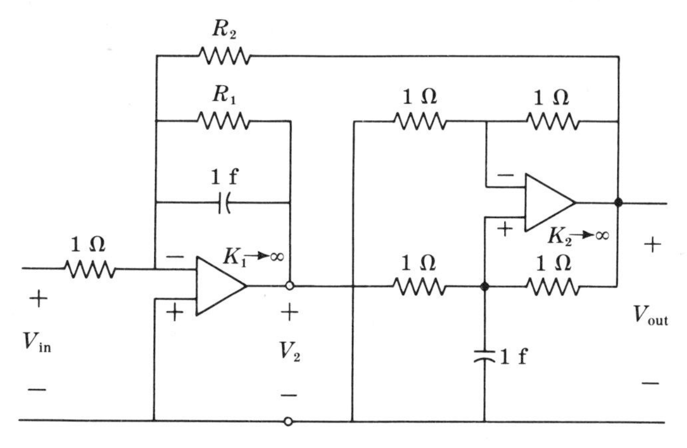

FIG. 2.66 Alternate low-pass second-order function synthesis using positive- and negative-gain integrators.

first integrator, in this case ($R_6 = 0$, $R_7 = \infty$), the p term in the numerator is zero and the zeros are on the $j\omega$ axis without regard to component tolerances or balancing of any kind. Of course, the location depends on the values of R_4 and R_5 or on the potentiometer setting.

The requirement of proper ratios of the resistors used with the positive-gain integrator is an added complication that produces a sensitivity problem at high Q. This increased sensitivity makes the negative-gain integrator much more attractive under high-Q conditions. However, some applications requiring moderate Q may be best served by the use of the positive-gain integrator realizations.

Summary. The extreme stability and ease of adjustment of both Q and frequency, of both poles and zeros, make the use of integrators a very

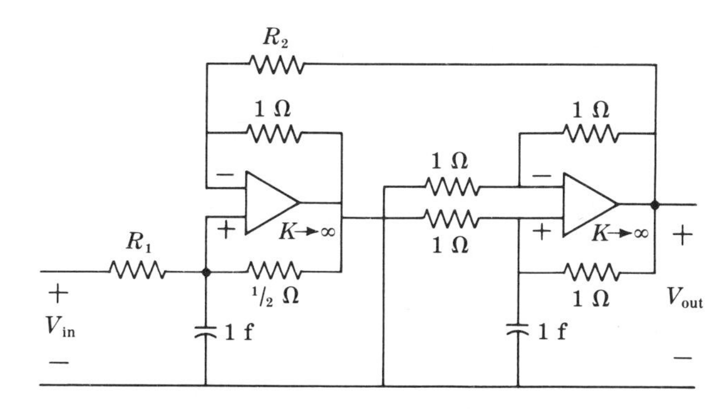

FIG. 2.67 Alternate low-pass second-order function synthesis using positive-gain integrators.

attractive method of active RC synthesis. However, the use of integrators and/or summers requires additional power due to the large number of active elements as compared to the active RC synthesis methods discussed in Sec. 2.2. Those methods are therefore more appropriate for low-Q applications. In addition, the operational amplifier gain required (at least $2Q$) limits the high-frequency performance of this synthesis method. At the present time, frequencies above 100 kHz are difficult to achieve and therefore other methods must be used for the higher-frequency regions. The complex-phantom-zero or limited-phase-shift approach (negative-sigma-axis phantom zeros) may be the best candidates for high Q for frequencies above 100 kHz.

2.4 Positive-gain VCVS Synthesis Using Distributed Elements

The positive-gain structures described earlier (Sec. 2.1) can be simplified and the gain required can be reduced if distributed RC elements are used (Happ and Riddle 1961). Figure 2.68 compares the structure of a two-pole network using lumped and distributed elements. In the case of the distributed line, Fig. 2.68*b*, the gain required is always less than 1! Oscillation occurs at a voltage gain of 0.9206. A further advantage of this realization is that since a single R and a single C are used, the system Q is independent of both R and C. Q is determined by gain alone. The peak frequency, however, is determined by the RC product and the gain.

The synthesis of distributed networks of this type is complicated by the irrational nature of the transfer function (Castro and Happ 1960; Happ and Castro 1961; Thompson 1964). If we are to compare these two

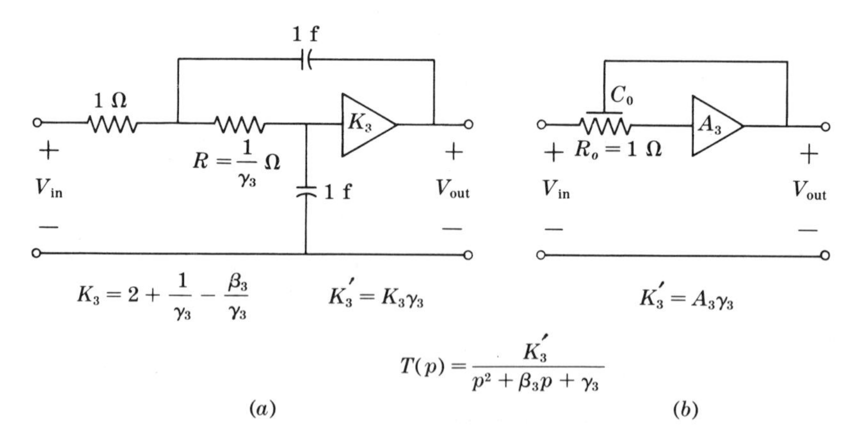

Fig. 2.68 Lumped and distributed active RC networks. (*a*) Lumped; (*b*) distributed.

structures, Fig. 2.68, we must do it on the basis of some sort of approximation, or modification, of the distributed network to obtain a rational transfer function (Heizer 1962; Barker 1965). A number of methods for describing passive distributed networks have been proposed (Kelly and Ghausi, June 1965; Wyndrum 1963). These are based on a dominant-pole or effective-dominant-pole idea. If we use an approach of this type, an equivalent pair of poles can be defined which produce approximately the same amplitude response as is obtained from the distributed *RC* active network (Kerwin 1967). The problem in this case then reduces to an analytical one of determining the amplitude response of the distributed network and then matching this response by means of a pair of equivalent poles. A digital computer program suited to the analysis of networks containing distributed, lumped, and active elements (DLA networks) is available (Kerwin 1967; Huelsman and Kerwin 1968), including the use of tapered lines (Kelly and Ghausi, March 1965).

In this case, Fig. 2.68*b*, however, the network is sufficiently simple that we can use conventional analytic methods. The uniform *RC* line of Fig. 2.69 has a Y matrix.

$$[Y] = \sqrt{\frac{Cp}{R}} \begin{bmatrix} \coth \sqrt{RCp} & -\operatorname{csch} \sqrt{RCp} \\ -\operatorname{csch} \sqrt{RCp} & \coth \sqrt{RCp} \end{bmatrix} \tag{2.175}$$

The addition of an amplifier as shown in Fig. 2.68*b* results in a system flowgraph as shown in Fig. 2.70 where the two transmissions $T_1(p)$ and $T_2(p)$ are determined by superposition; that is, $T_1(p)$ is the transmission from V_1 to V_2 with $V_3 = 0$, and $T_2(p)$ is the transmission from V_3 to V_2 with $V_1 = 0$, Fig. 2.69. The indefinite admittance matrix for this three-terminal network, Fig. 2.69, is

$$[Y] = \sqrt{\frac{Cp}{R}} \begin{bmatrix} \coth \sqrt{RCp} & -\operatorname{csch} \sqrt{RCp} & \operatorname{csch} \sqrt{RCp} - \coth \sqrt{RCp} \\ -\operatorname{csch} \sqrt{RCp} & \coth \sqrt{RCp} & \operatorname{csch} \sqrt{RCp} - \coth \sqrt{RCp} \\ \operatorname{csch} \sqrt{RCp} - \coth \sqrt{RCp} & \operatorname{csch} \sqrt{RCp} - \coth \sqrt{RCp} & 2 \coth \sqrt{RCp} - 2 \operatorname{csch} \sqrt{RCp} \end{bmatrix} \tag{2.176}$$

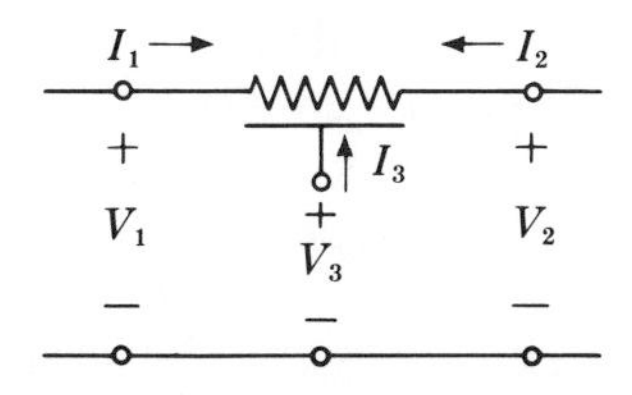

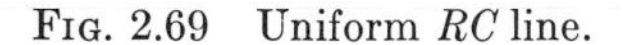

FIG. 2.69 Uniform *RC* line.

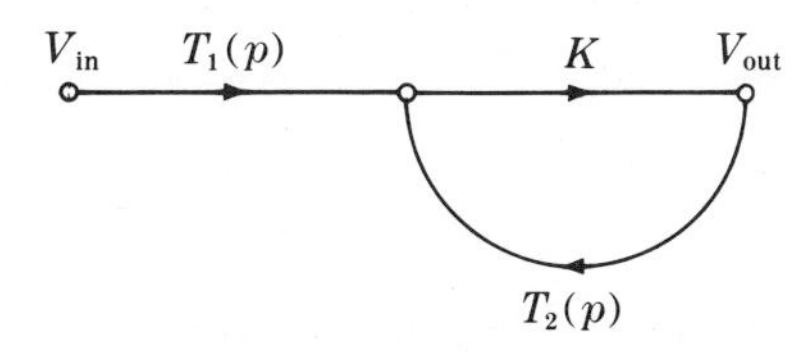

FIG. 2.70 Flowgraph of the network of Fig. 2.68*b*.

Setting $V_3 = 0$ and $I_2 = 0$, we obtain the open-circuit voltage transfer function

$$T_1(p) = \left.\frac{V_2}{V_1}\right]_{V_3=0} = \operatorname{sech}\sqrt{RCp} \tag{2.177}$$

and for $V_1 = 0$, $I_2 = 0$

$$T_2(p) = \left.\frac{V_2}{V_3}\right]_{V_1=0} \tag{2.178}$$

and since, for any three-terminal network of this kind,

$$T_2(p) = 1 - T_1(p) \tag{2.179}$$

we obtain

$$T_2(p) = \left.\frac{V_2}{V_3}\right]_{V_1=0} = 1 - \operatorname{sech}\sqrt{RCp} \tag{2.180}$$

and since from the flowgraph, Fig. 2.70,

$$\frac{V_{\text{out}}}{V_{\text{in}}} = \frac{KT_1}{1 - KT_2} \tag{2.181}$$

we obtain

$$\frac{V_{\text{out}}}{V_{\text{in}}} = \frac{K \operatorname{sech}\sqrt{RCp}}{1 - K + K\operatorname{sech}\sqrt{RCp}} = \frac{K/(1-K)}{K/(1-K) + \cosh\sqrt{RCp}} \tag{2.182}$$

From (2.182) the poles are located at

$$\frac{K}{1-K} + \cosh\sqrt{RCp} = 0 \tag{2.183}$$

and we find that oscillation occurs at a frequency $\omega = \sqrt{3RC}$ at a gain $K = 0.9206$.

In order to determine the equivalent-pole positions (2.182) is used to calculate the amplitude response. The result of a series of amplitude-response calculations and amplitude matchings is shown in Fig. 2.71, in which the equivalent-pole positions are plotted as a function of the line capacity C_0 and the amplifier gain K. The design procedure consists of entering Fig. 2.71 at the normalized value of pole position $p_j = \sigma_j + j\omega_j$ desired, reading off the value of C_0 and K required, and then scaling the network elements R_0 and C_0 to whatever impedance and frequency level are required. An important consideration is, of course, the degree of approximation involved. Since the matching is in the passband, any difference between the distributed network response and that of the two-pole function occurs in the stop-band region. A typical comparison is shown in Fig. 2.72. The values are $\sigma = -0.05$, $\omega = 0.86$ rad/sec.

The matching of this network response to a two-pole function is entirely arbitrary, but is very convenient for a synthesis procedure based on

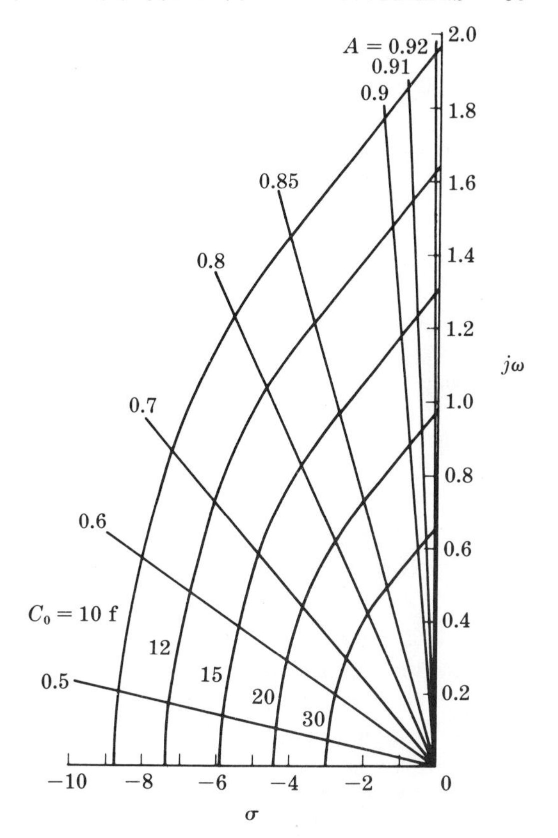

FIG. 2.71 Equivalent-pole positions for the network of Fig. 2.68*b*.

cascade connection of networks realizing first- and second-order factors. The second-order portion of a third-order Butterworth low-pass filter is approximated when $K = 0.76$. This gain can be conveniently set with a resistive divider at the output of an emitter follower, as indicated in Fig. 2.73. An advantage in this configuration is that the maximum gain is directly limited by the divider so that transistor parameter changes cannot increase the gain above $R_2/(R_1 + R_2)$. This is only important when the poles are of high Q as oscillation would occur if the voltage gain exceeded 0.92. R_1 must be kept small compared to R_0 since we have assumed a VCVS. Actually, R_1/R_0 should be less than 0.0001, or an emitter follower should be used to isolate the resistive divider and provide a voltage drive to the *RC* line as shown in Fig. 2.73. The amplitude response of the distributed, lumped, active *RC* network of Fig. 2.73 is compared with that of a third-order Butterworth function in Fig. 2.74.

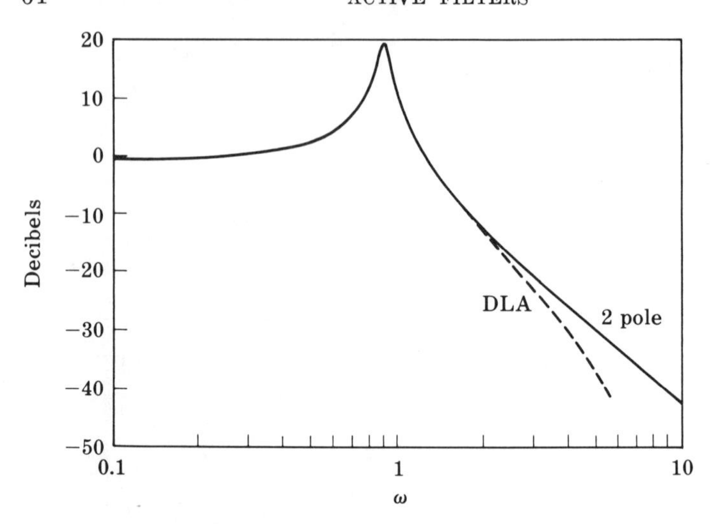

FIG. 2.72 Comparison of the response of the network of Fig. 2.68*b* and a two-pole function.

The use of this network as an oscillator is attractive because of its extreme simplicity. Another interesting feature is the independence of Q to changes in either R_0 or C_0; Q is determined by the gain K only. This allows trimming of either R_0 or C_0 to set the desired frequency without affecting the Q. This is advantageous since trimming both R_0 and C_0 in an RC distributed line is difficult. The Q sensitivity to gain change

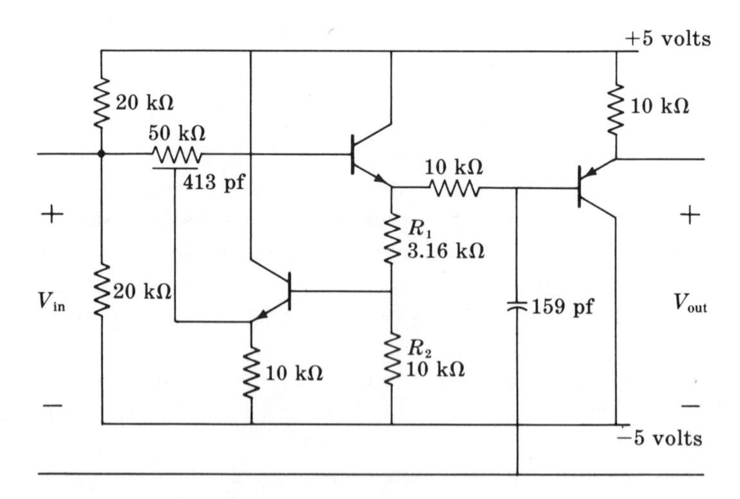

Fig. 2.73 Approximation of a third-order Butterworth low-pass filter function using a DLA network.

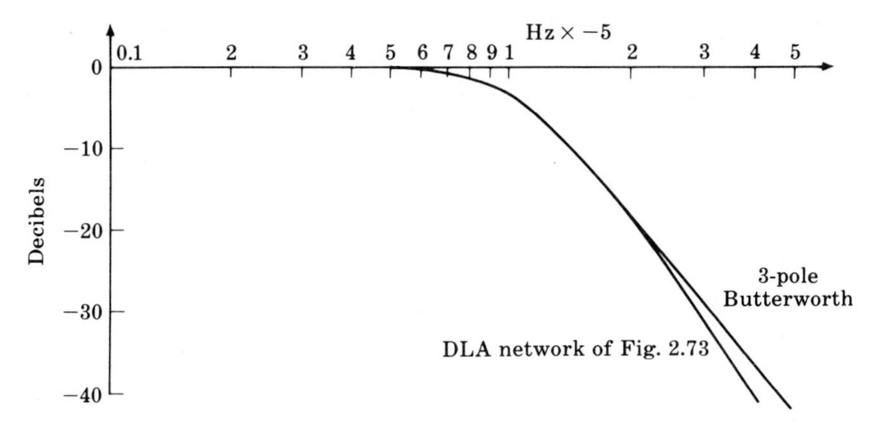

FIG. 2.74 Response of the network of Fig. 2.73 compared to the third-order Butterworth response.

is the primary disadvantage of this network. We cannot conveniently use the pole-position sensitivity in this case, but we can use an equivalent-pole-position sensitivity. The curves of Fig. 2.71 can be used to find the equivalent sensitivity since they show the equivalent-pole positions. The change in position of the pole for constant R_0 and C_0 at any gain K is indicated by the constant C_0 lines and a comparison of pole position for any two values of K establishes the percent change in sigma and the percent change in omega for the particular percent change in K. This can be seen from the figure; for example,

$$S_K{}^Q\Big]_{K=0.91} \approx 75 \qquad \text{for a } Q \text{ of } 10$$

The networks of Fig. 2.10*a*, *b*, and *c* all have zeros on the $j\omega$ axis that can be adjusted in position independently of the pole positions, and the poles can be made complex by adjustment of the amplifier gain without any effect on the zeros. These networks, however, use three or four capacitors and four or three resistors respectively to achieve a second-order function, and the gain required is near 2. The use of distributed elements allows a simplification in the number of elements and a significant reduction in the gain required.

If we consider first the passive, distributed, lumped network of Fig. 2.75

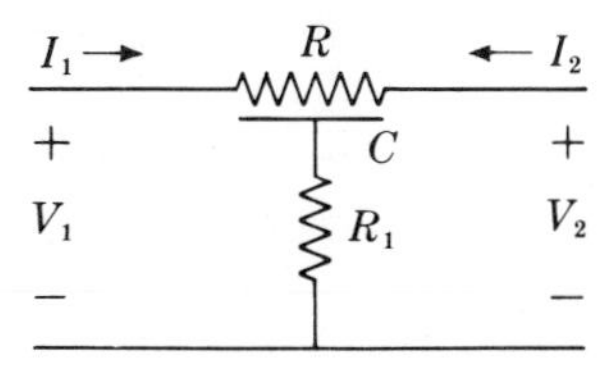

FIG. 2.75 Distributed, lumped RC network.

where R and C constitute a uniform RC line, we find that the Z matrix is

$$[Z] = \sqrt{\frac{R}{Cp}} \begin{bmatrix} \coth\sqrt{RCp} + \frac{R_1}{R}\sqrt{RCp} & \operatorname{csch}\sqrt{RCp} + \frac{R_1}{R}\sqrt{RCp} \\ \operatorname{csch}\sqrt{RCp} + \frac{R_1}{R}\sqrt{RCp} & \coth\sqrt{RCp} + \frac{R_1}{R}\sqrt{RCp} \end{bmatrix} \tag{2.184}$$

and therefore that

$$\frac{V_2}{V_1} = \frac{z_{21}}{z_{11}} = \frac{\operatorname{csch}\sqrt{RCp} + (R_1/R)\sqrt{RCp}}{\coth\sqrt{RCp} + (R_1/R)\sqrt{RCp}} \tag{2.185}$$

Expansion of (2.185), letting $p = j\omega$, and equating real and imaginary terms in the numerator to zero gives the conditions for zero transmission at a real frequency,

$$\omega = \frac{11.187}{RC} \tag{2.186}$$

$$R_1 = \frac{R}{17.798} \tag{2.187}$$

These are the conditions for the first $j\omega$-axis zero. Many others exist; however, the R_1/R ratio is different, so that only one pair of $j\omega$-axis zeros exists when $R = 17.798R_1$. Of course, there are also zeros off the $j\omega$ axis as well, and it is interesting to note that for an R_1/R ratio of 17.798, five pairs of complex zeros are present! The other four pairs of complex zeros are well removed from the $j\omega$ axis and are essentially no different in effect from the remaining negative-sigma-axis zeros. The steady-state frequency response of this network, Fig. 2.75, is described in detail by Kaufman and Garrett (1962), and the response is similar to that of the lumped-element twin-T. A comparison of the twin-T and the distributed, lumped, RC notch network is shown in Fig. 2.76*a* and *b*.

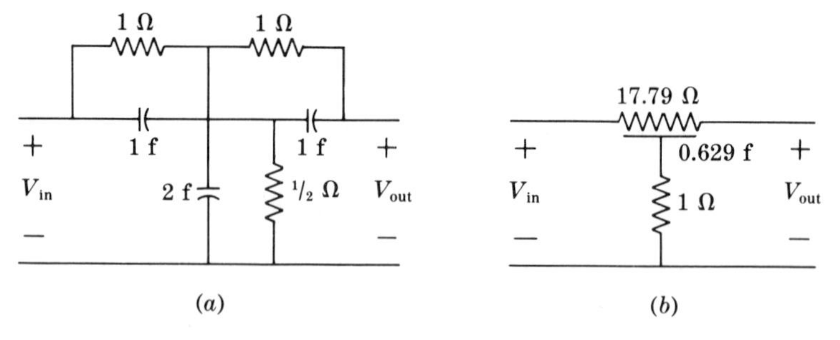

FIG. 2.76 Lumped RC and distributed RC notch networks.

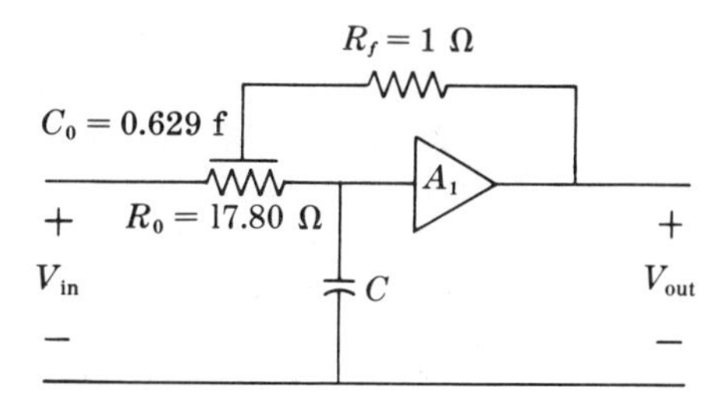

FIG. 2.77 Distributed, lumped, active *RC* network.

Figure 2.76*a* shows the conventional lumped-element twin-T network having $j\omega$-axis zeros at $\omega = \pm 1$, and Fig. 2.76*b* shows a distributed, lumped network also producing $j\omega$-axis zeros at $\omega = \pm 1$. The obvious reduction in elements is attractive, particularly the use of a single capacitor in one case. Since the total *RC* products of the two networks are similar and since the major parts of the resistance and all of the capacitance are in the distributed *RC* line, and thereby the *R* and *C* exist in the same physical location, the distributed network reduces the chip area occupied by nearly a factor of 2. If we apply feedback by means of a VCVS in a similar manner to that used in Fig. 2.10*c* to the network of Fig. 2.76*b* as shown in Fig. 2.77, we obtain a similar result to that of the lumped-element network, namely, the amplitude response is essentially the same as that of a two-complex-pole, two-$j\omega$-axis-zero rational function. The shunt capacitor in Fig. 2.77 is added to split the poles and zeros apart as was done with the lumped-element network of Fig. 2.10*c*. A significant difference, however, is the fact that the gain required for a given amplitude response is reduced by about a factor of 2. This is particularly worthwhile since most of the pole positions needed can thereby be obtained with VCVS gains near unity which is a very stable condition for the VCVS. In addition, for those cases where the gain is less than 1, the operational amplifier VCVS can be replaced with an emitter-follower VCVS. The resulting poles and zeros of the distributed, lumped, active (DLA) network of Fig. 2.77 are not as simply described as are those of Fig. 2.10*c*, but by matching the amplitude response of one to the other, we can obtain an equivalent-pole position to go with the known zeros at $\omega = \pm 1$. A computer program has been developed which allows the determination of the amplitude response of this network for various values of gain *A* and shunt capacity *C* (Huelsman and Kerwin 1968). The matching of this response to a two-pole, two-$j\omega$-axis-zero function then determines the equivalent-pole positions for many different values of *A* and *C*. A typical amplitude comparison between the DLA network response and the matched two-pole, two-zero response is shown in Fig. 2.78. The result of many such matchings produces the design chart of Fig. 2.79. In this figure, the zeros are located at $\omega = \pm 1$ and the gain A and capacity C are read directly from the figure for any particular equivalent-pole position desired. Since the zeros are located at unity to normalize the chart for use with any pole-zero combination, the transfer function specified must be transformed so that the zeros occur at $\omega = \pm 1$ before determining the pole positions. For example, consider the fifth-order elliptic low-pass filter function (five poles, four zeros) having 0.5-db passband ripple and 39.3-db stop-band ripple, shown below. This function is taken from Skwirzynski (1965). This is a particularly convenient source of filter functions in factored form for use in active RC synthesis.

$$T(p) = \frac{0.168(0.359p^2 + 1)(0.738p^2 + 1)}{(p^2 + 0.488p + 0.501)(p^2 + 0.114p + 0.808)(p + 0.416)} \tag{2.188}$$

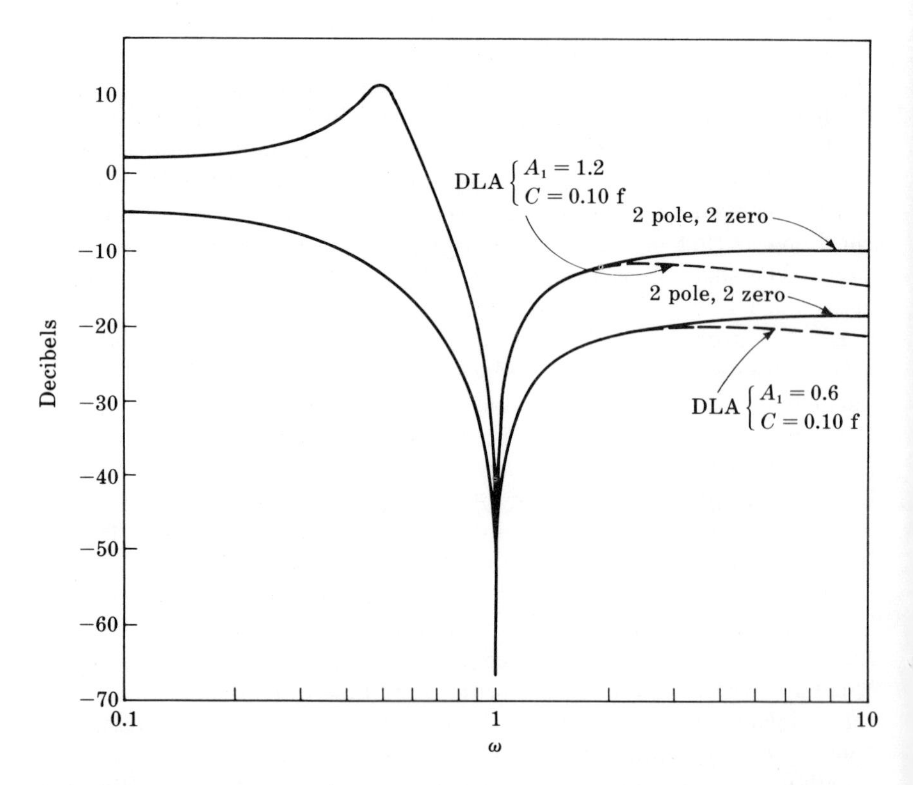

Fig. 2.78 Response of the DLA network of Fig. 2.77 compared to a two-pole, two-zero response.

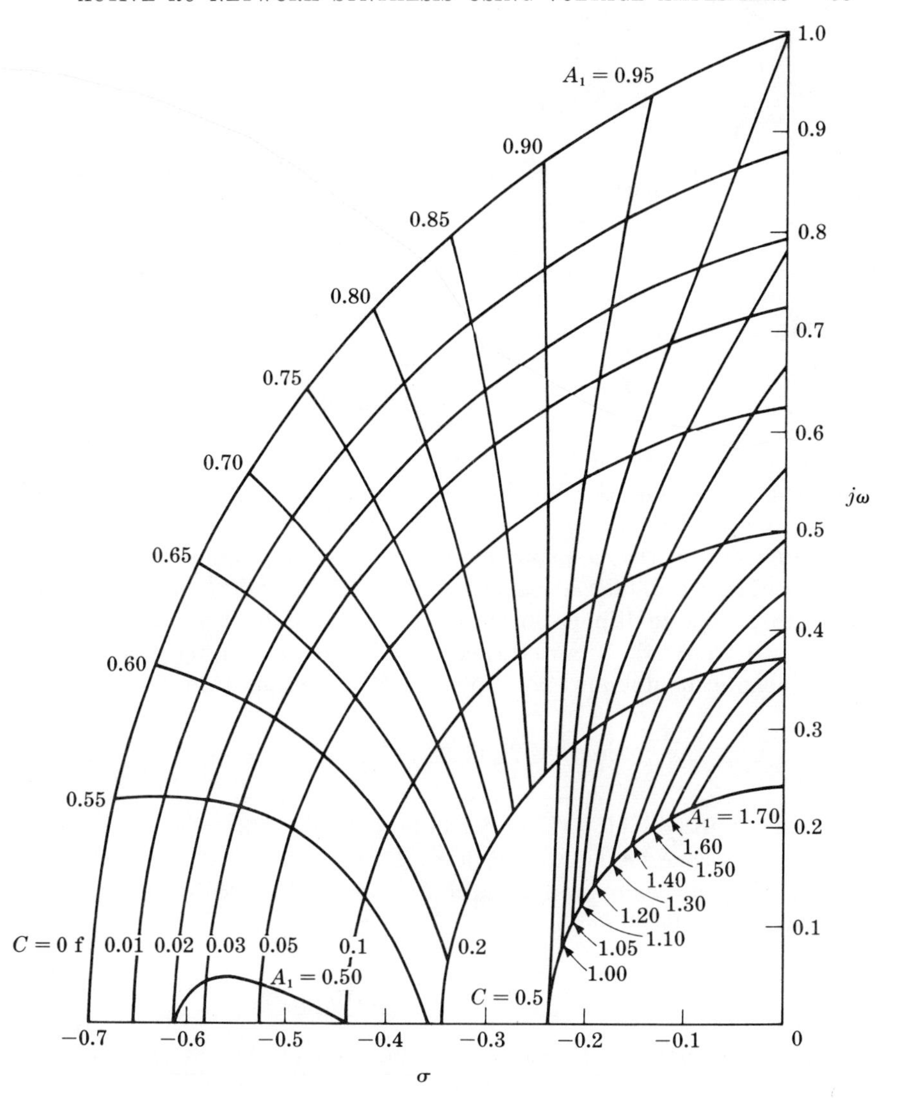

FIG. 2.79 Equivalent-pole positions of the DLA network of Fig. 2.77.

The first factor is

$$T_a(p) = \frac{0.359p^2 + 1}{p^2 + 0.488p + 0.501} \tag{2.189}$$

Transforming this so that the zeros are at $\omega = \pm 1$, we substitute

$$s^2 = 0.359p^2 \tag{2.190}$$

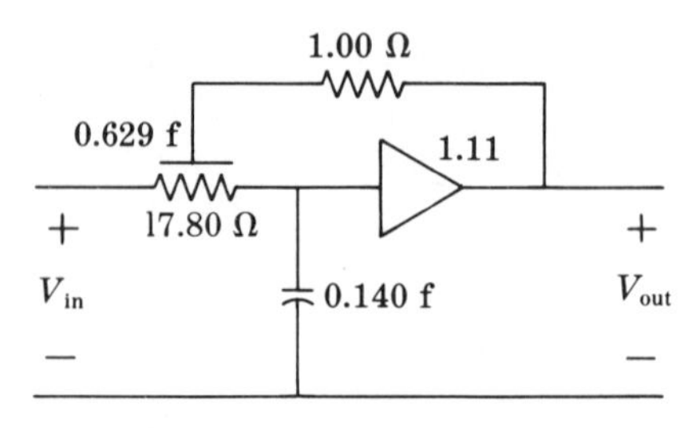

FIG. 2.80 DLA network approximating (2.191).

to obtain

$$T'_a(s) = \frac{s^2 + 1}{2.785s^2 + 0.814s + 0.501} \tag{2.191}$$

The poles are now located at $s = -0.146 \pm j0.398$. Entering Fig. 2.79 at this position, we find that a gain $A = 1.11$ and a capacity $C = 0.14$ farad are required. The network is shown in Fig. 2.80.

Returning to the p plane by using the s-to-p relation of (2.190) ($p = 1.67s$), we divide all capacities by 1.67 to obtain the network of Fig. 2.81. The multiplier obtained can easily be determined by finding $T(0)$ for the network. Since the feedback is zero at dc, $T(0) = A = 1.11$. If we denote the unknown multiplier as H, we obtain from $T_a(p)$

$$T_a(0) = \frac{H}{0.501} \tag{2.192}$$

and since $T(0) = 1.11 = T_a(0)$, we obtain $H = 0.556$. We can similarly realize the second biquadratic $T_b(p)$ of (2.188) and by cascade connection obtain the four zeros and four of the poles. The third factor, $T_c(p)$ of (2.188), can be chosen as

$$T_c(p) = \frac{0.416}{p + 0.416} \tag{2.193}$$

since we are only concerned with realization of the transfer function within a constant multiplier. $T_c(p)$ is easily realized by the network of

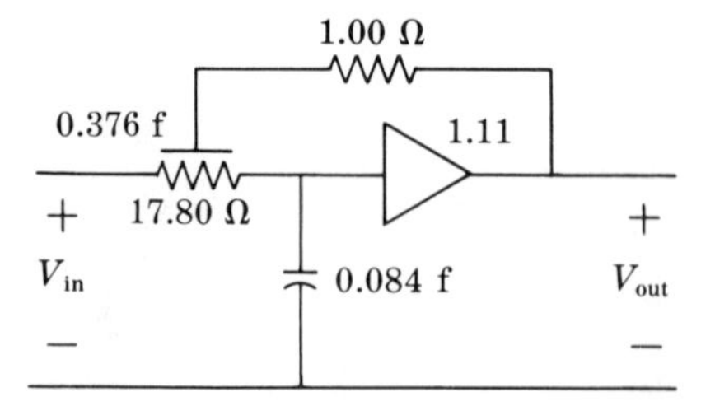

FIG. 2.81 DLA network approximating (2.189).

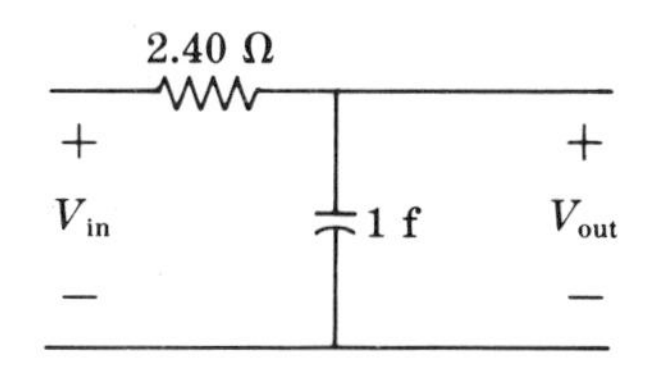

FIG. 2.82 Passive *RC* network realizing (2.193).

Fig. 2.82. The complete realization of the original function is obtained by cascade connection of the three networks and is shown in Fig. 2.83. A plot of the three theoretical factors of this function, their sum, the actual response of the two DLA networks, and the sum of the passive *RC* plus the two DLA network responses is shown in Fig. 2.84. In this case, there is no appreciable deviation from the equal-ripple specification in either the passband or the stop band, but the DLA network shows a greater attenuation beyond the second stop-band peak than the theoretical function of (2.188).

The two DLA networks described so far allow the synthesis of functions of any degree, having $j\omega$-axis zeros, if the zeros are located beyond the poles. If the function required is of odd degree, a passive *RC* network is added to produce the necessary first-order factor. For those functions in which the poles are located beyond the zeros, such as high-pass functions and certain bandpass functions, a new DLA network is needed. This is shown in Fig. 2.85 and is essentially similar to the lumped network of Fig. 2.10*b*. If the poles are not high-Q and are close to the zeros in radial distance from the origin, it may be that this new DLA network, Fig. 2.85, will be required even when the zeros are slightly beyond the poles. This dividing line is the curve $C = 0$ in Fig. 2.79.

Calculation of the amplitude response of this network for many different values of gain A and resistance R is followed by the determination of the equivalent-pole positions giving the best amplitude match using a two-pole, two-$j\omega$-axis-zero rational function. A typical comparison is shown

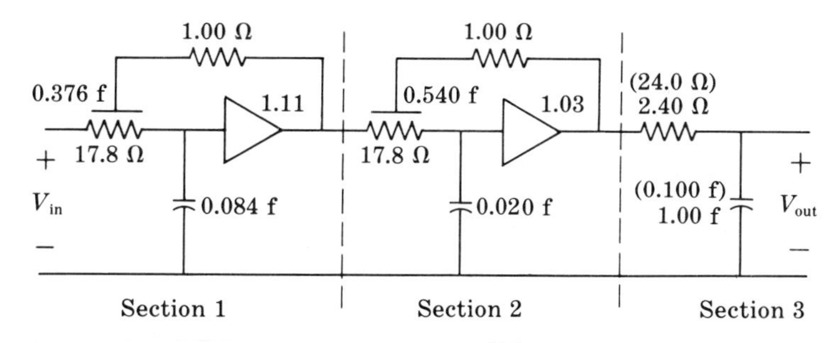

FIG. 2.83 DLA network approximating (2.188).

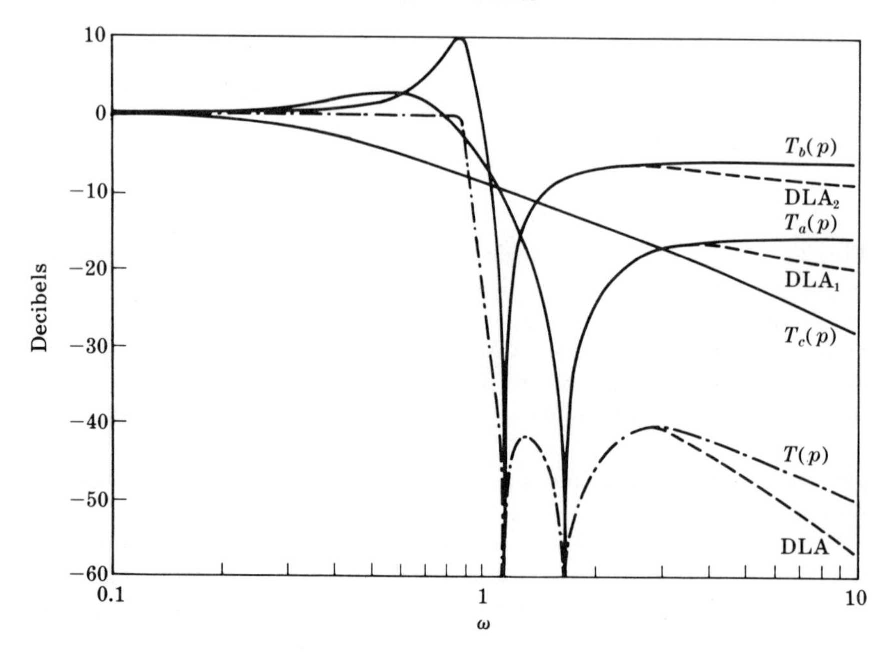

FIG. 2.84 Five-pole, four-$j\omega$-axis-zero low-pass elliptic-function filter response.

in Fig. 2.86 where the amplitude response of the DLA network of Fig. 2.85 is compared to a matched two-pole, two-$j\omega$-axis zero function. The equivalent-pole positions as a function of gain A_2 and resistance R which result from many such matchings are shown in Fig. 2.87. The zeros are normalized to $\omega = \pm 1$ on this design chart. If we choose, for example, the three-pole, two-$j\omega$-axis-zero maximally-flat high-pass function

$$T(p) = \frac{p(p^2 + 0.25)}{p^3 + 1.634p^2 + 1.585p + 0.750} \tag{2.194}$$

(-3 db at $\omega = 1.0$ rad/sec and zeros at $\omega = \pm\frac{1}{2}$ rad/sec) and factor

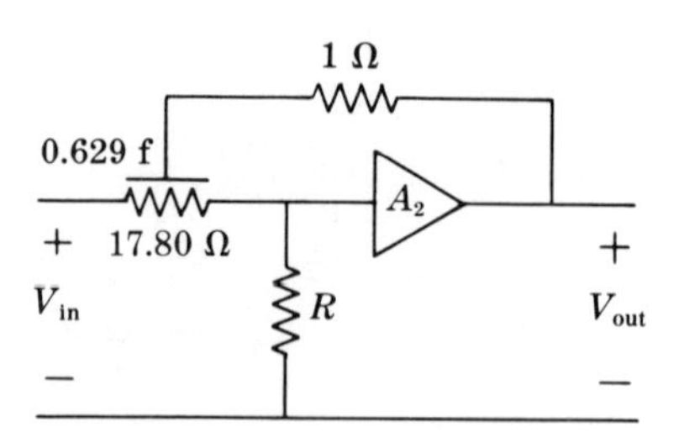

FIG. 2.85 DLA network for poles beyond the zeros.

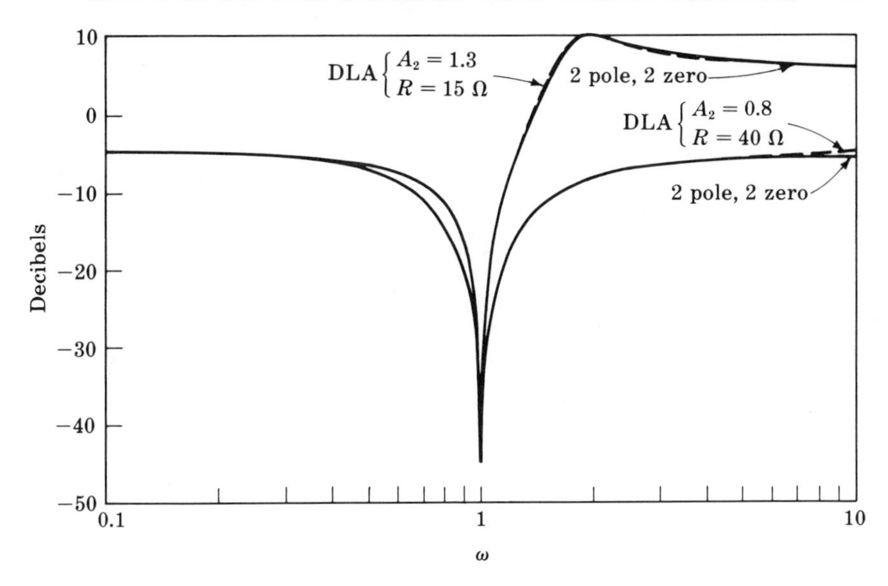

FIG. 2.86 Responses of the DLA network of Fig. 2.85 compared to a two-pole, two-zero function for two values of A_2 and R.

this transfer function, we obtain

$$T_1(p) = \frac{p^2 + 0.25}{p^2 + 0.82p + 0.92} \tag{2.195}$$

and

$$T_2(p) = \frac{p}{p + 0.82} \tag{2.196}$$

Normalization of (2.195) by substituting

$$p^2 = 0.25s^2 \tag{2.197}$$

places the zeros at $\omega = \pm 1$ and we have

$$T_1'(s) = \frac{0.25(s^2 + 1)}{0.25s^2 + 0.41s + 0.92} \tag{2.198}$$

The poles are now at $s = -0.82 \pm j1.73$ and entering Fig. 2.87 at this pole position, we find that $A_2 = 1.31$ and $R = 11.5$ ohms. Because of the simple nature of this transfer function (relatively mild cutoff slope), the zeros and poles are separated to a maximum extent. The pole position is therefore near the edge of Fig. 2.87. The network realizing $T_1'(s)$ is shown in Fig. 2.88. The dc gain of this network is $(11.5)(1.31)/29.3 = 0.514$ and therefore the transfer function actually realized is [note that $T_1''(0) = 0.473/0.92 = 0.514$]

$$T_1''(s) = \frac{0.473(s^2 + 1)}{0.25s^2 + 0.41s + 0.92} \tag{2.199}$$

We now transform this network to the p plane by dividing all capacitors by the scaling factor 0.5 [since $p = 0.5s$ in (2.197)] to obtain the network corresponding to $T_1(p)$ within a constant multiplier. This network is

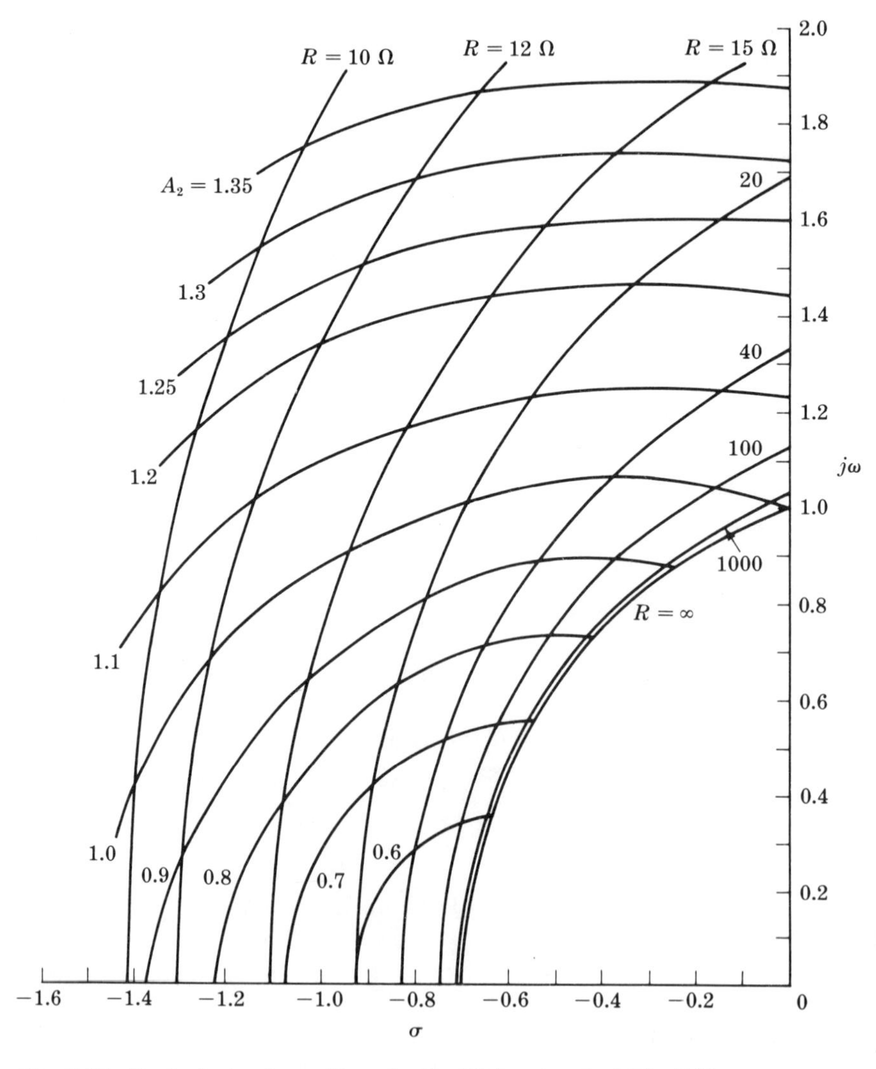

FIG. 2.87 Equivalent-pole positions for the DLA network of Fig. 2.85.

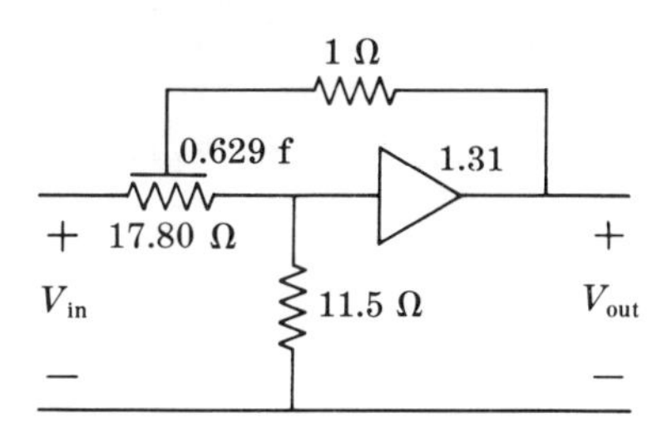

FIG. 2.88 DLA network approximating (2.198).

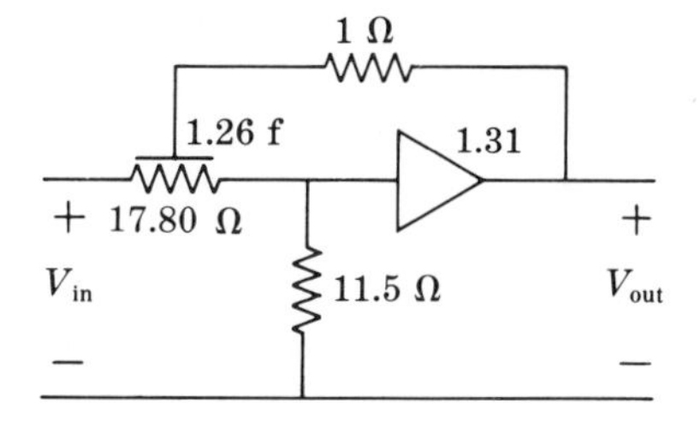

FIG. 2.89 DLA network approximating (2.195).

shown in Fig. 2.89. $T_2(p)$ in (2.196) is easily realized by a passive *RC* network and this network is then cascaded with the network of Fig. 2.89 to complete the synthesis. The values shown can be modified by independent impedance scaling of the two sections due to the isolation provided by the VCVS. This is a convenient feature of this method of synthesis and can be used to provide equal capacitor values or to minimize the total capacity, for example, and is applicable to all synthesis procedures in which a VCVS is used. The resulting network is shown in Fig. 2.90. The final transfer function approximated by the network of Fig. 2.90 is

$$T(p) = \frac{1.892(p^2 + 0.25)}{p^2 + 0.82p + 0.92} \frac{p}{p + 0.82} \tag{2.200}$$

which is (2.194) within a constant multiplier.

Summary. The use of distributed *RC* elements allows a considerable reduction in the number of components required for the realization of a given transfer function and a reduction in the VCVS gain required. In fact, fewer capacitors are required than the number of poles being realized in some of the distributed *RC* networks! This reduction in gain provides greater bandwidth capability and increased stability of the VCVS. In addition, the distributed elements provide approximately

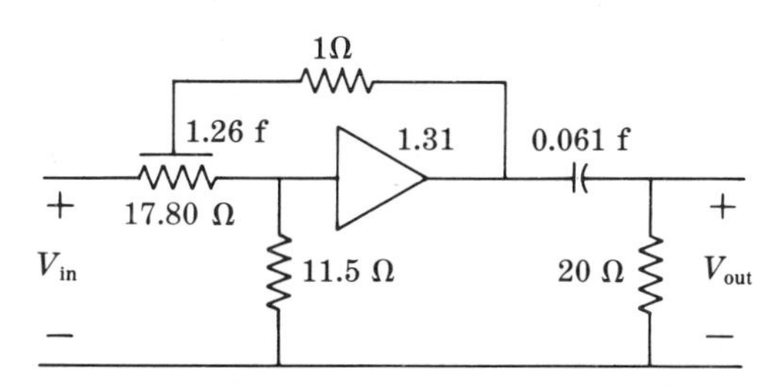

FIG. 2.90 DLA network approximating (2.200).

twice the RC product for a given chip area than lumped-element realizations. The increased difficulty in trimming the RC lines to a specific value of resistance and capacitance as well as the high equivalent-pole-position sensitivity to variations in the VCVS gain are the primary disadvantages of this method of synthesis. From a practical standpoint the pole Q should probably be less than 10 if this method is to be used.

Recently, some work has been done on multiloop DLA networks similar to the lumped-element multiloop networks described in Sec. 2.2 (Holt 1966). This would appear to be a very attractive way of extending these networks to higher Q with, of course, the possibility of use at high frequencies.

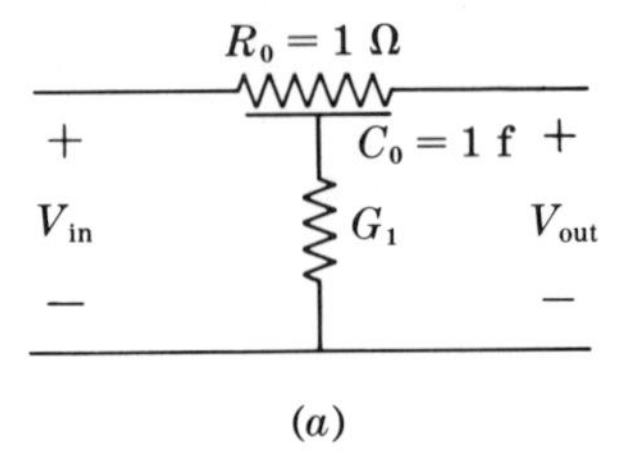

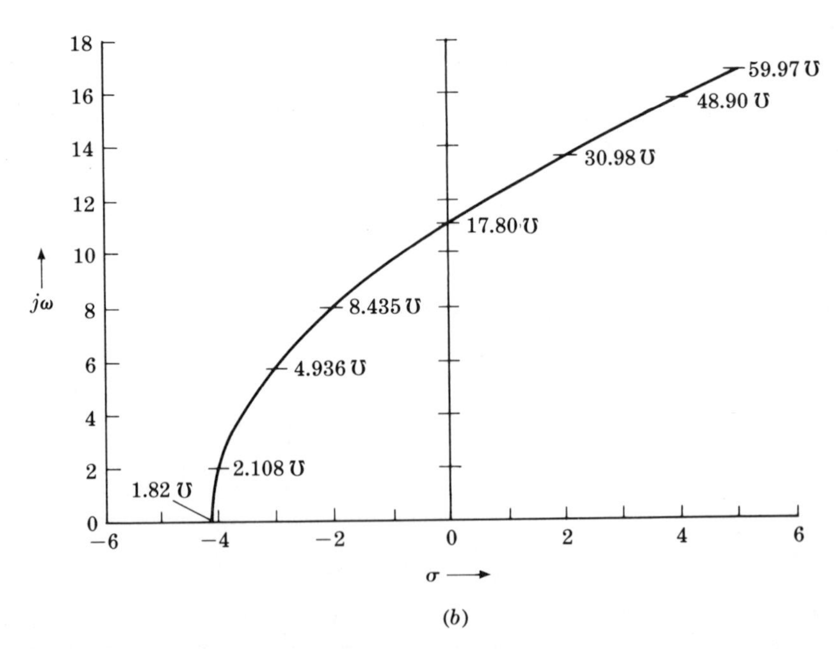

FIG. 2.91 Distributed, lumped RC network having complex zeros. (a) Schematic; (b) root locus of the dominant zeros as a function of G_1.

2.5 Negative-gain VCVS Synthesis Using Distributed Elements

The phantom-zero methods previously described using lumped *RC* elements had the disadvantage that a considerable number of *RC* elements were required for a single pair of high-Q poles and they had rather high sensitivity to passive-element variation. The use of distributed *RC* elements can greatly simplify these networks, at the expense, however, of a considerably more intractable transfer function. In addition, the distributed network may have a greatly reduced Q sensitivity to changes in the passive elements.

A suitable distributed *RC* network is shown in Fig. 2.91*a* (Kerwin and Shaffer 1968), normalized so that the two dominant zeros are located at $\omega = \pm 1$ when $G_1 = 17.79$ mhos. The position of the dominant zeros is determined by the value of G_1 as indicated by the root locus shown in Fig. 2.91*b*. The real part of the zero position, sigma, is related to R_1 (where $R_1 = 1/G_1$) as

$$R_1 \approx \frac{1}{17.80 + 5.40\sigma + 0.529\sigma^2} \tag{2.201}$$

This relation is accurate to within ± 0.4 percent from $\sigma = 0$ to $\sigma = 5.0$.

This network is then combined with a VCVS as shown in Fig. 2.92. The configuration is chosen to produce a zero at the origin and one at infinity so that a true second-order bandpass function is obtained. Since the transfer function of this distributed active *RC* network is irrational (the number of poles and zeros are infinite), we will determine the Q from the calculated amplitude response by means of

$$Q = \frac{\omega_{max}}{\omega_2 - \omega_1} \tag{2.202}$$

where ω_2 and ω_1 are the upper and lower 3-db points respectively. The

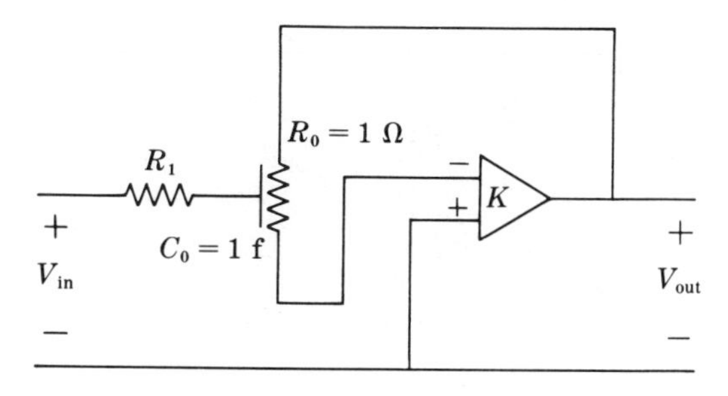

FIG. 2.92 Bandpass DLA network with phantom zeros in the right-half plane.

relation between Q, amplifier gain K, and the real part of the zero position, sigma, is shown in Fig. 2.93. The Q sensitivity $S_K{}^Q$ to change in gain K as a function of sigma is also shown in Fig. 2.93 and is determined by calculating ΔQ for incremental changes in K. The determination of Q by (2.202) allows an equivalent-pole position to be determined (Kerwin 1967) so that the two dominant zeros and two equivalent poles reasonably approximate the amplitude response of the network.

Using the data of Fig. 2.93, the amplifier gain K and the value of sigma can be determined for any specified Q and Q sensitivity. The resulting trade off between gain required and Q sensitivity is clearly shown. The advantage of positive values of sigma in increasing the Q for a given gain is also apparent. The frequency of maximum response ω_{max} as a function of gain K and sigma is shown in Fig. 2.94. The overall system gain as a function of K and sigma is shown in Fig. 2.95.

As a design example we will assume that a Q of 50 is required with a Q sensitivity $S_K{}^Q \leq 10$. From Fig. 2.93 we find that $\sigma = 0.8$ and $K = 42$. Solving (2.201) we obtain $R_1 = 0.045$ ohm for this value of σ. From Fig. 2.94 we find that $\omega_{max} = 12$ rad/sec and from Fig. 2.95 that the system gain at ω_{max} is 51 db.

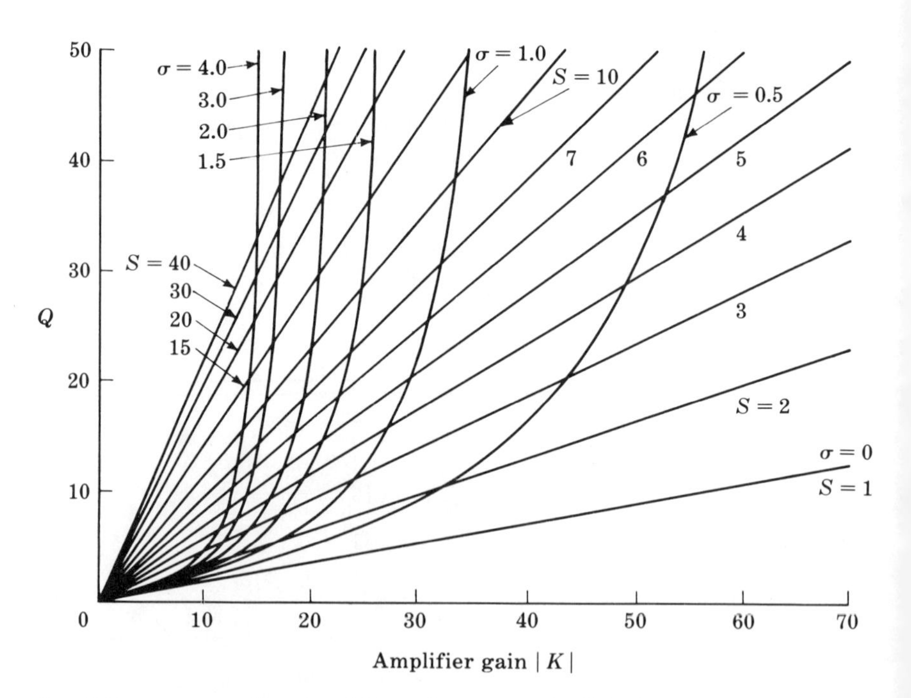

FIG. 2.93 Q vs. amplifier gain $|K|$ as a function of the right-half-plane phantom-zero position (σ) for the network of Fig. 2.92.

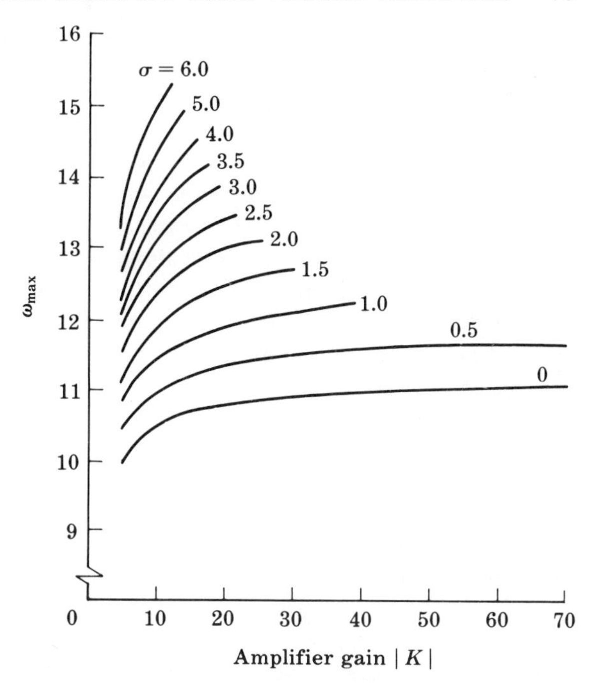

FIG. 2.94 Maximum-response frequency vs. amplifier gain as a function of the phantom-zero positions.

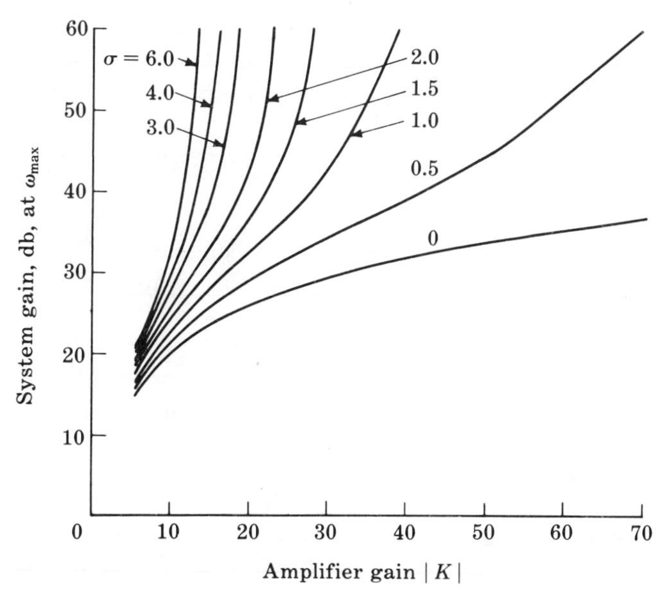

FIG. 2.95 Maximum system gain vs. amplifier gain as a function of the phantom-zero positions.

Summary. The use of right-half-plane phantom zeros allows a considerable reduction in amplifier gain for a given Q at the expense of increased Q sensitivity to amplifier-gain change, and therefore increased high-frequency capability. The use of distributed elements simplifies the network and in addition greatly reduces the Q sensitivity to the passive elements. Since only a single capacitor is used, $S_C{}^Q = 0$, and if the *ratio* between R_0 and R_1 remains constant with time and temperature, the Q sensitivity to the resistors is also zero. This is also true in the case of the previously discussed lumped RC twin-T network; however, all three capacitors must track exactly and so must all three resistors, a more difficult requirement to meet. The frequency of maximum response is inversely proportional to the R_0C_0 product and so no special problems exist here.

2.6 Conclusions

In this chapter, the use of one type of active element, the operational amplifier, in the realization of various transfer functions has been discussed. These amplifiers have been used either as a VCVS, as an integrator, or as a summer.

VCVS realizations have several advantages. First of all, the gain of the VCVS appears as a parameter of the transfer function in such a way that various characteristics of the transfer function may be easily adjusted by varying the gain. For example, a bandpass realization using a positive-gain VCVS may have its Q readily varied by changing the gain K. Another advantage of the VCVS class of active RC circuits in which the output is taken from the VCVS is that such realizations may be cascaded without interaction occurring between the networks. Thus, complicated network functions of relatively high degree may be realized by a cascade connection of the relatively simple structures described in this chapter.

Realization procedures may be found in the literature describing the use of other types of controlled sources and networks than those discussed here in realizing various network functions. In general, these procedures are of varying degrees of practicality. For example, some of them may be built with a single transistor, obviously an advantage. On the other hand, the synthesis procedures for determining the passive networks for such realizations are considerably more complicated than the ones described in this chapter, and in general the networks are a great deal more complex (Pande and Shukla 1965; Gorski-Popiel 1965; Holt and Sewell 1965). The large quantity of such realization techniques which have appeared in the literature precludes any detailed coverage of them. Many of the techniques attempt to minimize the number of active elements at the expense of an increased number of passive elements (Holt 1966; Hogin 1966; Hakim, May 1965a). In the case of the use of the voltage follower (a VCVS with unity gain) (Hogin 1966), the increased stability of gain allows the use of structures which would otherwise be much too sensitive to gain change. Hogin has suggested that the pole-position sensitivity to gain change be with respect to the open-loop gain in the case of the voltage follower. He has described synthesis procedures for network functions up to sixth order using a single voltage follower (Hogin 1968), and a second-order bandpass function at 10 MHz (Hogin 1967). A representative bibliography of references to such realization procedures, however, has been provided for the reader who wishes to pursue the subject further. Sensitivity studies using the pole-position sensitivity for comparison with the sensitivities given herein may be made of such realizations to establish a basis for comparison between the different techniques.

In summarizing the various techniques, the simplest networks suited to low-Q requirements are the distributed, lumped, active networks using a positive-gain VCVS, as described in Sec. 2.4. The practical Q limit depends on the acceptable amplitude-response tolerances, but is probably below 10. Of course, use of these networks is dependent on facilities suited to the construction of distributed *RC* elements. In the absence of such facilities, the lumped *RC* networks which use a positive-gain VCVS of Sec. 2.2 are the appropriate choice for low-Q applications (generally less than 10). Both of these methods are well suited to a wide range of frequencies due to the very low VCVS gains required. The upper limit depends primarily on the gain-bandwidth available, but 10 MHz should pose no great problem with present operational amplifiers even allowing for a reasonable amount of negative feedback to stabilize the gain, but with the condition that $Q \leq 10$.

The commercial availability of two-pole, two-zero building blocks which use the state-variable synthesis method of Sec. 2.3 coupled with their projected low price will probably make this method the most practical for all high-Q applications up to 250–500 kHz. At higher frequencies than this when high Q is needed, the multiloop, or phantom-zero methods of Sec. 2.2 should be considered.

In any particular synthesis problem we may well find that the best solution does not consist of the use of any particular synthesis method previously described, but rather in a combination of these methods in which the low-Q poles are realized with a single positive-gain VCVS and the moderate-Q poles by the multiloop or phantom-zero methods and the highest-Q poles with integrators. In this way, each section of the filter is realized in the simplest possible way and the power required is minimized.

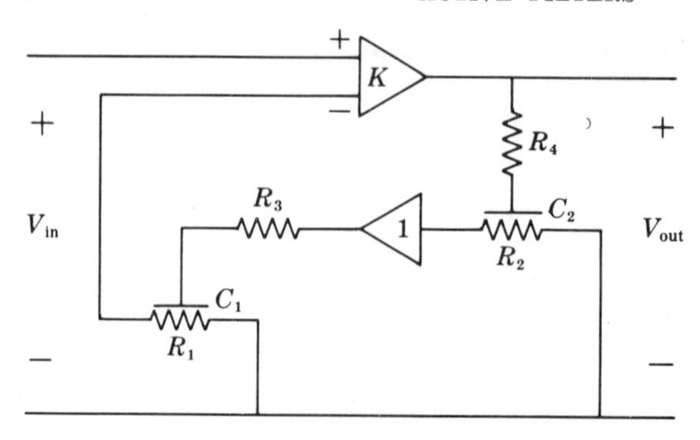

FIG. 2.96 Active *RC* network approximating a fourth-order bandpass function.

The most important problem at present seems to be the development of methods for the synthesis of high-Q systems at frequencies above 500 kHz having even lower sensitivities. The use of right-half-plane phantom zeros with a negative-gain VCVS as described in Sec. 2.5 is a step in that direction. If this can be done so that a fourth-order bandpass function can be easily obtained, it would be particularly desirable, since this would give an equivalent to the double-tuned interstage network. A possible approach is shown in Fig. 2.96.

References

Antoniou, A.: New RC-active Network Synthesis Procedures Using Negative-impedance Converters, *Proc. IEE*, vol. 114, pp. 894–902, July, 1967.

Armstrong, D. B., and F. M. Reza: Synthesis of Transfer Functions by Active RC Networks with Feedback Loops, *IRE Trans. Circuit Theory*, vol. CT-1, no. 2, pp. 8–17, June, 1954.

Bachmann, A. E.: Transistor Active Filters Using Twin-T Rejection Networks, *Proc. IEE*, vol. 106, pt. B, pp. 170–174, March, 1959.

Balabanian, N.: "Sensitivity Considerations in Active Network Synthesis," Syracuse University Res. Inst., New York, April, 1963.

——— and B. Patel: Active Realization of Complex Zeros, *IEEE Trans. Circuit Theory*, vol. CT-10, no. 2, pp. 299–300, June, 1963.

——— and I. Cinkilie: Expansion of an Active Synthesis Technique, *IEEE Trans. Circuit Theory*, vol. CT-10, no. 2, pp. 290–298, June, 1963.

Barker, D. G.: Synthesis of Active Filters Employing Thin Film Distributed Parameter Networks, *IEEE Intern. Conv. Record*, pt. 7, pp. 119–126, 1965.

Bennett, W. R.: Synthesis of Active Networks, *Proc. Symp. Modern Network Synthesis*, Polytechnic Institute of Brooklyn, vol. 5, pp. 45–61, 1956.

Bialko, M.: Two-Stage Selective RC Amplifiers with Small S^Q Sensitivities, *Electron. Letters*, vol. 1, no. 3, pp. 102–104, March, 1967.

Biswas, R. N., and E. S. Kuh: Optimum Synthesis of a Class of Multi-loop Feedback Systems, *Electron. Res. Lab. Memo. ERL-M218*, University of California, Berkeley, July 11, 1967.

Blackburn, H., D. S. Campbell, and A. J. Muir: Low-pass Active Filters, *Radio & Electron. Engr.*, vol. 34, pp. 90–96, August, 1967.

Blecher, F. H.: Application of Synthesis Techniques to Electronic Circuit Design, *IRE Trans. Circuit Theory*, vol. CT-7, special supplement, pp. 79–91, August, 1960.

Bloodworth, G. G., and N. R. S. Nesbitt: Active Filters Operating below 1 Hz, *IEEE Trans. Instr. Meas.*, vol. IM-16, no. 2, pp. 115–120, June, 1967.

Bobrow, L. S., and S. L. Hakim: A Note on Active RC Realization of Voltage Transfer Functions, *IEEE Trans. Circuit Theory*, vol. CT-11, no. 4, pp. 493–494, December, 1964.

Bodner, H. A., and S. K. Mitra: Active All-pass Network, *Electron. Letters*, vol. 1, no. 3, p. 98, May, 1965.

Bongiorno, J. J.: Synthesis of Active RC Single-tuned Bandpass Filters, *IRE Natl. Conv. Record*, 1958, pt. 2, pp. 30–41.

Bown, G. C.: Sensitivity in Active RC Filters, *Electron. Letters*, vol. 3, no. 7, pp. 298–299, July, 1967.

Boyce, A. H.: A Theoretical and Practical Study of Active Filters, *Marconi Rev.*, vol. 30, 2d qtr., pp. 68–97, 1967.

Brennan, R., and A. Bridgman: Simulation of Transfer Functions Using Only One Operational Amplifier, *IRE WESCON Conv. Record*, pt. 4, pp. 273–278, August, 1957.

Brugler, J. S.: RC Synthesis with Differential-input Operational Amplifiers, in Papers on Integrated Circuit Synthesis (compiled by R. W. Newcomb and T. N. Rao), *Stanford University Center for Systems Research Tech. Rept.* 6560–4, pp. 115–130, June, 1966.

Burr-Brown Research Corporation: "Handbook of Operational Amplifier Applications," Tucson, Ariz., 1963.

Butler, E.: The Operational Amplifier as a Network Element, *Cornell Res. Rept. EERL*-40, *Tech. Rept.* 86, June, 1965.

Butler, E. M., and S. K. Mitra: An Equivalent Circuit for the Operational Amplifier, *Proc. Allerton Conf. Circuit and System Theory*, 3d, p. 845, 1965.

Calahan, D. A.: Sensitivity Minimization in Active RC Synthesis, *IRE Trans. Circuit Theory*, vol. CT-9, no. 1, pp. 38–42, March, 1962.

Castro, P. S.: RC Distributed Parameter Network with Circular Geometry, *Proc. Natl. Electron. Conf.*, vol. 19, pp. 98–106, 1963.

——— and W. W. Happ: Distributed Parameter Circuits and Microsystem Electronics, *Proc. Natl. Electron. Conf.*, vol. 16, pp. 448–460, 1960.

Chirlian, P. M.: "Integrated Active Network Analysis and Synthesis," Prentice-Hall, Inc., Englewood Cliffs, N.J., 1967.

Cooper, R. E., and C. O. Harbourt: Sensitivity Reduction in Amplifier-RC Synthesis, *Proc. IEEE*, vol. 54, no. 11, pp. 1578–1579, November, 1966.

Deboo, G. J.: A Novel Integrator Results by Grounding Its Capacitor, *Electron. Design*, vol. 15, no. 12, p. 90, June 7, 1967.

Deliyanis, T.: Sensitivity Study of Five RC Active Networks Using the Method of Single Inversion, *Intern. J. Electron.*, vol. 22, pp. 197–213, March, 1967.

de Pian, L.: "Linear Active Network Theory," Prentice-Hall, Inc., Englewood Cliffs, N.J., 1962.

Dutta-Roy, S. C.: On Some Three-terminal Lumped and Distributed RC Null Networks, *IEEE Trans. Circuit Theory*, vol. CT-11, no. 1, pp. 98–103, March, 1964.

Edson, W. A.: Tapered Distributed RC Lines for Phase-shift Oscillators, *Proc. IRE*, vol. 49, no. 6, pp. 1021–1024, June, 1961.

Electro-technology staff: Packaged Operational Amplifiers, *Electro-technol.*, vol. 79, no. 1, pp. 73–77, January, 1967.

Fialkow, A. D., D. Hazony, P. Kodali, and T. Zeren: Transfer Function Realizability of Multi-port RLC Transformerless Grounded Networks, *IEEE Trans. Circuit Theory*, vol. CT-11, no. 1, pp. 73–80, March, 1964.

Field, R. K.: The Tiny, Exploding World of Linear Micro-circuits, *Electron. Design*, vol. 15, no. 15, pp. 49–66 and 68–72, July 19, 1967.

Ford, R. L., and F. E. J. Girling: Active Filters and Oscillators Using Simulated Inductance, *Electron. Letters*, vol. 2, no. 2, p. 52, February, 1966.

Foss, R. C., and B. J. Green: Qfactor, Qstability, and Gain in Active Filters, *Electron. Letters*, vol. 2, no. 3, pp. 99–100, March, 1966.

Foster, E. J.: Active Low-pass Filter Design, *IEEE Trans. Audio*, vol. AU-13, no. 5, pp. 104–111, September/October, 1965.

Fowler, W. J.: A Look at Linear Integrated Circuits, *Electron. Ind.*, vol. 24, pp. 64–69 and 194, September, 1965.

Fuller, W. D., and P. S. Castro: A Microsystems Bandpass Amplifier, *Proc. Natl. Electron. Conf.*, vol. 16, pp. 139–151, 1960.

Geffe, P.: Make a Filter Out of an Oscillator, *Electron. Design*, vol. 15, no. 10, pp. 56–58, May 10, 1967.

Ghausi, M. S.: "Principles and Design of Linear Active Circuits," McGraw-Hill Book Company, New York, 1965.

——— and G. J. Herskowitz: The Effect of Shaping and Loading on Distributed RC Networks, *J. Franklin Inst.*, vol. 278, no. 2, pp. 108–123, August, 1964.

——— and D. O. Pederson: A New Design Approach for Feedback Amplifiers, *IRE Trans. Circuit Theory*, vol. PGCT-8, no. 3, pp. 274–284, September, 1961.

——— and ———: A New Feedback Broadbanding Technique for Transistor Amplifiers, *Proc. Natl. Electron. Conf. Chicago*, vol. 18, p. 325, 1962.

Giles, J. N. (ed.): "Linear Integrated Circuits Applications Handbook," Fairchild Semiconductor, Mountain View, Calif., 1967.

Good, E. F.: A Two-phase Low Frequency Oscillator, *Electron. Eng.*, pp. 164–169, April, 1957.

Gorski-Popiel, J.: RC Active Networks, *Electron. Letters*, vol. 1, no. 10, pp. 288–289, December, 1965.

———: Reduction of Network Sensitivity through the Use of Higher Order Approximating Functions, *Electron. Letters*, vol. 3, no. 8, pp. 365–366, August, 1967.

——— and A. J. Drew: RC Active Ladder Networks, *Proc. IEE*, vol. 112, no. 12, pp. 2213–2219, December, 1965.

Grabel, Arvin: The Use of Sensitivity as a Criterion in the Synthesis of Active Networks, *New York University AL TDR* 64 165, July, 1964.

Hakim, S. S.: Feedback Network Synthesis, *Intern. J. Control*, vol. 1, no. 1, pp. 47–54, January, 1965.

———: Internal Feedback in Active Two-port Networks, *Intern. J. Control*, vol. 1, no. 5, pp. 455–460, May, 1965b.

———: RC-active Circuit Synthesis Using an Operational Amplifier, *Intern. J. Control*, vol. 1, no. 5, pp. 433–445, May, 1965a.

———: RC Active Filters Using Amplifiers as Active Elements, *Proc. IEE*, vol. 112, no. 5, pp. 901–912, May, 1965c.

———: Synthesis of RC Active Filters with Prescribed Pole Sensitivity, *Proc. IEE*, vol. 112, no. 12, pp. 2235–2242, December, 1965.

Happ, W. W., and P. S. Castro: Distributed Parameter Circuit Design Techniques, *Proc. Natl. Electron. Conf.*, vol. 17, pp. 45–70, 1961.

——— and G. C. Riddle: Limitations of Film Type Circuits Consisting of Resistive and Capacitive Layers, *IRE Intern. Conv. Record*, vol. 9, pt. 5, pp. 141–165, 1961.

Hatley, W. T., Jr.: High-frequency Tuning and Filtering in Integrated Circuits, in

Papers on Integrated Circuit Synthesis (compiled by R. W. Newcomb and T. N. Rao), *Stanford University Center for Systems Research Tech. Rept.* 6560-4, pp. 93–113, June, 1966.

Hazony, D.: Grounded RC–Unity Gain Amplifier Transfer Vector Synthesis, *IEEE Trans. Circuit Theory,* vol. CT-14, no. 1, pp. 75–76, March, 1967.

——— and R. D. Joseph: Transfer Matrix Synthesis with Active RC Networks, *J. SIAM Appl. Math.*, vol. 14, pp. 739–761, July, 1966.

——— and R. Lagerlof: Canonical Active Networks, *Chalmers Institute of Technology, Division of Network Theory TR* 6602, Gothenberg, Sweden, June 30, 1966.

Heizer, K. W.: Distributed RC Networks with Rational Transfer Functions, *IRE Trans. Circuit Theory,* vol. CT-9, no. 4, pp. 356–362, December, 1962.

———: Rational Parameters with Distributed Networks, *IEEE Trans. Circuit Theory,* vol. CT-10, no. 4, pp. 531–532, December, 1963.

Hellstrom, M. J.: Equivalent Distributed RC Networks or Transmission Lines, *IRE Trans. Circuit Theory,* vol. CT-9, no. 3, pp. 247–251, September, 1962.

———: Symmetrical RC Distributed Networks, *Proc. IRE,* vol. 50, pp. 97–98, January, 1962.

Hesselberth, C. A.: Synthesis of Some Distributed RC Networks, *University of Illinois Rept. R*-164, August, 1963.

Hogin, J. L.: "Active RC Networks Utilizing the Voltage Follower," Ph.D. dissertation, Montana State University, Bozeman, March, 1966.

———: Development of an Active Low Pass 6-Pole Butterworth Filter, *ERL Rept.* 815-*I*1-0368, Montana State University, Bozeman, March, 1968.

———: Study and Development of an Active RC Tuned Network, *ERL Report* 902-*F*2-0367, Montana State University, Bozeman, March, 1967.

Holland, L.: "Thin Film Microelectronics," John Wiley & Sons, Inc., New York, 1965.

Holt, A. G. J.: RC Active Circuits, *University of Newcastle-on-Tyne Conf. Electron. Networks*, September, 1966.

——— and R. Linggard: Active Chebyshev Filters, *Electron. Letters,* vol. 1, no. 5, pp. 130–131, July, 1965.

——— and ———: RC Active Synthesis Procedure for Polynomial Filters, *Proc. IEE,* vol. 113, no. 5, pp. 777–782, May, 1966.

——— and J. I. Sewell: Active RC Filters Employing a Single Operational Amplifier to Obtain Biquadratic Response, *Proc. IEE,* vol. 112, no. 12, pp. 2227–2234, December, 1965.

——— and F. W. Stephenson: Specialization of Yanagisawa Synthesis Procedure to Obtain RC Active Networks Having Optimum Pole Sensitivity, *Radio Electron. Engr.*, vol. 29, no. 6, pp. 362–368, June, 1965.

Horowitz, I. M.: Active Network Synthesis, *IRE Natl. Conv. Record,* pt. 2, pp. 38–45, 1956.

———: Active RC Synthesis, *IEEE Trans. Circuit Theory,* vol. CT-13, no. 1, pp. 101–102, March, 1966.

———: Exact Design of Transistor RC Band-pass Filters with Prescribed Active Parameter Insensitivity, *IRE Trans. Circuit Theory,* vol. CT-7, no. 3, pp. 313–320, September, 1960.

———: RC Transistor Network Synthesis, *Proc. Natl. Electron. Conf.*, vol. 12, pp. 818–829, October, 1956.

———: Synthesis of Active RC Transfer Functions, *Microwave Res. Inst. Rept. R*507-56, Polytechnic Institute of Brooklyn, November, 1956.

——— and G. R. Branner: A Unified Survey of Active RC Synthesis Techniques, *Proc. Natl. Electron. Conf.*, vol. 23, pp. 257–261, 1967.

Huelsman, L. P.: "Handbook of Operational Amplifier Active RC Networks," Burr-Brown Research Corp., Tucson, Ariz., 1966.

———: Stability Criteria for Active RC Synthesis Techniques, *Proc. Natl. Electron. Conf.*, vol. 20, pp. 731–736, 1964.

———: "Theory and Design of Active RC Circuits," McGraw-Hill Book Company, New York, 1968.

——— and W. J. Kerwin: Digital Computer Analysis of Distributed Lumped Active Networks, *IEEE J. Solid State Circuits*, vol. SC-3, no. 1, pp. 26–29, March, 1968.

Jagoda, N. H.: An Active Realization of the Elliptic Function Approximation, *IRE Trans. Circuit Theory*, vol. CT-9, no. 4, p. 423, December, 1962.

Joseph, R. D., and D. Hilberman: Immittance Matrix Synthesis with Active Networks, *IEEE Trans. Circuit Theory*, vol. CT-13, no. 3, p. 324, September, 1966.

Joyce, M. V., and K. K. Clarke, "Transistor Circuit Analysis," Addison-Wesley Publishing Company, Inc., Reading, Mass., 1961.

Kalman, R. E.: Mathematical Description of Linear Dynamical Systems, *J. SIAM Control.*, ser. A, vol. 1, no. 2, pp. 162–192, 1963.

Kaufman, W. M., and S. J. Garrett: Tapered Distributed Filters, *IRE Trans. Circuit Theory*, vol. CT-8, no. 4, pp. 329–336, December, 1962.

Keller, J. P.: Linear IC's: Part 3—Differential Amplifiers at Work, *Electronics*, vol. 40, no. 19, pp. 96–105, Sept. 18, 1967.

Kelly, J. J.: Analysis and Design Considerations for Distributed RC Networks with Arbitrary Taper, *New York University Rept. TR*-400-115 *SR*-18, August, 1965.

———: Analysis of Drift Transistor Current Gain with Application to Distributed RC Networks, *New York University Rept. TR*-400-98, July, 1964.

——— and M. S. Ghausi: Network Properties of Distributed RC Networks with Arbitrary Geometric Shapes, *New York University Rept. TR*-400-107 *SR*-15, March, 1965.

——— and ———: On the Effective Dominant Pole of the Distributed RC Network, *J. Franklin Inst.*, vol. 297, no. 6, pp. 417–429, June, 1965.

———, ———, and J. H. Mulligan, Jr.: On the Analysis of Composite Lumped-distributed Systems, *IEEE Intern. Conv. Record*, vol. 14, pt. 7, pp. 308–318, 1966.

Kerwin, W. J.: An Active RC Elliptic Function Filter, *IEEE Region Six Conf. Record, Tucson, Ariz.*, April, 1966.

———: Analysis and Synthesis of Active RC Networks Containing Distributed and Lumped Elements, *Stanford University Tech. Rept.* 6560-14, August, 1967.

———: RC Active Networks, in Papers on Integrated Circuit Synthesis (compiled by R. W. Newcomb and T. N. Rao), *Stanford University Center for Systems Research Tech. Rept.* 6560-4, pp. 153–170, June, 1966.

——— and L. P. Huelsman: The Design of High-performance Active RC Band-pass Filters, *Proc. IEEE Intern. Conv. Record, New York*, pt. 10, pp. 74–80, March, 1966.

——— and C. V. Shaffer: An Integrable IF Amplifier of High Stability, *Proc. Midwest Symp. Circuit Theory, 11th, Notre Dame, Ind.*, pp. 78–88, 1968.

———, L. P. Huelsman, and R. W. Newcomb: State Variable Synthesis for Insensitive Integrated Circuit Transfer Functions, *IEEE J. Solid State Circuits*, vol. SC-2, no. 3, pp. 87–92, September, 1967.

———, ———, and ———: State Variable Synthesis for Insensitive Integrated Circuit Transfer Functions, *Stanford University Tech. Rept.* 6560-10, December, 1966.

Kinariwala, B. K.: Synthesis of Active RC Networks, *Bell Syst. Tech. J.*, vol. 38, pp. 1269–1316, September, 1959.

Kuh, E. S.: Transfer Function Synthesis of Active RC Networks, *IRE Intern. Conv. Record*, vol. 8, pt. 2, pp. 134–138, 1960.

Kuo, F. F.: Sensitivity of Transmission Zeros in RC Network Synthesis, *IRE Natl. Conv. Record*, pt. 2, pp. 18–22, 1959.

———: Transfer Function Synthesis with Active Elements, *Proc. Natl. Electron. Conf.*, vol. 13, pp. 1049–1056, 1957.

Landee, R. W., D. C. Davis, and A. P. Albrecht: "Electronic Designers' Handbook," McGraw-Hill Book Company, 1957.

Lee, S. C.: Sensitivity Minimization in Active RC Integrated-circuit Design, *Proc. Ann. Allerton Conf. Circuit and Syst. Theory, 4th, Urbana, Ill.*, pp. 269–281, 1966.

Leeds, M. B.: Linear Integrated Circuit Fundamentals, *RCA Tech. Ser.* IC-40, Harrison, N.J., 1966.

Lindgren, A. G.: Transfer Characteristics of a Class of Distributed RC Networks, *Proc. IEEE*, vol. 53, no. 6, pp. 625–626, June, 1965.

Lovering, W. P.: Analog Computer Simulation of Transfer Functions, *Proc. IEEE*, vol. 53, no. 3, pp. 306–307, March, 1965.

Lynn, Meyer, Hamilton (Motorola): "Analysis and Design of Integrated Circuits," McGraw-Hill Book Company, New York, 1967.

Manolescu, A.: Single Transistor Oscillators with Distributed RC Networks, *Electron. Letters*, vol. 2, no. 4, pp. 151–152, April, 1966.

Margolis, S. G.: On the Design of Active Filters with Butterworth Characteristics, *IRE Trans. Circuit Theory*, vol. CT-3, no. 3, p. 202, September, 1956.

McVey, P. J.: An Active RC Filter Using Cathode-followers, *Electron. Eng.*, vol. 34, pp. 458–463, July, 1962.

———: Sensitivity in Some Simple RC Active Networks, *Proc. IEE*, vol. 112, no. 7, pp. 1263–1269, July, 1965.

Mitra, S. K.: Active RC Filters Employing a Single Operational Amplifier as the Active Element, *Proc. Hawaii Intern. Conf. Syst. Sci., Honolulu*, January, 1968.

———: A New Approach to Active RC Network Synthesis, *J. Franklin Inst.*, vol. 274, no. 3, pp. 185–197, September, 1962.

———: Synthesis of Active RC Networks, University of California, Berkeley, *IER Ser.* 60, issue no. 442, Mar. 20, 1962.

———: Transfer Function Realization Using RC One-ports and Two Grounded Voltage Amplifiers, *Proc. Ann. Princeton Conf. Inform. Sci. and Syst., 1st, Princeton University, N.J.*, pp. 18–23, March, 1967.

———: Transfer Function Synthesis Using a Single Operational Amplifier, *Electron. Letters*, vol. 3, no. 7, pp. 332–333, July, 1967.

Morse, A. S.: The Use of Operational Amplifiers in Active Network Theory, *Proc. Natl. Electron. Conf.*, vol. 20, pp. 748–752, 1964.

Moschytz, G. S.: Miniaturized Filter Building Blocks Using Frequency Emphasizing Networks, *Proc. Natl. Electron. Conf.*, vol. 23, pp. 364–369, 1967.

Myers, B. R.: Transistor-RC Network Synthesis, *IRE WESCON Conv. Record*, pt. 2, pp. 65–74, 1959.

Newell, W. E.: Pole Sensitivity Relationships, *Westinghouse Elec. Corp. Sci. Paper* 116-3000-*B*2, Aug. 31, 1961.

———: Tuned Integrated Circuits—A State-of-the-art Survey, *Proc. IEEE*, vol. 52, no. 12, pp. 1603–1608, December, 1964.

O'Neill, J. F., and M. S. Ghausi: Approximation in Gain and Phase of Two-ports Using Lattice of Paralleled Resonators, *Proc. Hawaii Intern. Conf. Syst. Sci., Honolulu*, January, 1968.

——— and ———: Transfer Function Synthesis Using Resonators and Some Active RC Resonators, *Proc. Ann. Princeton Conf. Inform. Sci. and Syst., Princeton, N.J.*, pp. 150–154, March, 1967.

O'Shea, R. P.: Synthesis Using Distributed RC Networks, *IEEE Trans. Circuit Theory*, vol. CT-12, no. 4, pp. 546–554, December, 1965.

Pande, H. C., and R. S. Shukla: Synthesis of Transfer Functions Using an Operational Amplifier, *Proc. IEE*, vol. 112, no. 12, pp. 2208–2212, December, 1965.

Paul, R. J. A.: Active Network Synthesis Using One-port RC Networks, *Proc. IEE*, vol. 113, no. 1, pp. 83–86, January, 1966.

———: Simulation of Rational Transfer Functions with Adjustable Coefficients, *Proc. IEE*, vol. 110, no. 4, pp. 671–679, April, 1963.

Piercy, R. N. G.: Synthesis of Active RC Filter Networks, *A.T.E. J.*, vol. 21, no. 2, pp. 61–75, April, 1965.

Prabhaker, A.: An A. C. Integrator, *Electron. Eng.*, vol. 35, no. 7, pp. 444–448, July, 1963.

Radiation, Inc.: "Integrated Operational Amplifiers, Technical Information and Applications—Microelectronic Division," 1st ed., Melbourne, Fla., December, 1966.

Rajasikharan, P. K., and S. N. Rao: Synthesis of Active RC Networks Using a Differential Amplifier, *Intern. J. Electron.*, vol. 22, pp. 297–305, April, 1967.

RCA: Linear Integrated Circuit Fundamentals, *RCA Tech. Ser.* IC-40, Harrison, N.J., 1966.

Richards, P. I.: Synthesizing Transfer Functions with Two Grounded Pentodes, *Proc. IEEE*, vol. 55, no. 4, pp. 552–553, April, 1967.

Rohrer, R. A., J. A. Resh, and R. A. Hoyt: Distributed Network Synthesis and Approximation in the Time Domain, *University of Illinois CSL-R*-222, July, 1964.

Saito, M.: Sensitivity in Active RC Networks, *Electron. Commun. Japan*, vol. 47, pp. 54–62, February, 1964.

Sallen, R. P., and E. L. Key: A Practical Method of Designing RC Active Filters, *IRE Trans. Circuit Theory*, vol. CT-2, no. 1, pp. 74–85, March, 1955.

Sandberg, I. W.: Active RC Networks, *Polytechnic Institute of Brooklyn, Microwave Res. Inst. Res. Rept. R*-662-58, *PIB* 590, May, 1958.

———: Synthesis of N-port Active RC Networks, *Bell Syst. Tech. J.*, vol. 40, pp. 329–347, January, 1961.

Saraga, W.: Sensitivity of 2nd-order Sallen-Key Active RC Filters, *Electron. Letters*, vol. 3, no. 10, pp. 442–443, October, 1967.

Shenoi, B. A.: A New Technique for Twin-T RC Network Synthesis, *IEEE Trans. Circuit Theory*, vol. CT-11, no. 3, pp. 435–436, September, 1964.

Sipress, J. M.: Synthesis of Active RC Networks, *IRE Trans. Circuit Theory*, vol. CT-8, no. 3, pp. 260–269, September, 1961.

———: Synthesis of Multiparameter Active RC Networks, *NEREM Record*, November, 1960.

Skwirzynski, J. K.: "Design Theory and Data for Electrical Filters," D. Van Nostrand Company, Inc., Princeton, N.J., 1965.

Somerville, M. J., and G. H. Tomlinson: Filter Synthesis Using Active RC Networks, *J. Electron. Control*, vol. 12, no. 5, pp. 401–420, May, 1962.

Stata, R.: Operational Amplifiers, Part III—Survey of Commercially Available Operational Amplifiers, *Electromech. Design*, vol. 10, no. 2, pp. 40–51, February, 1966.

Stuart, A. G., and D. G. Lampard: Bridge Networks Incorporating Active Elements and Application to Network Synthesis, *IEEE Trans. Circuit Theory*, vol. CT-10, no. 3, pp. 357–362, September, 1963.

Su, K. L.: "Active Network Synthesis," McGraw-Hill Book Company, New York, 1965.

Sutcliffe, H.: Tunable Filter for Low Frequencies Using Operational Amplifiers, *Electron. Eng.*, vol. 36, no. 6, pp. 399–403, June, 1964.

Taylor, P. L.: Flexible Design Method for Active RC Two-ports, *Proc. IEE*, vol. 110, no. 9, pp. 1607–1616, September, 1963.

Thompson, C.: The Transfer and Impedance Functions for the Case of 2-port Thin Film R-C Type Networks, *Arch. Übertragung*, pp. 565–568, September, 1964.

Thornton, R. D.: Active RC Networks, *IRE Trans. Circuit Theory*, vol. CT-4, no. 3, pp. 78–89, September, 1957.

Thorp, J.: Realization of Variable Active Networks, *IEEE Trans. Circuit Theory*, vol. CT-12, no. 4, pp. 511–514, December, 1965.

Truxal, J. G.: "Control System Synthesis," McGraw-Hill Book Company, New York, 1955.

——— and I. M. Horowitz: Sensitivity Considerations in Active Network Synthesis, *Proc. Midwest Symp. Circuit Theory, 2d*, pp. 6-1 to 6-11, 1956.

Uzunoglu, V. N.: "Semiconductor Analysis and Design," McGraw-Hill Book Company, New York, 1964.

Van Valkenburg, M. E.: Recent Advances in Active Network Synthesis, *Proc. Midwest Symp. Circuit Theory, 4th, Marquette University, Milwaukee, Wis.*, vol. 4, pp. N1–N13, December, 1959.

Vidal, J. J., and S. M. Bawin: Frequency Analysis of Truncation Errors in RC Networks, *IEEE Trans. Electron. Computers*, vol. EC-14, no. 2, pp. 229–233, April, 1965.

Vlach, J., and J. Bendik: Active Filters with Low Sensitivity to Element Changes, *Radio Electron. Engr.*, vol. 33, pp. 305–316, May, 1967.

von der Pfordten, Hans J.: Substitution of Inductances in Integrated Circuits, in Papers on Integrated Circuit Synthesis (compiled by R. W. Newcomb and T. N. Rao), *Stanford University Center for Systems Research, Tech. Rept.* 6560-4, pp. 171–203, June, 1966.

Wadha, L. K.: Simulation of Third Order Systems by a Single Operational Amplifier, *IEEE Trans. Electron. Computers*, vol. EC-13, no. 2, p. 128, April, 1964; *Proc. IRE*, pp. 201–202, February, 1962; *Proc. IRE*, p. 465, April, 1962; *Proc. IRE*, pp. 1538–1539, June, 1962.

Warner, R. M., Jr., and J. N. Fordemwalt (eds.): "Integrated Circuits, Design Principles and Fabrication," McGraw-Hill Book Company, New York, 1965.

Widlar, R. J.: A Unique Circuit Design for a High Performance Operational Amplifier Especially Suited to Monolithic Construction, *Proc. Natl. Electron. Conf.*, vol. 21, pp. 169–174, 1965.

Wilson, B. L. H., and R. B. Wilson: Shaping of Distributed RC Networks, *Proc. IRE*, vol. 49, no. 8, pp. 1330–1331, August, 1961.

Wolff, Dieter: Two Structures to Minimize Sensitivity in Active RC Filters, *Electron. Letters*, vol. 2, no. 4, pp. 152–153, April, 1966.

Woo, B. B., and R. G. Hove: Synthesis of Rational Transfer Functions with Thin Film Distributed-parameter RC Active Networks, *Proc. Natl. Electron. Conf.*, vol. 21, pp. 241–246, 1965.

Wyndrum, R. W., Jr.: The Exact Synthesis of Distributed RC Networks, *New York University Laboratory for Electroscience, TR*-400-76, May, 1963.

Youla, D. C.: The Synthesis of Networks Containing Lumped and Distributed Elements, Part 1, *Proc. Polytechnic Symp. Generalized Networks*, 1966.

Zai, Y. F.: RC-Active Filters Using Unity-gain Amplifiers, *Electron. Letters*, vol. 3, no. 10, p. 461, October, 1967.

CHAPTER 3

GYRATOR CIRCUITS

H. J. Orchard
Lenkurt Electric Company
San Carlos, Calif.

In this chapter we shall be concerned with one specialized approach to the general problem of duplicating or improving the performance of conventional passive *RLC* networks by means of circuits that contain no inductors. The wish to solve this problem may arise from one or more of several engineering needs. Foremost among these is size reduction. When amplifiers and related electronic equipment were made with vacuum tubes, the associated passive networks were relatively small and there was little incentive to make them much smaller. But the advent of integrated circuits has so reduced the size of electronic equipment that, by comparison, the conventional *RLC* network is quite enormous and the need for a significant size reduction in order to take full advantage of the integrated circuits becomes urgent.

With thin-film and thick-film technology it is possible to reduce the size of resistors and capacitors by an acceptable factor, but for the inductor no such treatment appears possible. A good-quality inductor still requires a ferrite core and a winding, and, compared to a thin-film resistor or capacitor, it is still a very large and heavy component. Hence the need for inductorless networks.

In principle it is possible to duplicate the performance of any *RLC* network, to any desired accuracy, with a *passive RC* network; this rather remarkable result was due to Guillemin.[1] Unfortunately the price of this approach is very high. A passive *RC* network must have real and simple natural frequencies, even though its transmission zeros may be complex, and it is a very inefficient tool for simulating the behavior of an *RLC* network with lightly damped, complex natural frequencies. Not only does it require the use of an excessively large number of components, but it also results in a very high flat loss that must be offset by tandem-connected amplifiers.

The restriction to real, simple natural frequencies in the *RC* network can be removed by using active devices, basically amplifiers, in the network. One can look upon these active devices as a means of supplying power to the *RC* network, to overcome the power loss occurring in the resistors and so to reduce the damping on the natural frequencies. By supplying sufficient power to the network the natural frequencies can be made complex and as lightly damped as desired, even, as the neophyte experimenter soon discovers to his dismay, to the extent of having no damping at all. For this reason the driving-point and transfer impedances of an *RC* active network are even less restricted than those of a passive *RLC* network and theoretically one can reproduce exactly any impedance function of any *RLC* network. The practical problem thus reduces to one of discovering an optimum circuit arrangement and method of design. To this end an enormous effort has been devoted over the last 15 years.

For passive *RLC* networks the optimum circuit arrangement is usually synonymous, or nearly so, with the one that contains the minimum number of components; and the choice between equivalent circuits that contain approximately equal numbers of components is then made on the basis of sensitivity to component tolerances. One common example is the choice made between ladder and lattice filter circuits. However, it is rare that there is a great disparity between the sensitivities of different equivalent circuits to component tolerances, or that much improvement is possible by using more than the minimum number of components, although one may use extra components in order to obtain a better distribution of values.

With *RC* active circuits, on the other hand, the situation appears to be quite different. Counting each active device, say each amplifier, as one element (and this is fairly generous in view of its price and complexity) it is certainly not true for any circuits so far discovered that the one with the minimum number of components is also the one with the lowest sensitivity, or even remotely close to it. Typical of these circuits which use a minimum of components is the one originally suggested by Linvill,[2] shown in Fig. 3.1. The negative impedance converter can be made with one high-gain differential amplifier and is counted as one active element. The natural frequencies of the system are produced by the interaction of one *RC* network with the negative of the other, as seen through the negative impedance converter, and the large amount of

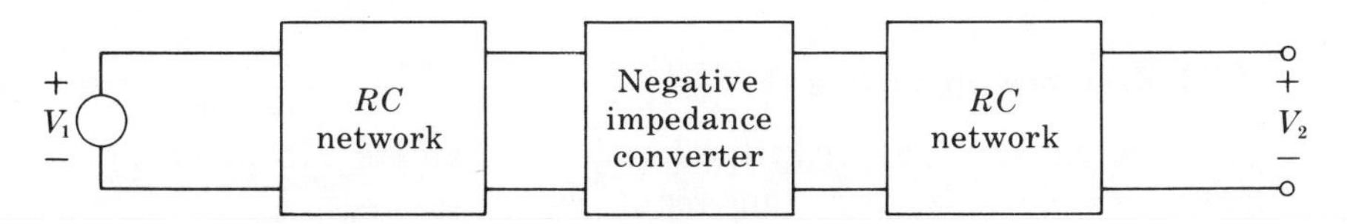

FIG. 3.1 *RC* active circuit proposed by Linvill.

cancellation which occurs between positive and negative components at the interface makes the performance very sensitive to tolerances on the components. So sensitive is this circuit in fact that it is virtually unusable except for the simplest type of network function.

In the case of transfer functions it is possible to reduce the sensitivity by factoring the transfer function into its simplest possible bilinear or biquadratic factors, providing a Linvill-type circuit for each such factor and then tandem-connecting these circuits without allowing any interaction, e.g., by using isolating amplifiers. A considerable literature exists describing many varieties of circuits in this class and discussing this dominating aspect of sensitivity. They all illustrate the process of buying reduced sensitivity at the cost of more active elements.

However, none of these active circuits ever provides a sensitivity to component tolerances which is lower than can be obtained with a well-designed passive *RLC* network, and usually it is much higher. This suggests that, in seeking inductorless circuits with acceptably low component sensitivity, one should perhaps consider those which include, as far as possible, the particular physical properties of passive *RLC* networks that are responsible for their superior quality.

One quite obvious approach to this which, until recently, has been almost completely neglected is to use exactly the same circuit as one would if inductors were available and then simply replace each inductor by some *RC* active, two-terminal device that behaves like an inductor. Assuming that such a device can be made and that the equivalent inductance has the same order of stability as that of a real inductor, then one should be able to duplicate the performance of the *RLC* network with perfectly acceptable engineering tolerances.

The great advantage of this approach is that one can still make use of the many decades of experience in designing and manufacturing *RLC* networks which are available. For almost every possible network application there exists ample documentation of what has been found to be the best circuit arrangement for economic manufacture and it seems unlikely that significantly better *RC* active circuits of any different kind could be found or could exist.

In the remainder of this chapter we shall investigate various ways of making *RC* active circuits to simulate inductors and comment on their merits and failings.

3.1 Inductance Simulation

The voltage and current at the terminals of an ideal inductor satisfy the simple, first-order, linear differential equation:

$$V = L\frac{dI}{dt} \tag{3.1}$$

where L is the inductance in henrys. Any two-terminal device that satisfies (3.1) will qualify as a replacement for an inductor.

One aspect of (3.1) is that the inductor is an energy-storage element; the energy is stored in the magnetic field and has a magnitude:

$$LI^2/2 \qquad \text{joules} \tag{3.2}$$

Of the various elements available to us in an RC active circuit, only the capacitor has the ability to provide a derivative relation between voltage and current, and also to store energy. It is therefore essential that the inductance-simulating circuit contain at least one capacitor.

On the other hand, an RC circuit containing n capacitors will, in general, cause the voltage and current at any terminal pair to satisfy an nth-order differential equation. Only by introducing some form of degeneracy into the network, equivalent to cancelable factors in the impedance function, is it possible to reduce the order of the differential equation. This is normally undesirable because it increases both the difficulty of adjustment and the sensitivity to component errors. We conclude therefore that the optimum inductance-simulating circuit should contain just one capacitor.

In this event the simulating circuit reduces to the general form shown in Fig. 3.2. It consists of a frequency-independent two-port network, containing only resistors and active devices, terminated at port 2 by a capacitor and presenting the desired inductive behavior at port 1. This two-port network proves to be the vital ingredient in our present approach to the construction of inductorless filters; we shall examine its properties in some detail.

The voltage and current at port 1 are related by:

$$V_1 = L\frac{dI_1}{dt} \tag{3.3a}$$

while at port 2 the capacitor imposes the condition:

$$I_2 = -C\frac{dV_2}{dt} \tag{3.3b}$$

These two voltages and currents are also related algebraically by the

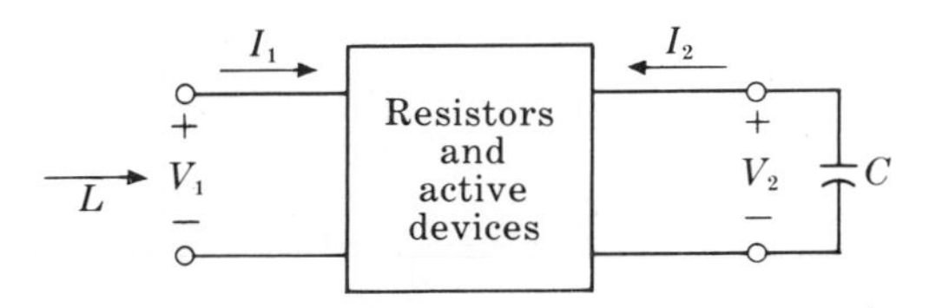

Fig. 3.2 General inductance-simulating circuit.

two-port network equations:

$$\begin{aligned} V_1 &= R_{11}I_1 + R_{12}I_2 \\ V_2 &= R_{21}I_1 + R_{22}I_2 \end{aligned} \tag{3.4}$$

Differentiating the second of these equations and substituting, via (3.3*b*), for I_2 in the first gives:

$$V_1 = R_{11}I_1 - CR_{12}R_{21}\frac{dI_1}{dt} - CR_{12}R_{22}\frac{dI_2}{dt} \tag{3.5}$$

Comparing this with (3.3*a*) we see that the two-port network must satisfy the conditions:

$$R_{11} = R_{22} = 0 \tag{3.6a}$$

$$R_{12}R_{21} = -\frac{L}{C} \tag{3.6b}$$

Assuming that we wish to produce a positive inductance from a positive capacitor, (3.6*b*) indicates that one of the two factors R_{12} and R_{21} has to be negative and the other positive. Consequently R_{12} and R_{21} cannot be equal and it follows that the network must be nonreciprocal.

To simplify the notation, let the port numbering be such that R_{12} is the negative term† and then put

$$R_1 = -R_{12} \qquad R_2 = R_{21}$$

i.e., so that R_1 and R_2 are both positive. The matrix form of (3.4) is then:

$$\begin{bmatrix} V_1 \\ V_2 \end{bmatrix} = \begin{bmatrix} 0 & -R_1 \\ R_2 & 0 \end{bmatrix} \begin{bmatrix} I_1 \\ I_2 \end{bmatrix} \tag{3.7}$$

Such a network has the quite general property of impedance inversion illustrated in Fig. 3.3. If one port is terminated with an impedance Z the input impedance at the other port becomes R_1R_2/Z, as may readily be checked from (3.7). The transformation of a capacitive reactance

† There is no special significance in this choice of the signs; the other possible choice will occur naturally in some of the circuits we shall examine and all the results derived are equally valid for both cases.

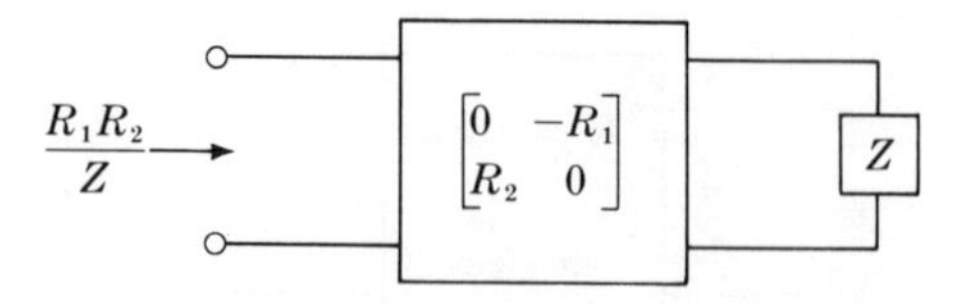

FIG. 3.3 Impedance-inverting circuit.

into an inductive reactance is merely a special case of this impedance inversion.

In addition to being nonreciprocal this network is also, in general, an active one as may be seen by examining the expression for the total power absorbed at the two ports, namely,

$$V_1I_1 + V_2I_2 = I_1I_2(R_2 - R_1) \tag{3.8}$$

If $R_2 - R_1$ is not zero, one can always choose the signs for I_1 and I_2 so that this power is negative. In other words there will always be circumstances under which the network can *supply* power to the equipment to which it is connected.

When the network is used in the simple inductance-simulating circuit of Fig. 3.2 it is of little consequence that it is an active device, because the voltages and currents at the ports are so constrained that the power flow *into* one port is always equal to the power flow *out* of the other port. In some other applications of the network, however, it is desirable to suppress the active property, and we now examine the special conditions under which this can occur.

3.2 The Gyrator

The exceptional case in (3.8) occurs with $R_1 = R_2$ when the expression for the total power vanishes for all values of I_1 and I_2. The network can then neither supply nor absorb any power and it reduces to a passive, nonreciprocal, nondissipative two-port network.

This special network was first conceived by Tellegen[3] as an additional ideal circuit element, to complement the ideal transformer, inductor, capacitor, and resistor, and intended as a means by which synthesis techniques could be extended to nonreciprocal circuits. He named it the *gyrator* and devised for it the symbol shown in Fig. 3.4. The name was selected to reflect the similarity between the descriptive equations for the network and those of a mechanical gyrostatic system; so to speak, the voltages at the ports are gyrated with the currents.

The common value of R_1 and R_2 is usually referred to as the *gyration resistance.* As mentioned at the end of the previous section, it matters little for most of our applications whether R_1 and R_2 are exactly equal or not; it is primarily the product R_1R_2 which concerns us. For this reason

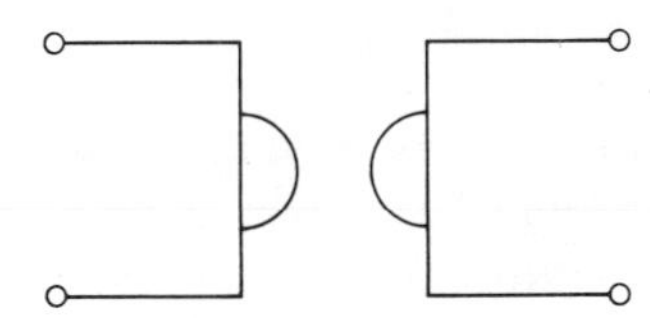

FIG. 3.4 Symbol for a gyrator.

we shall refer to the network with the impedance matrix of (3.7), where R_1 is not necessarily equal to R_2, as a gyrator, even though, properly speaking, the name should be reserved for the special case when $R_1 = R_2$. Some people have described the more general case as an *active gyrator* but we shall not draw this distinction.

Although, in the original paper describing the gyrator, Tellegen gave some consideration to possible methods of construction, it was more in the nature of a theoretical justification and there was no serious thought at the time of making the gyrator as a practical circuit element. Nevertheless, the attractive possibilities opened up by the concept of the gyrator spurred people to seek practical ways of building the device.

The prime property of the gyrator is its nonreciprocity, and to effect a practical construction one needs nonreciprocal building blocks of some kind. Passive, nonreciprocal devices, using the Hall effect, are available but they have the drawback of also being very lossy. The most commonly available nonreciprocal devices are vacuum tubes and transistors, which, in contrast to the Hall-effect device, are very active. By suitable interconnections of transistors and resistors it is possible to retain the essential nonreciprocity of the former while at the same time so diluting their active nature that the overall result is a gyrator. Many writers have described ingenious ways of doing this.

Before discussing the details of some of these circuits we shall first examine the two principal imperfections that can exist in practical gyrators and see how they influence the ability of the gyrator to provide an inductance from a capacitor. Ideally, the gyrator should have an impedance matrix:

$$\begin{bmatrix} 0 & -R \\ R & 0 \end{bmatrix} \tag{3.9}$$

or, by inverting (3.9), an admittance matrix:

$$\begin{bmatrix} 0 & G \\ -G & 0 \end{bmatrix} \tag{3.10}$$

where $G = 1/R$.

The first and most obvious imperfection is that the elements on the principal diagonal of these matrices will not be exactly zero. Assume that the impedance matrix is

$$\begin{bmatrix} \epsilon R & -R \\ R & \epsilon R \end{bmatrix} = \begin{bmatrix} 0 & -R \\ R & 0 \end{bmatrix} + \begin{bmatrix} \epsilon R & 0 \\ 0 & \epsilon R \end{bmatrix} \tag{3.11}$$

where ϵ is dimensionless and small compared to unity. It is evident from (3.11) that the ϵR terms simply represent resistances in series with the ports of a perfect gyrator. The equivalent circuit of such an imperfect gyrator, terminated at one port by a capacitor C, is shown in Fig. 3.5.

The impedance seen looking into the left-hand port consists of the resistance ϵR at that port in series with the inverse, as produced by the ideal gyrator, of the capacitor and series resistance ϵR at the right-hand port. The net result is an inductance R^2C with a shunt resistance R/ϵ and a series resistance ϵR. In this respect it is similar to a normal iron-cored inductor in which core and winding losses are represented by shunt and series resistances respectively. In both cases the Q factor varies with frequency in the same way, reaching a maximum at the point where the power loss in the series resistance equals the power loss in the shunt resistance.

Neglecting terms in ϵ^2 and assuming a lossless capacitor, the Q factor of the simulated inductance is

$$Q = \frac{1}{\epsilon} \frac{\omega CR}{1 + (\omega CR)^2} \tag{3.12}$$

The maximum value of this is $(2\epsilon)^{-1}$ and occurs when $\omega CR = 1$. An exactly dual argument, starting from an imperfect admittance matrix

$$\begin{bmatrix} \epsilon G & G \\ -G & \epsilon G \end{bmatrix} \tag{3.13}$$

leads to the equivalent circuit of Fig. 3.6 and the same expression $(2\epsilon)^{-1}$ as the maximum obtainable Q factor.

Nonzero values in the principal diagonal of either matrix thus correspond almost exactly to dissipation in a real inductor and limit, by a very simply computed amount, the maximum achievable Q factor in any simulated inductance.

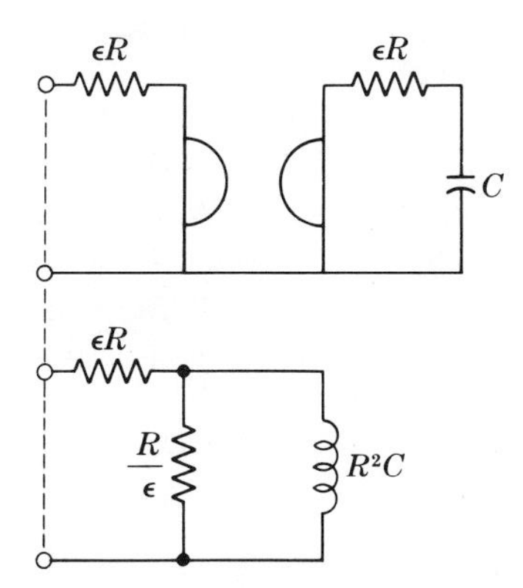

Fig. 3.5 Equivalent circuit of a gyrator having nonzero terms ϵR on the principal diagonal of its impedance matrix.

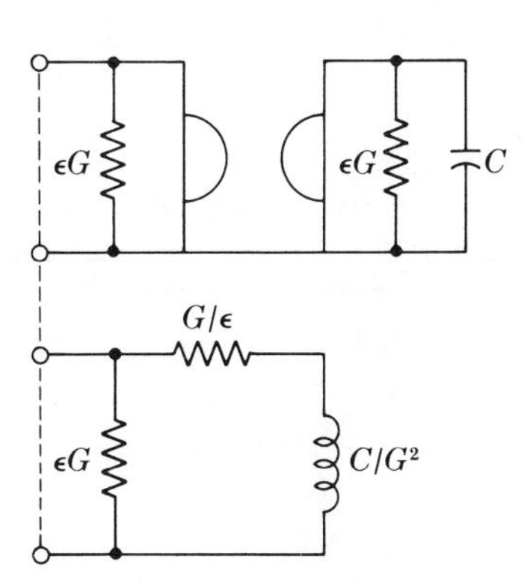

Fig. 3.6 Equivalent circuit of a gyrator having nonzero terms ϵG on the principal diagonal of its admittance matrix.

The second main imperfection that can occur in a gyrator is that the gyration resistances are not purely resistive but have a small phase angle ϕ. The impedance matrix can then be written

$$\begin{bmatrix} 0 & -\mathrm{R}e^{j\phi} \\ \mathrm{R}e^{j\phi} & 0 \end{bmatrix} \tag{3.14}$$

and the simulated inductance obtained from a perfect capacitor has an impedance

$$j\omega CR^2 e^{j2\phi} \tag{3.15}$$

When ϕ is small the exponential can be replaced by the first two terms of its Taylor series and (3.15) becomes:

$$-\omega CR^2 2\phi + j\omega CR^2 \tag{3.16}$$

A *positive* phase angle in the gyration resistances thus produces a *negative* series resistance, and hence a negative Q factor of magnitude $-(2\phi)^{-1}$.

By an exactly dual argument one finds that a *negative* phase angle ϕ in each gyration conductance G in (3.10) produces a negative shunt conductance across the simulated inductance and hence a negative Q factor of the same magnitude, $-(2\phi)^{-1}$.

When both imperfections are present simultaneously the net effect upon the Q, found by computing the total effective series resistance (or shunt conductance), is quite simply

$$Q = \frac{1}{2(\epsilon - \phi)} \tag{3.17}$$

The cancellation which can occur in this expression when ϵ and ϕ are both positive proves to be of great practical importance. In striving for maximum Q in the simulated inductance one naturally attempts to keep both ϵ and ϕ separately as small as possible, but it is inevitable that neither can be made exactly zero, and so the possibility of achieving nominally infinite Q by slightly increasing the smaller to make $\epsilon = \phi$ is quite valuable. In practice, of course, the capacitor which terminates the gyrator will have some dissipation and this will augment slightly the ϵ term; the phase angle of the gyration resistance can help to cancel this dissipation as well.

In the event that the gyration resistances are not equal it is convenient to define the principal diagonal terms in (3.11) as

$$\begin{bmatrix} \epsilon_1 \sqrt{R_1 R_2} & -R_1 \\ R_2 & \epsilon_2 \sqrt{R_1 R_2} \end{bmatrix} \tag{3.18}$$

The expression for the maximum obtainable Q factor is then simply $1/(2\sqrt{\epsilon_1 \epsilon_2})$. Likewise, if the phase angles of R_1 and R_2 are ϕ_1 and ϕ_2

respectively, (3.17) extends to

$$Q = \frac{1}{2\sqrt{\epsilon_1\epsilon_2} - \phi_1 - \phi_2} \tag{3.19}$$

3.3 Some Gyrator Realizations

In this section we shall consider a few typical examples of the many circuits which either have been used or suggested for use as practical gyrators. To make a gyrator one has to use at least one component which is inherently nonreciprocal, and of these the most commonly available are vacuum tubes and transistors. When operated in the common-cathode or common-emitter mode these devices can be regarded as grounded two-port networks with an admittance matrix, which, to a very crude approximation, is of the form

$$y = \begin{bmatrix} 0 & 0 \\ g & 0 \end{bmatrix} \tag{3.20}$$

The pentode is the device most accurately represented by (3.20) and the g is its transconductance. The transistor, on the other hand, has values for y_{11} and y_{12} which initially are too large to be neglected, but which can be substantially reduced by local feedback.

Possibly the earliest circuit used for a gyrator, although not explicitly recognized as such at the time, was the so-called reactance tube[4] which was invented in the 1930s for use with automatic frequency control. The circuit is shown in Fig. 3.7 and consists of a two-port network formed by the grid, plate, and cathode of a pentode, bridged by a resistor R; a capacitor C normally terminates port 1 and the desired inductive reactance is presented at port 2. The admittance matrix of the bridging

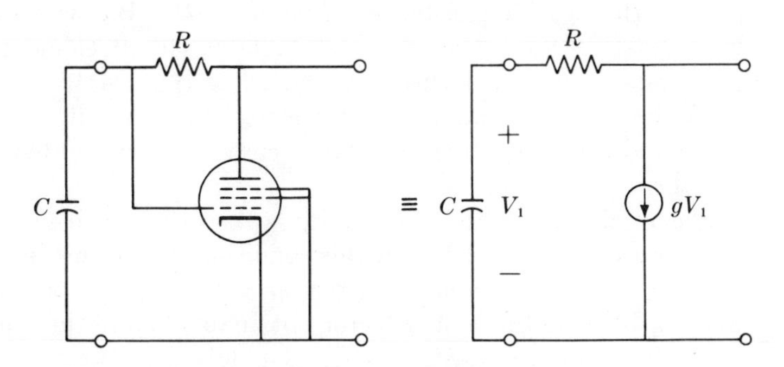

FIG. 3.7 Reactance tube circuit.

resistor is

$$y = \begin{bmatrix} \frac{1}{R} & -\frac{1}{R} \\ -\frac{1}{R} & \frac{1}{R} \end{bmatrix} \tag{3.21}$$

and, adding this to (3.20), we obtain

$$y = \begin{bmatrix} \frac{1}{R} & -\frac{1}{R} \\ g - \frac{1}{R} & \frac{1}{R} \end{bmatrix} \tag{3.22}$$

as the matrix of the reactance-tube gyrator.

Applying the results of the previous section one finds the value of the simulated inductance and its maximum Q factor to be

$$L = \frac{CR^2}{Rg - 1} \qquad Q = \frac{\sqrt{Rg - 1}}{2} \tag{3.23}$$

R is usually chosen so that $Rg \gg 1$ and then, very nearly, $L = CR/g$ and $Q = \frac{1}{2}\sqrt{Rg}$. But even if Rg is made as large as 100, for example, by having $R = 20$ kΩ and $g = 5$ mA/volt, the maximum Q factor is still only 5. The circuit is thus not capable of producing very high-Q inductances. In its original application this feature hardly mattered because the dissipation was removed by the oscillator circuit to which the reactance tube was connected. The important point was that the simulated inductance could easily be controlled by varying the transconductance of the tube by means of the grid bias.

Virtually the same circuit was suggested, much later, by Shekel[5] in a deliberate attempt to show how a gyrator might be constructed. He realized that the nonzero principal diagonal terms in (3.22) were undesirable and arranged to remove them simply by shunting both gyrator ports with negative resistances of values $-R$. It was assumed by then that negative resistances were available as circuit elements; in practice one would have had to use something like the negative-resistance characteristic of the dynatron or transitron[6] tube circuit, or else the input impedance of a negative impedance converter terminated by a positive resistor.

Shekel also suggested making $Rg = 2$ so that both gyration conductances were equal. Although this condition does result in a true, passive gyrator, it is undesirable from the special viewpoint of inductance simulation, because the best Q factor obtainable before the ports are loaded with the negative resistances is then only $\frac{1}{2}$. Achieving a Q factor of several hundred would thus require a matching of positive and negative resistances to an accuracy of a few parts in a thousand. Making $Rg = 100$ would ease this matching problem by a factor of 10.

The elimination of nonzero terms on the principal diagonal of the y matrix by cancellation with negative resistances shunting the ports is a temptingly easy artifice which has been used by many gyrator designers, but if the gyrator is to be used for the rather special purpose of simulating a high-Q inductance it is a temptation which must be resisted if one is to avoid a prohibitively difficult matching problem. However, it is not necessary for the negative resistances to appear explicitly across the ports for this matching problem to exist. This feature is illustrated by the gyrator circuit of Fig. 3.8 which was first given, implicitly, by Carlin[7] and later discussed by Harrison.[8]

The T network formed by the three resistors R_1, $-R_2$, and R_3 has the impedance matrix

$$Z = \begin{bmatrix} R_1 - R_2 & -R_2 \\ -R_2 & R_3 - R_2 \end{bmatrix} \tag{3.24}$$

The negative impedance converter reverses the direction of I_2 while preserving the polarity of V_2 and this has the effect of changing the sign of the elements in the right-hand column of (3.24). The impedance matrix of the tandem connection is therefore

$$Z = \begin{bmatrix} R_1 - R_2 & +R_2 \\ -R_2 & R_2 - R_3 \end{bmatrix} \tag{3.25}$$

For the circuit to act as a gyrator we must evidently have $R_1 = R_2 = R_3$ in order that the principal diagonal terms should vanish. Here again we see that these required zero terms are obtained by a simple cancellation between resistances equal in magnitude to the gyration resistance. It means, for example, that to achieve a Q factor of 500 from a simulated inductance, both R_1 and R_3 must match R_2 to within 0.1 percent.

One way of avoiding this cancellation, and the consequent difficulties of component matching, is to use in the construction only such devices as will naturally make negligible contribution to the y_{11} and y_{22} (or z_{11} and z_{22}) terms. For example, the admittance matrix of the gyrator

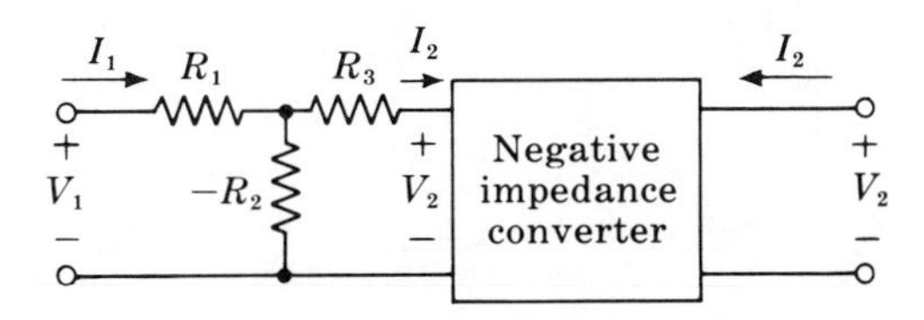

FIG. 3.8 Gyrator circuit due to Carlin.

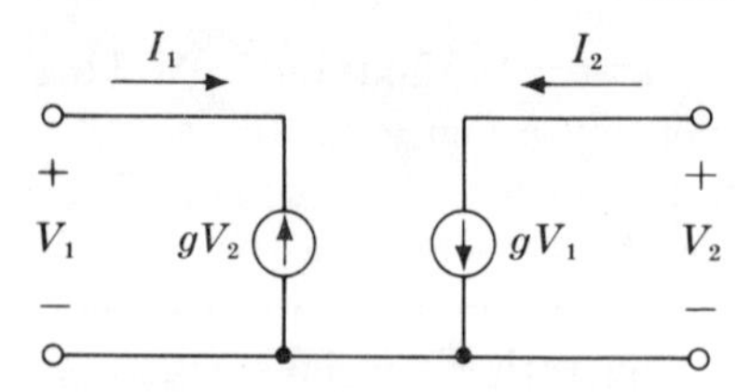

FIG. 3.9 Gyrator constructed from two voltage-controlled current sources.

can be split up thus

$$\begin{bmatrix} 0 & -g \\ g & 0 \end{bmatrix} = \begin{bmatrix} 0 & 0 \\ g & 0 \end{bmatrix} + \begin{bmatrix} 0 & -g \\ 0 & 0 \end{bmatrix} \tag{3.26}$$

where the two matrices on the right-hand side represent ideal voltage-controlled current sources, i.e., amplifiers with infinite input and output impedances and controlled transconductance $\pm g = I_{\text{out}}/V_{\text{in}}$. Two such controlled sources, of opposite polarity and connected back to back as in Fig. 3.9, will provide the required gyrator matrix with no cancellation. A single pentode is a controlled source with a positive value of g in its y matrix, and a pentode preceded by a phase-inverting stage forms a controlled source with a negative value of g. Thus, in principle at least, the circuit of Fig. 3.10 will act as a gyrator. In practice the presence of dc power-feeding paths to the output tubes would contribute something to the main diagonal terms of the y matrix and so limit the maximum Q of a simulated inductance. A practical version of this circuit, but using transistors, was described by Shenoi.[9]

The need for the phase-inverting stage can be eliminated if one considers a balanced-to-ground version of this gyrator, for then the phase inversion can be accomplished merely by a cross connection. This arrangement, shown in skeleton form in Fig. 3.11, was described by Sharpe.[10] As with the grounded version, however, the quality of this gyrator would also suffer because of the shunting of the ports by the inevitable dc power-feeding resistors.

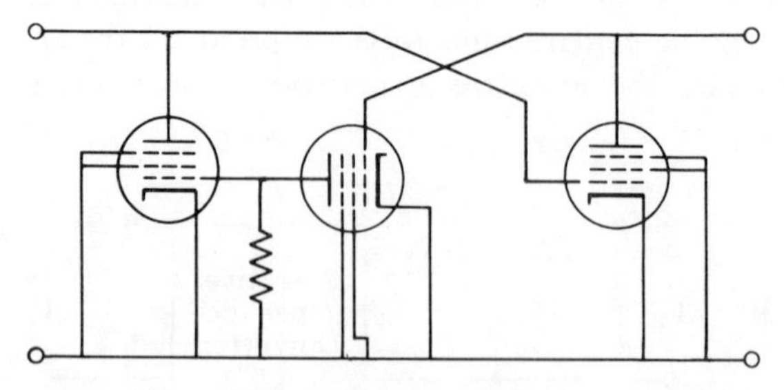

FIG. 3.10 Elements of a vacuum-tube circuit to realize the controlled-source gyrator of Fig. 3.9.

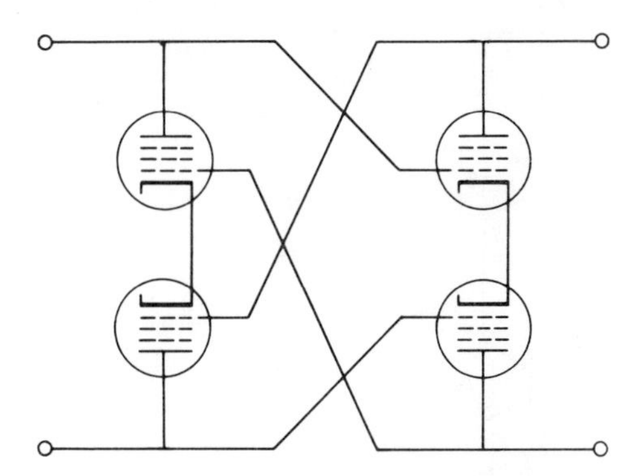

FIG. 3.11 Gyrator circuit due to Sharpe.

If one wishes to exploit this approach to produce a gyrator which can give stable Q factors of several hundred in a simulated inductance it is necessary to use rather more elaborate circuits for the voltage-controlled current sources. The major limitation of the simple circuits considered so far lies in the input and output impedances. Even though the basic nonreciprocal device itself may have very high impedances the latter will necessarily be reduced by simple power-feeding and bias arrangements. In using a gyrator to simulate an inductor in a filter it will usually happen that the gyrator faces a capacitor at both ports and there is then no means of feeding power supplies to it from outside; from a dc point of view it must therefore be completely self-contained. It is also highly desirable that it be dc coupled throughout, partly so that any inductance simulation will hold down to zero frequency and partly to eliminate bulky blocking capacitors.

An output circuit which satisfies these requirements can be made with two complementary bipolar transistors as in Fig. 3.12. Each transistor has a naturally high differential resistance at its collector and this is increased by the series feedback from the resistors in the emitter leads. With a perfectly balanced circuit the dc output voltage will be just half the supply voltage. The total output impedance is the parallel combination of the two transistor circuits, and in practice can reach several megohms. The drive to the circuit can be applied to either base separately or to both bases in parallel.

The input impedance at the base of a bipolar transistor is intrinsically quite low and must be increased by a large amount of feedback in order to approach the high output impedance obtainable from the circuit of Fig. 3.12. For modest applications it may suffice simply to use a fairly large resistor in the emitter lead, as in Fig. 3.13a. Under normal circumstances the bias current for the base of this transistor would be

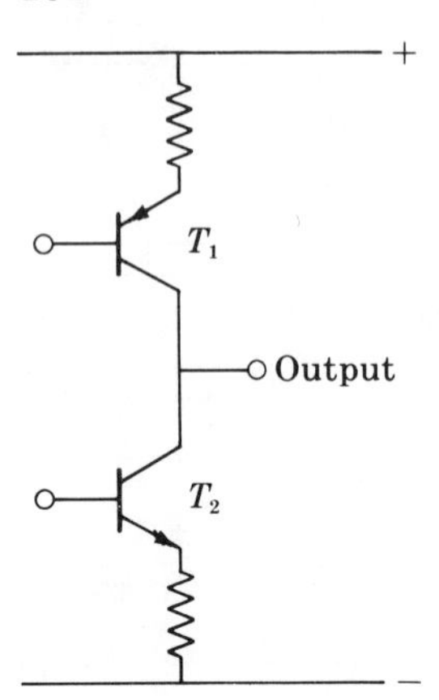

FIG. 3.12 High-impedance output circuit for a voltage-controlled current source.

derived from a potentiometer across the supply voltage, but using this method here would shunt the input impedance quite seriously. If both controlled sources of the gyrator use similar input and output circuits it is possible, in this special application, to supply the bias current for each input from the output of the other controlled source. This will slightly unbalance the currents in the two complementary transistors of each output circuit but the extent is negligible.

For this arrangement to work the circuit must be designed so that the input and output terminals of each controlled source are nominally at

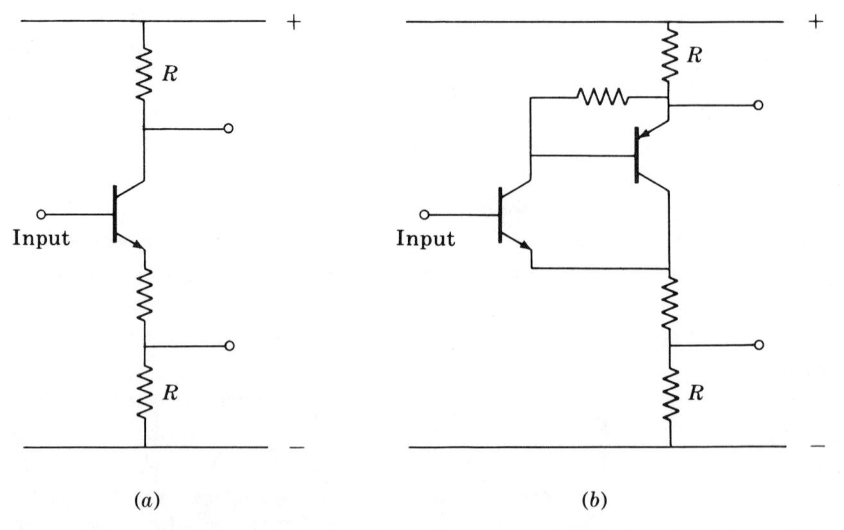

FIG. 3.13 (a) Simple, single-transistor input circuit; (b) two-transistor, high-impedance input circuit for voltage-controlled current source.

the same potential, i.e., the base of the transistor in Fig. 3.13 must be at the same potential as the two collectors in Fig. 3.12. This in turn dictates to some extent how the input and output circuits can be coupled together. A complete simple gyrator built from these units is shown in Fig. 3.14. The difference of polarity between the two controlled sources is achieved by taking the drive to the output circuit from the input *collector* in one case, and from a tap on the input *emitter* resistance in the other. This circuit was given, in essence, by Rao and Newcomb.[11]

Around the loop formed by the two controlled sources there is a net reversal of polarity and, with no external loading on the ports, a very high gain. In the notation of (3.13) this loop gain is

$$20 \log_{10} \frac{1}{\epsilon^2} \qquad \text{dB} \tag{3.27}$$

An ϵ of only 0.01 thus leads to a loop gain at 80 dB. All this gain appears as negative feedback to stabilize the dc operating conditions which are consequently quite rigidly controlled. The problem of keeping such a large amount of negative feedback stable is, in most cases, solved quite simply when the gyrator is loaded by a capacitor to simulate an inductor. This capacitor causes the loop gain to decrease steadily at 20 dB/decade, starting at a frequency of a few hertz. Only if the gyrator is required to work at several hundred kilohertz will difficulties of instability from this cause arise.

With careful design this type of gyrator will produce Q factors of up to perhaps 200, the limitation being caused by the relatively low input

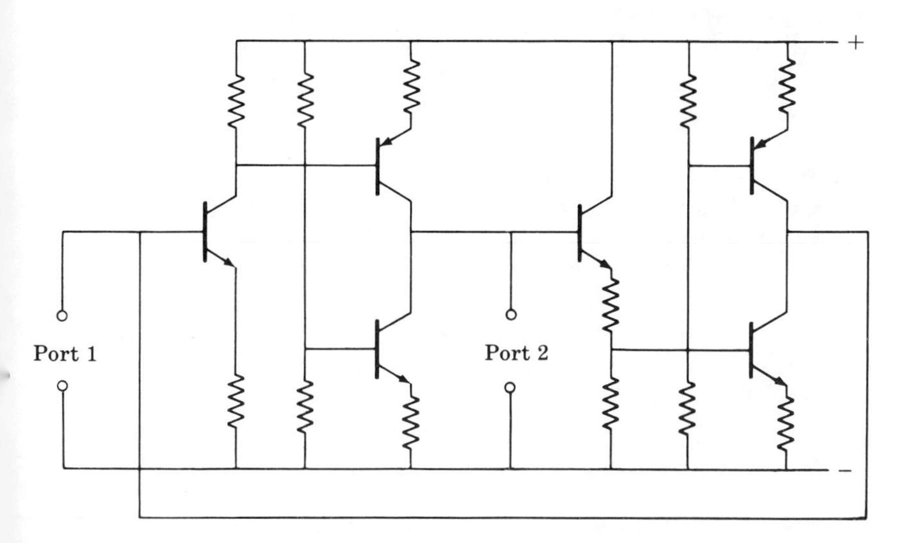

Fig. 3.14 Gyrator built from the circuits of Figs. 3.12 and 3.13*a*.

impedance of the controlled sources. The latter may be increased either by replacing the input bipolar transistor with a metal-oxide-semiconductor field-effect transistor (MOSFET) or else by using more elaborate circuitry. As an example of the latter one can replace the single-transistor input circuit of Fig. 3.13*a* with the two-transistor feedback circuit of Fig. 3.13*b*. A gyrator using this arrangement, combined with an MOS input transistor, was described by Sheahan and Orchard.[12] A very similar gyrator, but using a differential input stage, was published by Holmes, Gruetzmann, and Heinlein.[13]

With gyrators of this class the phase shift which occurs in the voltage-controlled current sources, due to the natural band limitation of the amplifiers, results in gyration conductances with a negative phase angle. This causes an increase in the Q factor of a simulated inductance for the reasons given in Sec. 3.2. By adding capacitors of a few picofarads across appropriate resistors it is possible to change this phase shift slightly and hence to control the Q factor. However, if Q factors of several hundred are required the total phase shift from all causes has to be kept down to about a thousandth of a radian. This means that the 3-dB frequency of the amplifiers must be about three decades higher than the operating frequency.

This situation would be easier to achieve and control if it were possible to apply feedback, directly from output to input, over each controlled source. But the voltage fed back to the input stage has to be derived from, and directly proportional to, the output current from the controlled source, and herein lies the difficulty. The only convenient location for a resistor, across which the output current can develop a feedback voltage, is in the emitter lead of that output transistor which is being driven. The polarity of the voltage across a resistor there is correct for negative feedback only when the transconductance of the controlled source is positive; even then the dc potential is not correct. In spite of these remarks it may nevertheless be possible, with sufficient ingenuity, to apply overall feedback to both halves of the gyrator, although no such circuits have yet been published.

On account of these difficulties, and the fact that monolithic integration of complementary bipolar transistors has so far not been practical, interest has been directed toward gyrator circuits using readily available amplifiers that are integrable. The most common such device is the operational amplifier.

3.4 Gyrators Made with Operational Amplifiers

Operational amplifiers are high-gain, dc-coupled differential amplifiers. Originally they were made with vacuum tubes and were primarily

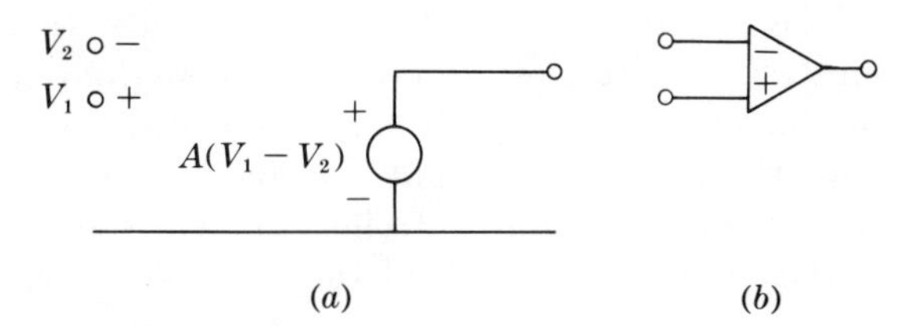

FIG. 3.15 (*a*) Equivalent circuit of and (*b*) symbol for an ideal operational amplifier.

intended for use in analog computers, but since then they have become widely available at low cost in integrated form and they are now accepted as active devices of almost universal application. The idealized equivalent circuit is shown in Fig. 3.15. The output is a grounded, zero-impedance generator whose emf is A times the difference between the potentials applied to the two differential input terminals. The input impedance at each of these terminals is infinite and the gain factor A is assumed to approach infinity.

Practical amplifiers typically have output impedances of a few hundred ohms and input impedances of a few tens of thousands of ohms. The common-mode input impedance, i.e., with both inputs paralleled, on the other hand, is usually very high, in the order of several megohms. The gain factor A may range from 10^3 to 10^6; commonly it is of the order of 10^4. The input and output terminals are arranged to be at ground potential and the amplifiers consequently require both a positive and a negative power supply.

In most practical applications the operational amplifier has a large amount of negative feedback applied to it by the circuitry in which it is embedded, and in order to keep the whole system stable it is necessary to control the way in which the gain of the amplifier falls off with increasing frequency. This is done by adding external components, usually a small capacitor and resistor, across some part of the amplifier circuit to cause the gain to drop off at a uniform rate of 20 dB/decade from some chosen frequency up to the point where the gain has been reduced to zero. To a fair approximation the input/output characteristic of the amplifier can then be represented by

$$\frac{V_{\text{in}}}{V_{\text{out}}} = \frac{1}{A} = \delta\left(1 + \frac{p}{\omega_0}\right) \tag{3.28}$$

where ω_0 is the angular frequency at the 3-dB point on the response. In what follows we shall often omit the frequency-dependent factor $(1 + p/\omega_0)$ and use the symbol δ to represent the complete function. The specific meaning in each case should be clear from the context.

The negative feedback which is applied to the operational amplifier must, in its simplest form, be applied from the output terminal back to the negative input terminal as shown in Fig. 3-16*a*. This reduces both the input and output impedances to a very low value and forms essentially a current-controlled voltage source. If it were possible to float the input terminals of this controlled source off ground one might consider constructing a gyrator by simulating the component parts of the impedance matrix when split up thus

$$\begin{bmatrix} 0 & -R \\ R & 0 \end{bmatrix} = \begin{bmatrix} 0 & -R \\ 0 & 0 \end{bmatrix} + \begin{bmatrix} 0 & 0 \\ R & 0 \end{bmatrix} \tag{3.29}$$

by analogy with (3.26) but there appears to be no way of doing this. If the operational amplifier, with its negative feedback, is driven instead at the positive input terminal as in Fig. 3.16*b*, the output impedance remains low but the input impedance now becomes very high and the circuit forms a voltage-controlled voltage source.

Any gyrator built from operational amplifiers can be regarded as an interconnection of these two types of controlled sources. We are thus unable to produce by simple feedback arrangements the voltage-controlled current sources required for providing the two right-hand-side terms in (3.26). Instead we are forced to adopt some form of cancellation in order to achieve the high output impedance from the controlled sources which is necessary for giving the desired zeros in the principal diagonal of the admittance matrix. The cancellation can be done in two main ways. The first, a relatively obvious one, involves a cancellation between positive and negative resistances and demands, in practice, a critical match between component values. The second operates by a cancellation of emfs which in some realizations, though not all, may be

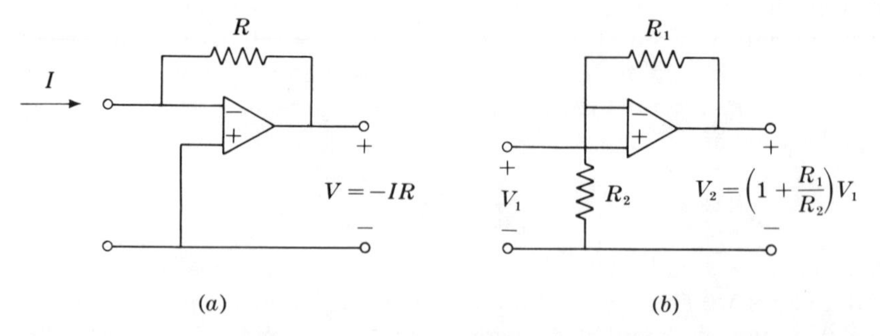

Fig. 3.16 Simple controlled sources made with operational amplifiers. (*a*) Current-controlled voltage source; (*b*) voltage-controlled voltage source.

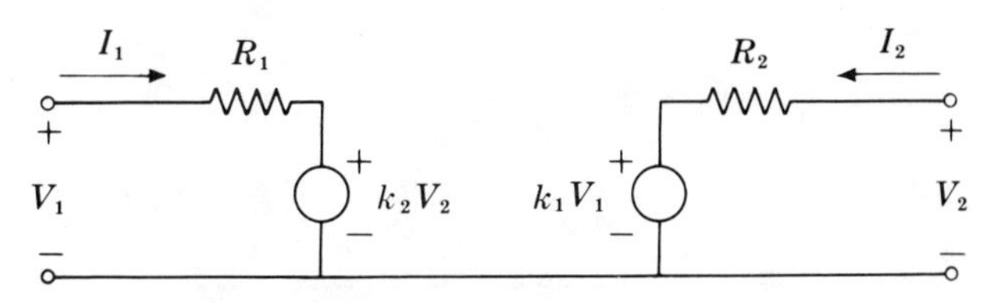

FIG. 3.17 First stage in the development of a gyrator using voltage-controlled voltage sources.

achieved without any critical matching. The latter appears to offer most promise for the design of high-grade gyrators for inductance simulation. We shall examine both approaches.

Consider the simple two-port circuit of Fig. 3.17 which consists of two grounded voltage generators, one proportional to V_2 and feeding through R_1 to port 1, and one proportional to V_1 feeding through R_2 to port 2. The current at port 1 is

$$I_1 = \frac{V_1 - k_2V_2}{R_1} \tag{3.30}$$

and at port 2

$$I_2 = \frac{V_2 - k_1V_1}{R_2} \tag{3.31}$$

Hence the admittance matrix is

$$\begin{bmatrix} \dfrac{1}{R_1} & \dfrac{-k_2}{R_1} \\ \dfrac{-k_1}{R_2} & \dfrac{1}{R_2} \end{bmatrix} \tag{3.32}$$

The main diagonal terms can be reduced to zero by shunting port 1 with a negative resistance of value $-R_1$ and port 2 with a negative resistance $-R_2$. The off-diagonal terms can be adjusted for gyrator behavior by setting, for example, $k_1 = -1$, $k_2 = 1$, and the circuit then takes the form shown in Fig. 3.18. The admittance matrix becomes

$$\begin{bmatrix} 0 & \dfrac{-1}{R_1} \\ \dfrac{1}{R_2} & 0 \end{bmatrix} \tag{3.33}$$

The grounded voltage generators can be made with the controlled-source circuits of Fig. 3.16 and the negative resistances with the circuit

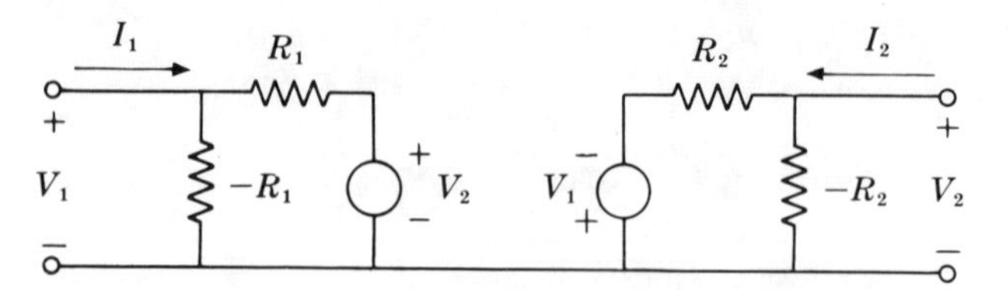

FIG. 3.18 Equivalent circuit of a gyrator using voltage-controlled voltage sources and resistance cancellation.

of Fig. 3.19. The latter is simply the voltage-controlled voltage source of Fig. 3.16*b* set for a gain of two and with a resistor R bridging from output to input; the input impedance is a short-circuit-stable, negative resistance $-R$. One can economize in amplifiers by incorporating the negative-resistance circuitry into that used for the grounded voltage generators. A complete gyrator of this type could, for example, take the form shown in Fig. 3.20. Here $k_1 = -2$, $k_2 = 2$, and $R_1 = R$. The admittance matrix is

$$\begin{bmatrix} 0 & \frac{-2}{R} \\ \frac{2}{R} & 0 \end{bmatrix} \tag{3.34}$$

Several variants of this gyrator circuit have been described in the literature. A relatively complicated one, due to Morse and Huelsman,[14] is shown in Fig. 3.21; the notation has been simplified slightly from the original publication. All four amplifiers operate as current-controlled voltage sources and have output voltages as indicated. Knowing that

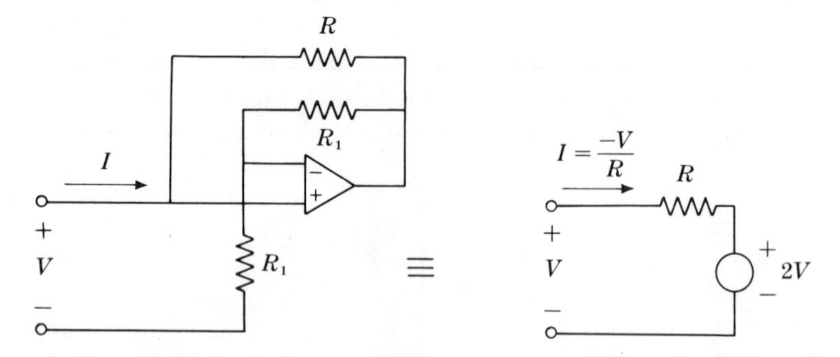

FIG. 3.19 Simple negative-resistance circuit using a voltage-controlled voltage source.

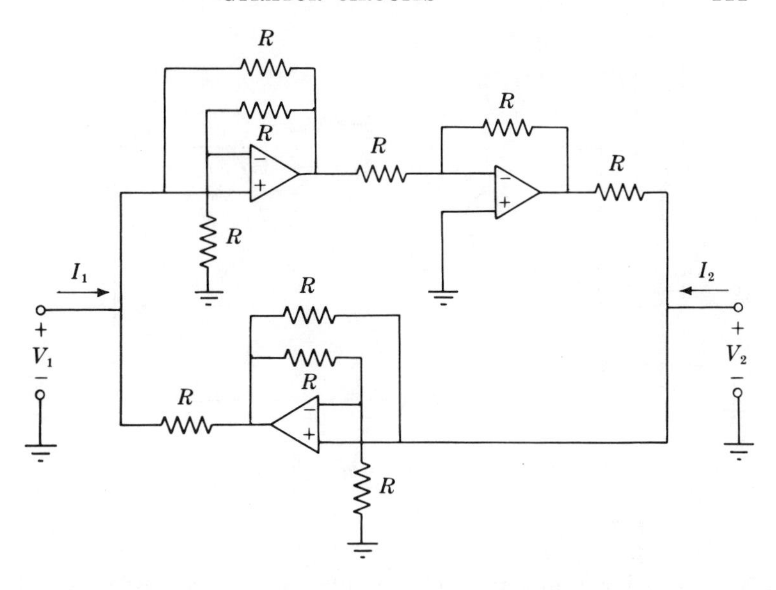

FIG. 3.20 A gyrator made with operational amplifiers and having the equivalent circuit of Fig. 3.18.

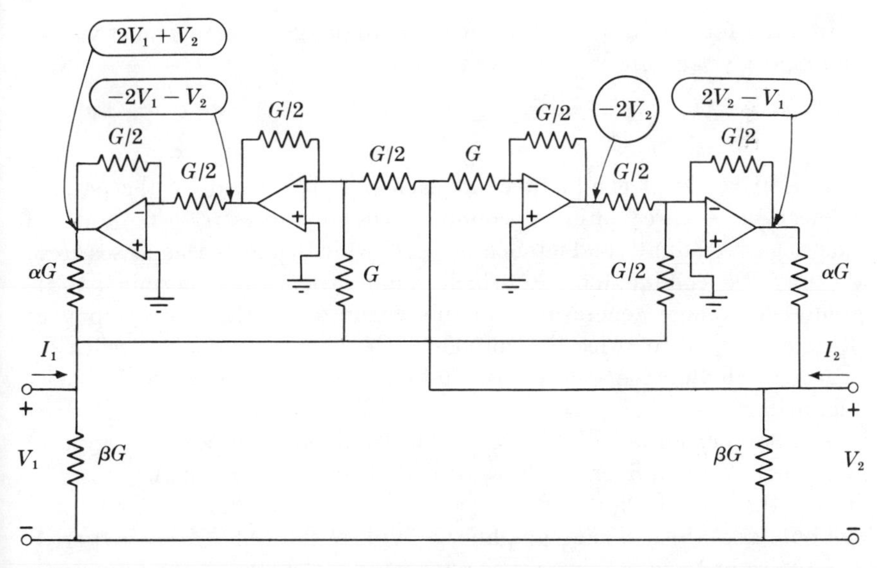

FIG. 3.21 A gyrator circuit due to Morse and Huelsman. Its mode of operation is basically that given by the equivalent circuit of Fig. 3.18.

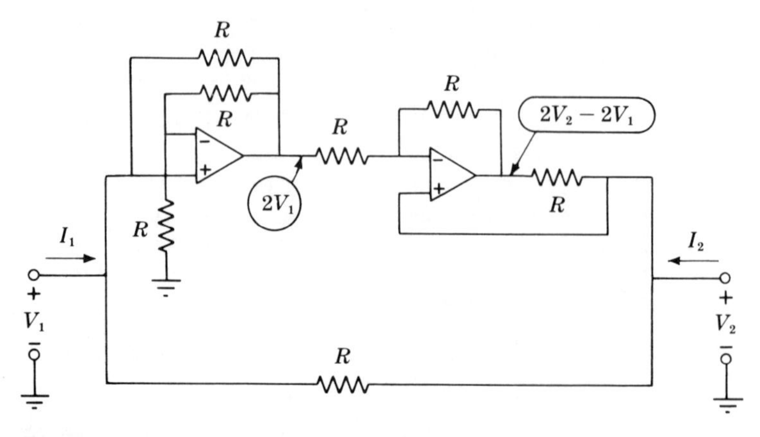

FIG. 3.22 A gyrator circuit due to Deboo.

the negative input to each amplifier is at ground potential one can directly calculate the input current at each port and obtain the admittance matrix:

$$\begin{bmatrix} \tfrac{3}{2} - \alpha + \beta & -\alpha \\ \alpha & \tfrac{3}{2} - \alpha + \beta \end{bmatrix} \tag{3.35}$$

In order for it to act as a gyrator the main diagonal terms must vanish and hence we require

$$\alpha = \tfrac{3}{2} + \beta \tag{3.36}$$

The positive shunt conductance at each port is formed by the parallel connection of three physical conductances of values G, $\tfrac{1}{2}G$, and βG. The negative shunt conductance of $-\alpha G$ which cancels this arises from the physical conductance αG which connects each port terminal to its grounded voltage generator. As this generator contains a component of its emf equal to twice the voltage at the port concerned, it reflects a negative conductance $-\alpha G$ across that port by the same mechanism as in Fig. 3.19.

The simpler circuit of Fig. 3.22, due to Deboo,[15] produces a gyrator with only two amplifiers. These provide transmission from left to right in just the same fashion as the top two amplifiers of Fig. 3.20, but here, not only does the left-hand amplifier provide a shunt of $-R$ across port 1, but the right-hand amplifier is also modified, by virtue of the connection to the positive input, so as to shunt $-R$ across port 2. The third amplifier of Fig. 3.20 is replaced by a resistor bridging the ports. This bridging resistor contributes the term

$$\begin{bmatrix} \frac{1}{R} & -\frac{1}{R} \\ -\frac{1}{R} & \frac{1}{R} \end{bmatrix} \tag{3.37}$$

to the admittance matrix while the two-amplifier part contributes

$$\begin{bmatrix} -\frac{1}{R} & 0 \\ \frac{2}{R} & -\frac{1}{R} \end{bmatrix} \tag{3.38}$$

In adding these two matrices we see that cancellation is occurring in the circuit not only to achieve the main diagonal zeros but also to obtain the correct sign in the y_{21} term.

These two examples are quite typical of most of the gyrator circuits that have been suggested for construction with operational amplifiers and they all suffer, in common with the arrangement of Fig. 3.8, from the same practical disadvantage. The quantities involved in the resistance cancellation are always derived from separate resistors which must consequently be adjusted and maintained to a very precise match if high and stable Q factors are to be achieved in a simulated inductance. In our analysis so far it has been unnecessary to examine the effects of amplifier imperfections because they are small compared with the effects of resistor tolerances. This is no longer true for the circuits we shall now describe which use the other form of cancellation.

Considering again our basic two-port circuit of Fig. 3.17 we see that an alternative way of modifying the main diagonal terms in its admittance matrix, (3.32), is to add additional emf components to the two grounded voltage generators, as in Fig. 3.23. Specifically we add k_3V_1 to the generator feeding into port 1 and k_4V_2 to the other. The currents entering the ports are then:

$$I_1 = \frac{V_1 - k_2V_2 - k_3V_1}{R_1} \qquad I_2 = \frac{V_2 - k_1V_1 - k_4V_2}{R_2} \tag{3.39}$$

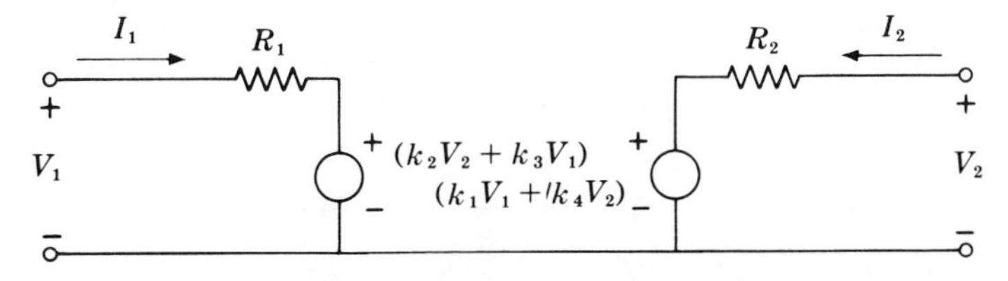

FIG. 3.23 A modification of the circuit of Fig. 3.17 in order to allow control of the y_{11} and y_{22} terms of the admittance matrix.

whence the admittance matrix is

$$\begin{bmatrix} \dfrac{1-k_3}{R_1} & \dfrac{-k_2}{R_1} \\ \dfrac{-k_1}{R_2} & \dfrac{1-k_4}{R_2} \end{bmatrix} \tag{3.40}$$

The main diagonal terms will vanish if $k_3 = k_4 = 1$. With this condition met, and k_1 and k_2 set as before, the circuit becomes that of Fig. 3.24. However, it is just as important and as critical here that k_3 and k_4 should equal unity as it was, in the previous cancellation scheme, that positive and negative resistances should be equal in magnitude. For the method to have any advantage over resistance matching it is essential to use some means of keeping k_3 and k_4 close to unity without relying on a match of component values.

One way of doing this is illustrated in the sketches in Fig. 3.25. The first, Fig. 3.25*a*, repeats the circuitry associated with port 1 of Fig. 3.24, but now shows the components of the generator emf as arising from two separate generators. The second merely interchanges the positions of the various items of the first sketch around the single loop. Finally, in the third sketch, the V_1 generator is replaced by the operational amplifier. Here we no longer assume an ideal amplifier as far as gain is concerned and show the voltage between the differential input terminals explicitly as δ times the output voltage V_0. Evidently $V_1 = V_0(1 + \delta)$ and then

$$I_1 = \frac{\delta V_0 - V_2}{R_1} = \frac{\delta V_1}{(1+\delta)R_1} - \frac{V_2}{R_1} \tag{3.41}$$

The input admittance at the port is, very nearly, δ/R_1 and is small just because the gain of the amplifier is high. The higher the gain, the lower will be the admittance and there is no critical cancellation, only a cancellation such as occurs at the input of a feedback system.

Two such circuits can be joined together to form a gyrator as in Fig. 3.26. The idealized version is shown in Fig. 3.26*a* and the finite-gain

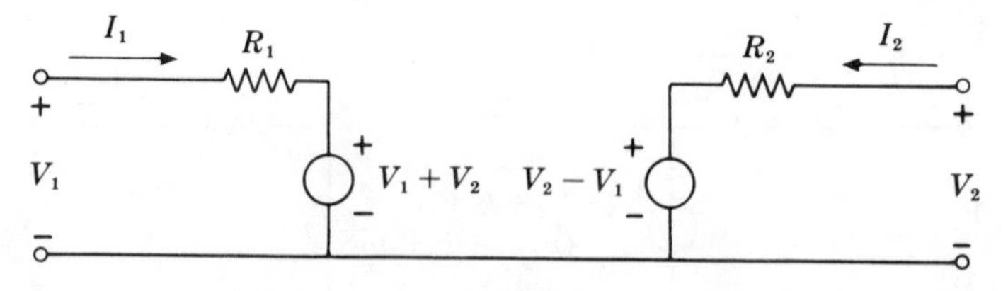

Fig. 3.24 The gyrator circuit which results when the parameters k_1, k_2, k_3, and k_4 in Fig. 3.23 are appropriately adjusted.

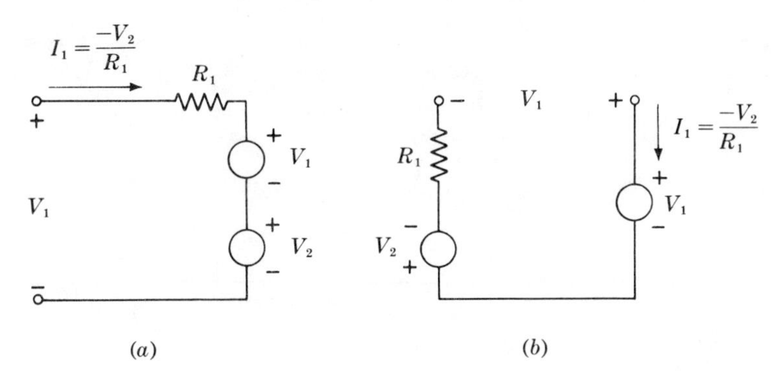

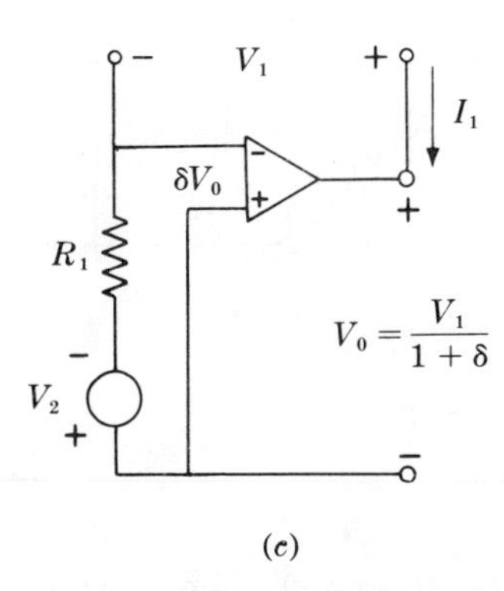

FIG. 3.25 Steps in the development of the circuitry associated with port 1 of Fig. 3.24. In (c) a high input impedance is presented at port 1 without requiring an accurate match of component values.

amplifier version in Fig. 3.26*b*. This still leaves a grounded voltage generator of $-V_1$ to be provided, but this can be supplied via an inverting amplifier from the output of amplifier 1. The final gyrator circuit appears in Fig. 3.26*c*. Assuming ideal amplifier impedances the port currents will be just the currents in R_1 and R_2; the voltages across these resistors are known and we can write the admittance matrix directly from these as

$$\begin{bmatrix} \dfrac{\delta_1}{(1+\delta_1)R_1} & \dfrac{-1}{(1+\delta_2)R_1} \\ \dfrac{R_3}{(1+\delta_1)(R_4+\delta_3 R_4+\delta_3 R_3)R_2} & \dfrac{\delta_2}{(1+\delta_2)R_2} \end{bmatrix} \tag{3.42}$$

(a)

(b)

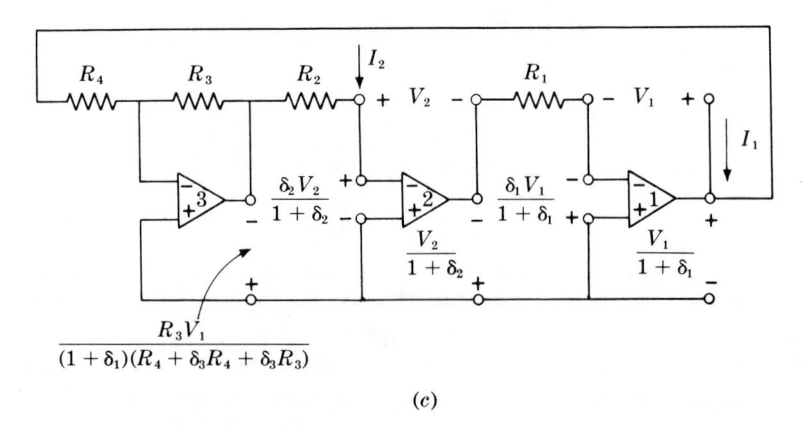

(c)

FIG. 3.26 Gyrator using the insensitive circuitry developed in Fig. 3.25. The idealized equivalent circuit appears in (a). The final, finite-gain amplifier version is shown in (c).

It would usually be convenient to make all four resistors equal, say to R, and then, assuming equal behavior in the three amplifiers, the matrix reduces, without any cancellations, to

$$\frac{1}{(1+\delta)R}\begin{bmatrix} \delta & -1 \\ \dfrac{1}{1+2\delta} & \delta \end{bmatrix} \tag{3.43}$$

Terminating each port with a capacitor C produces the equivalent of a tuned circuit of resonant frequency very nearly $(2\pi CR)^{-1}$. Using the results of Sec. 3.2 one can obtain the approximate expression for the Q factor

$$\frac{1}{2\delta(1-2\omega/\omega_0)} \tag{3.44}$$

By adjusting the 3-dB frequency of the amplifiers this Q factor can be made slightly negative and the circuit will then oscillate.

It was as an oscillator that this circuit was first described, by Good.[16] The same arrangement was later adapted by Geffe[17] and also by Kerwin, Huelsman, and Newcomb[18] as the heart of an insensitive means of realizing the response of a second-degree transfer function. In all this work it apparently passed unnoticed that the circuit was in fact a first-class gyrator. Unfortunately it is difficult to use this gyrator for simulating an inductor in a filter because neither port is either properly grounded or properly floating.

With a little rearrangement, however, it is possible to get *one* of the ports grounded and, incidentally, to save one amplifier. The loop associated with port 2 in Fig. 3.24 can be linked with the generator driving port 1 and the modified circuit arranged as in Fig. 3.27*a*. Replacing the generators by operational amplifiers then gives the circuit of Fig. 3.27*b*. The calculation of the amplifier output voltages is a quite straightforward step and from these the port currents can be found. The resulting admittance matrix is

$$\begin{bmatrix} \dfrac{\delta_1}{(1+\delta_1)R_1} & \dfrac{-1}{(1+\delta_1)R_1} \\ \dfrac{R_3+(\delta_1-\delta_2)(R_4+R_3)}{(1+\delta_1)(R_4+\delta_2R_4+\delta_2R_3)R_2} & \dfrac{\delta_1}{(1+\delta_1)R_2} \end{bmatrix} \tag{3.45}$$

which is almost identical to that of (3.42). By making all resistors equal to R and assuming similar amplifiers it does in fact reduce to the same simplified form of (3.43).

This circuit was discovered by Riordan[19] and, at the time of writing, appears to be one of the best gyrators known for the purpose of inductance simulation. Its use as a gyrator is restricted to such applications because port 2 can be connected only to some impedance, such as a capacitor, which can float off ground; it cannot be used as a general two-port gyrator like the circuit of Fig. 3.14.

Apart from the finite amplifier gains, whose effect on the Q is summarized by (3.44), the only other amplifier imperfections which are important are the common-mode input impedances which shunt R_2 and R_4. In the case of amplifier 2, its common-mode input impedance falls directly across R_4; the shunt *resistive* component of the input impedance merely changes the effective value of R_4 very slightly and is probably insignificant. However, the shunt *capacitance* will change the phase angle of R_4 slightly, and hence will contribute a positive phase angle to one gyration conductance. This has the effect of reducing the Q factor.

With all resistances equal, so that the voltage across R_2 becomes the same as the common-mode input voltage to amplifier 1, namely V_1, one can see that the effect on I_2 of the current flowing into this common-mode input impedance is the same as if, instead, the negative of this input impedance were shunted across R_2. In these circumstances the effects of the two amplifier input impedances will tend to cancel. The input impedances will normally be at least a hundred times greater than R_2 or

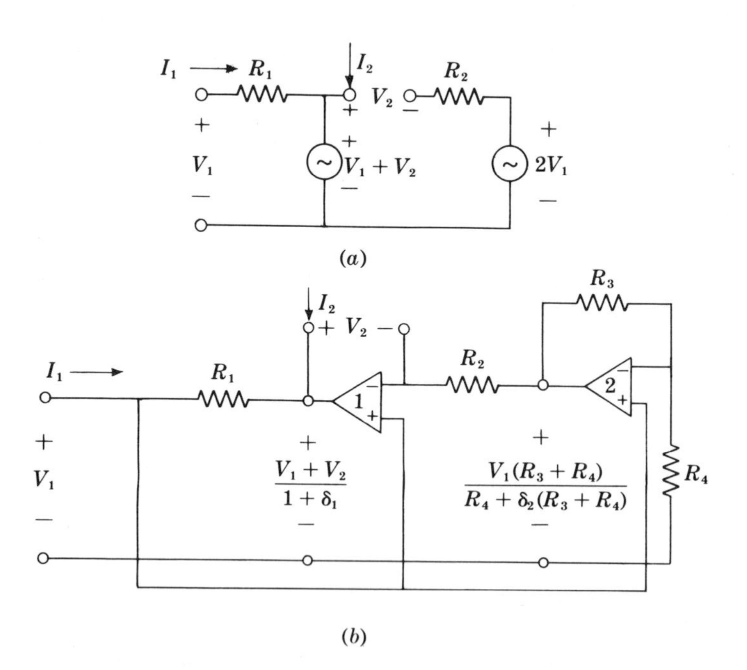

FIG. 3.27 Modification of the gyrator of Fig. 3.26*c* in order to get one port grounded. The idealized equivalent circuit is given in (*a*) and the finite-gain amplifier version in (*b*). The latter circuit is due to Riordan.

R_4, and if they match to within 10 percent the net effect is then very small. One useful by-product of this cancellation feature is that one can achieve a fine control over the Q of a simulated inductor by adding a very small capacitance to ground across the negative input to one or the other of the amplifiers.

At low frequencies the gyrator contribution to the total Q factor depends primarily upon the amplifier gain and, with a very high-quality terminating capacitor, Q factors exceeding 1,000 are not difficult to obtain. However, at frequencies below 100 Hz the practical limitation is more likely to be set by the capacitor Q than by the gyrator. At higher frequencies the effect of the increasing amplifier phase shift is to increase the Q factor, and in order to prevent oscillation it becomes necessary to add a small capacitance across R_4. The high-frequency limit is set by the difficulty of maintaining a sufficiently stable balance with temperature between all the various phase angles that affect the Q.

3.5 Simulation of Floating Inductors

All the gyrators described so far have been restricted in that the simulated inductances which can be obtained from them must have one terminal connected to ground. This arises primarily from the fact that the gyrators themselves are assumed to be connected to a common, grounded power supply. If each gyrator could have its own private power supply, isolated from ground, the simulated inductor would more nearly behave like a real inductor and could act as a replacement in almost all circuit applications. Unfortunately this is hardly practicable. Nevertheless, all low-pass filters and many bandpass filters do contain some inductors that are not connected directly to ground and, if the design process is to be of any general practical value, some solution must be found to the problem of simulating a floating inductor.

One approach, suggested by Holt and Taylor,[20] which has particular appeal because of its ingenious exploitation of the properties of the gyrator, is to use a shunt capacitor and two gyrators in the circuit of Fig. 3.28*a*. Both gyrators are assumed to have one terminal connected to ground at each port. If the left-hand and right-hand gyrators have admittance matrices, respectively:

$$\begin{bmatrix} 0 & g_1 \\ -g_2 & 0 \end{bmatrix} \quad \text{and} \quad \begin{bmatrix} 0 & g_3 \\ -g_4 & 0 \end{bmatrix} \tag{3.46}$$

then the equivalent circuit becomes that of Fig. 3.28*b*. The voltage V_3 across the capacitor is

$$V_3 = \frac{V_1 g_2}{Cp} - \frac{V_2 g_3}{Cp} \tag{3.47}$$

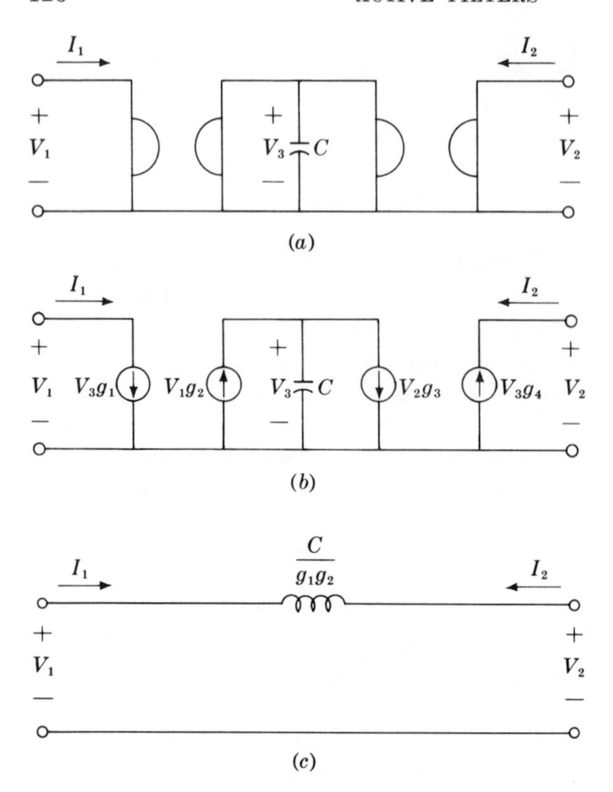

FIG. 3.28 Floating-inductor circuit suggested by Holt and Taylor.

and, since $I_1 = V_3g_1$ and $I_2 = -V_3g_4$, we can immediately write

$$I_1 = \frac{V_1g_1g_2}{Cp} - \frac{V_2g_2g_3}{Cp}$$
$$I_2 = -\frac{V_1g_2g_4}{Cp} + \frac{V_2g_3g_4}{Cp} \tag{3.48}$$

In order that this circuit shall represent a floating component connected to the nongrounded terminals of the ports we must arrange that, for all values of V_1 and V_2, the current entering one port is the negative of the current entering the other. In other words we require

$$I_1 + I_2 = \frac{V_1g_2(g_1 - g_4)}{Cp} - \frac{V_2g_3(g_1 - g_4)}{Cp} \tag{3.49}$$

to vanish for all V_1 and V_2. This is evidently satisfied by setting

$$g_1 = g_4 \tag{3.50}$$

The floating component so produced will be an inductor if the common current flowing through the pair of terminals is proportional to $V_1 - V_2$ and not merely a linear combination of V_1 and V_2. Examination of (3.48) shows that this condition will be satisfied if

$$g_2 = g_3 \tag{3.51}$$

With both $g_1 = g_4$ and $g_2 = g_3$ the circuit then becomes that of Fig. 3.28*c*, a perfect floating inductor of value C/g_1g_2.

In practice, of course, it is not possible to maintain an exact match between the conductances of the two separate gyrators and the resulting unbalance between I_1 and I_2 must be represented in Fig. 3.28*c* by additional components and controlled sources. There appears to be no simple equivalent circuit which allows the effect of the unbalance to be estimated easily, and the question whether, in any particular case, a given accuracy in the conductances is acceptable is probably best answered by a direct analysis of the circuit in which the floating inductor is to be used, including all the imperfections due to the unbalance. In the case of a bandpass filter of moderately high degree Sheahan[21] has demonstrated that the effect on the passband quality of a 2 percent mismatch between the conductances can be quite serious. In simple filters, on the other hand, such a mismatch may confer only minor blemishes to the performance and hence be perfectly acceptable.

The use of two gyrators to provide a floating inductor is not only troublesome because of the matching problem, but also rather expensive, and several writers have described special gyrator circuits for this purpose which are intended to be both easier to adjust and more economical. A good example is the circuit due to Holmes, Gruetzmann, and Heinlein,[22] which associates the matching requirements represented by (3.50) and (3.51) with separate units as shown in Fig. 3.29*a*.

In order to ensure that the current flowing into the capacitor is proportional to $V_1 - V_2$ the floating terminals are connected to the inputs of a differential amplifier whose output current is $(V_1 - V_2)/R_1$. The voltage V_3 which this current produces across the capacitor is applied to the input of a second amplifier whose balanced output is joined to the floating terminals. The extent to which the condition represented by (3.50) is satisfied is measured by the degree of balance achieved in the output currents of this second amplifier. All input and output impedances ideally are infinite.

Practical circuits for realizing these two amplifiers should preferably produce the desired behavior without at the same time requiring a critical match between resistors or other circuit parameters. The Holmes circuit for the differential amplifier, shown in Fig. 3.29*b*, does this in a very satisfactory way, but the balanced-output amplifier which they suggest,

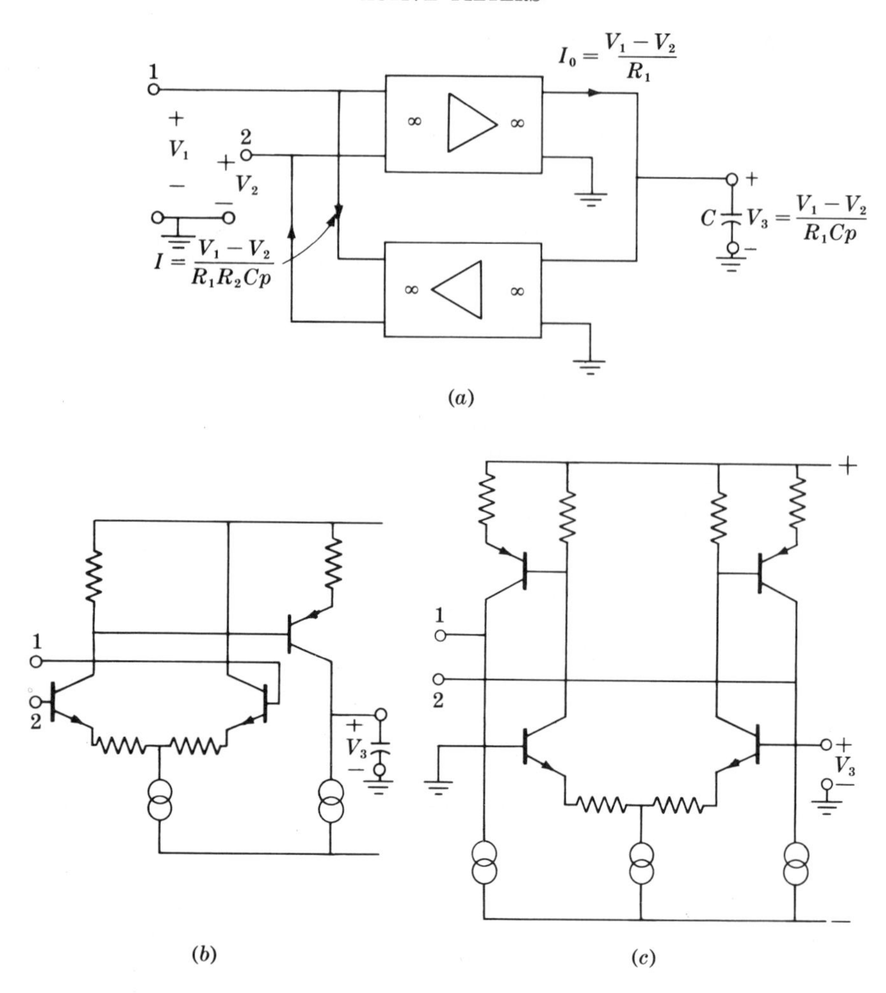

FIG. 3.29 Gyrator with a floating port as described by Holmes et al. The block form is shown in (*a*), the differential-input amplifier circuit in (*b*), and the balanced-output amplifier in (*c*).

shown in Fig. 3.29*c*, unfortunately still demands a match between the gains of the two separate output stages in order to achieve balanced-output currents. One possible solution is to use a balanced-output stage of the kind shown in Fig. 3.30.

Circuits of this type for producing floating inductors have the same disadvantage as those described in Sec. 3.3, namely, that it is difficult to apply enough overall feedback to the controlled sources. In view of the superior qualities of the Riordan gyrator using operational amplifiers it is natural to consider whether some variant of it could also provide a solution to the floating-inductor case. Riordan himself[19] suggested a balanced-to-ground version of his gyrator for this purpose, but insofar as it uses two capacitors it fails, strictly speaking, to qualify as a gyrator. The circuit is shown in Fig. 3.31.

Assuming ideal amplifiers and noting that the voltage drop across R_4 is $V_1 - V_2$, it is a straightforward calculation to obtain the expressions for the currents at the terminals:

$$I_1 = \frac{R_3(V_1 - V_2)}{R_1 R_2 R_4 Cp} \qquad I_2 = -\frac{R_3'(V_1 - V_2)}{R_1' R_2' R_4 C'p} \tag{3.52}$$

We see that, by virtue of the high gain of the amplifiers, the circuit automatically guarantees that the two currents are each proportional to $V_1 - V_2$. This is equivalent to satisfying the condition expressed by (3.51) for the Holt and Taylor circuit. However, a match between the resistance values is still called for in order to obtain equal magnitudes of the two currents and so to guarantee that the simulated inductor "floats." The principal merit of the circuit is that, if the required match of resistances can be obtained initially and then held over a period of time, the resulting floating inductor should have as good a quality as that

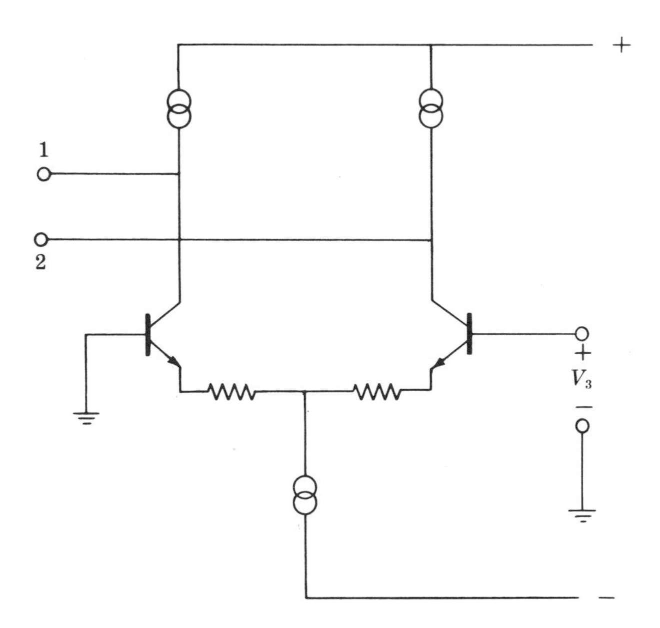

FIG. 3.30 Alternative circuit to that of Fig. 3.29c which avoids the need for a critical component match.

of the grounded version. But having to adjust simultaneously for this match and also for a given inductance value would be troublesome and is a poor reward for using four amplifiers and two capacitors.

The best practical solution to this problem seems to be that of simulating a private floating power supply for each gyrator that is to be used for replacing a nongrounded inductor. A simple way of doing this was described by Sheahan[21] and is shown in Fig. 3.32. The essential components are the two transistors T_1 and T_2 which, with a modest amount of series feedback from their emitter resistances, act as high-impedance current sources at their collectors. The current which flows from the collector of T_1 to the collector of T_2 also passes through the gyrator and constitutes its power supply.

If it were possible to bias the two transistors completely independently so that the current from T_1 *exactly* equaled the current into T_2, then the dc potential at the gyrator power-supply terminals would, under quiescent conditions, settle down halfway between the positive and negative rails. When a signal is applied to the circuit to which the floating gyrator is connected, the potential at *both* terminals of the gyrator port and hence also at the power-supply terminals will vary in sympathy. This it can do because of the high impedances presented by T_1 and T_2. As far as the circuit is concerned, the floating inductor has just the impedance of one

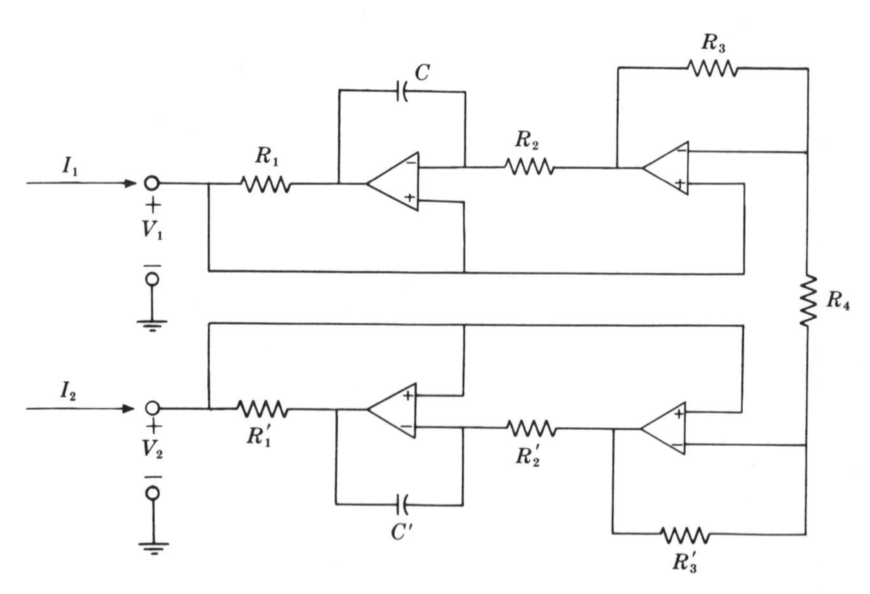

FIG. 3.31 Balanced-to-ground version of the gyrator of Fig. 3.27 for simulating floating inductors, due to Riordan.

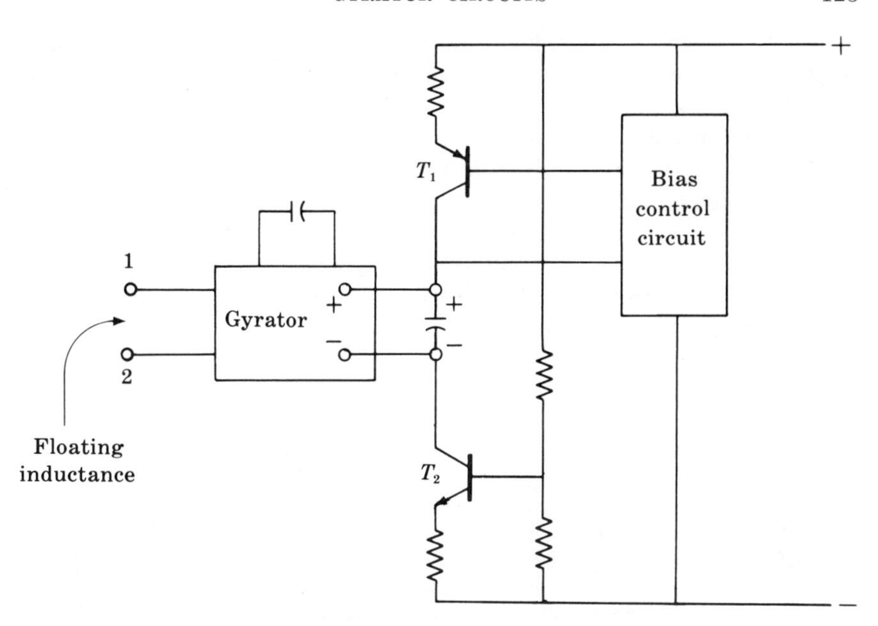

FIG. 3.32 Gyrator flotation circuit described by Sheahan.

collector connected from each terminal to ground. This impedance is so high that normally it has negligible effect upon the circuit.

The only practical difficulty is that it is not possible to bias the two transistors independently to give exactly equal currents. The slightest error between the two currents causes the potential at the gyrator power-supply terminals to either increase or decrease until T_1 or T_2 saturates. To solve this problem Sheahan added a control circuit to the bias supply for one of the transistors.

This control circuit monitors the dc potential of the gyrator power-supply terminals and adjusts the bias until the potential is at an appropriately central value. The monitoring terminal from the control circuit must itself present a high impedance so as not to degrade the high-impedance power supply and moreover, ideally, the control circuit should operate only at zero frequency. For if it operates at the signal frequency, the negative-feedback action will reduce the impedance of the gyrator power supply to a low value and so nullify the action of the flotation circuit. To prevent this it is necessary to put a simple RC low-pass filter into the control-circuit loop so that the feedback is effective only in the neighborhood of zero frequency and is removed at frequencies above a few hertz.

With all these conditions met the circuit works very well and with its help a 16th-degree channel bandpass filter for a frequency-division multiplex system has been successfully made in inductorless form.[23] Almost any gyrator can be used in the flotation circuit; the only condition is that the current consumption of the gyrator be constant, independent of signal drive. Unfortunately this rules out the use of such circuits as class *B*–type output stages in operational amplifiers. The other disadvantage of the flotation circuit is that it is expensive in power consumption; the voltage of the main power supply has to be at least twice that required by the gyrator itself, and so the total power dissipated is also at least twice that of the gyrator.

References

1. Guillemin, E. A.: Synthesis of RC-Networks, *J. Math. Phys.*, vol. 28, pp. 22–42, 1949.
2. Linvill, J. G.: RC Active Filters, *Proc. IRE*, vol. 42, pp. 555–564, 1954.
3. Tellegen, B. D. H.: The Gyrator, a New Electric Network Element, *Philips Res. Rept.*, vol. 3, no. 2, pp. 81–101, 1948.
4. Foster, D. E., and S. W. Seeley: Automatic Tuning, Simplified Circuits, and Design Practice, *Proc. IRE*, vol. 25, pp. 289–313, 1937.
5. Shekel, J.: The Gyrator As a Three-terminal Element, *Proc. IRE*, vol. 41, pp. 1014–1016, 1953.
6. Herold, E. W.: Negative Resistance and Devices for Obtaining It, *Proc. IRE*, vol. 23, pp. 1201–1223, 1935.
7. Carlin, H. J.: Synthesis of Non-reciprocal Networks, *Proc. Symp. Modern Network Synthesis*, Polytechnic Institute of Brooklyn, vol. 5, pp. 11–44, 1955.
8. Harrison, T. J.: A Gyrator Realization, *IEEE Trans. Circuit Theory*, vol. CT-10, p. 303, 1963.
9. Shenoi, B. A.: Practical Realization of a Gyrator Circuit and RC-gyrator Filters, *IEEE Trans. Circuit Theory*, vol. CT-12, pp. 374–380, 1965.
10. Sharpe, G. E.: The Pentode Gyrator, *IRE Trans. Circuit Theory*, vol. CT-4, pp. 321–323, 1957.
11. Rao, T. N., and R. W. Newcomb: Direct-coupled Gyrator Suitable for Integrated Circuits and Time Variation, *Electron. Letters*, vol. 2, pp. 250–251, 1966.
12. Sheahan, D. F., and H. J. Orchard: Integratable Gyrator Using M.O.S. and Bipolar Transistors, *Electron. Letters*, vol. 2, pp. 390–391, 1966.
13. Holmes, W. H., S. Gruetzmann, and W. E. Heinlein: High-performance Direct-coupled Gyrators, *Electron. Letters*, vol. 3, pp. 45–46, 1967.
14. Morse, A. S., and L. P. Huelsman: Gyrator Realization Using Operational Amplifiers, *IEEE Trans. Circuit Theory*, vol. CT-11, pp. 277–278, 1964.
15. Deboo, G. J.: Application of a Gyrator-type Circuit to Realize Ungrounded Inductors, *IEEE Trans. Circuit Theory*, vol. CT-14, pp. 101–102, 1967.
16. Good, E. F.: A Two-phase Low Frequency Oscillator, *Electron. Eng.*, pp. 164–169, 1957.
17. Geffe, P. R.: Make a Filter Out of an Oscillator, *Electron. Design*, pp. 56–58, May 10, 1967.
18. Kerwin, W. J., L. P. Huelsman, and R. W. Newcomb: State-variable Synthesis for Insensitive Integrated Circuit Transfer Functions, *IEEE J. Solid State Circuits*, vol. SC-2, pp. 87–92, 1967.

19. Riordan, R. H. S.: Simulated Inductors Using Differential Amplifiers, *Electron. Letters*, vol. 3, pp. 50–51, 1967.
20. Holt, A. G. J., and J. Taylor: Method of Replacing Ungrounded Inductances by Grounded Gyrators, *Electron. Letters*, vol. 1, p. 105, 1965.
21. Sheahan, D. F.: Gyrator-flotation Circuit, *Electron. Letters*, vol. 3, pp. 39–40, 1967.
22. Holmes, W. H., S. Gruetzmann, and W. E. Heinlein: Direct-coupled Gyrators with Floating Ports, *Electron. Letters*, vol. 3, pp. 46–47, 1967.
23. Sheahan, D. F., and H. J. Orchard: Bandpass Filter Realization Using Gyrators, *Electron. Letters*, vol. 3, pp. 40–42, 1967.

CHAPTER 4

ELECTRONIC-CIRCUIT ASPECTS OF ACTIVE FILTERS

Graham A. Rigby
University of California
Berkeley

4.1 Introduction

If the objective of active *RC* filter theory is to lead ultimately to practical realizations with good performance, then there are two important considerations which must follow the establishment of basic configurations. The first is an understanding of the effects of practical circuit and device limitations on the performance of a filter and the second involves the development of circuit-design methods which account for and minimize these limitations. These two aspects are closely related and are treated concurrently in this chapter. One further point, which will be brought out here, is that electronic-circuit considerations add extra criteria for judging basic active *RC* filter schemes, in terms of sensitivity and ease of realization, and these may alter some of the conclusions made on the basis of purely theoretical considerations.

Although the use of active *RC* techniques for filter synthesis was an important technique before the advent of integrated circuits, it has now assumed far wider importance. The absence of any appreciable physical inductance in microcircuits means that active *RC* techniques must be used at all frequencies up to about 100 MHz for the realization of higher-order transfer functions within the physical dimensions associated with integrated circuits. At frequencies above about 100 MHz, small inductors may be made by depositing helical conductor patterns on the surface of a circuit, for example, and enable *RLC* techniques to be used. At still higher frequencies, stripline methods may be used in the typical dimensions of microcircuits. Finally, there is a growing class of filters based on microminiature electromechanical and piezoelectric devices, but our attention here is concentrated on circuit approaches for monolithic realization and which are useful at frequencies below the VHF range.

In spite of the emphasis on integrated circuits, much of the following is equally applicable to discrete-component circuits. Some of the differences between the two media are as follows. The absolute accuracy of component parameters can be higher in discrete than in integrated circuits because of the possibility of preselecting these components, but a high degree of matching between elements in the latter case may be obtained without difficulty. The close thermal tracking of elements in an integrated circuit, compared with the discrete version, enables special design techniques to be used which particularly enhance stability of bias points. From the viewpoint of parasitics, linear integrated circuits are superior in most respects to discrete circuits, because of small geometries and high packing densities. In fact, it is difficult to simulate high-frequency integrated circuits in discrete form where wiring inductance and interlead capacitances are usually higher than in monolithic versions.

The economic advantages of integrated circuits over discrete circuits have been demonstrated far more dramatically in the case of digital circuits than linear. With some exceptions, the same trend is present in the latter, however. Since the processing yield of an IC has a strong influence on its cost, attention is given here to design techniques which place minimal requirements on process control, thereby contributing to higher yields. There are other cost-related factors over which the designer has some control. These include choices between monolithic construction and various hybrid methods involving thin-film capacitors and resistors, for example. Since many active *RC* techniques must lead to precise transfer functions, adjustment of the final circuit is often required and although this is usually a simpler matter in discrete circuits, some practical methods for adjusting integrated-circuit structures are described in this chapter.

After the properties and limitations of the basic integrated-circuit elements (resistors, capacitors, bipolar and field-effect transistors) are treated, the modeling of these elements is described. Two levels of analysis are then developed. The first is aimed at those calculations which can be done conveniently by hand and which give the circuit designer a sound, physically based understanding of what is happening in the circuit. Since the quantitative results of such analysis are only approximate, a higher-order analysis using the computer is then treated. The conversion of active *RC* schemes developed from network theory to forms suitable for electronic-circuit design is then considered and the chapter concludes with a treatment of some standard integrated-circuit building blocks.

4.2 Integrated-circuit Elements

The Fabrication Process. Out of the development of planar silicon diffusion technology has evolved a standard fabrication technique for integrated circuits[1,2] which is widely used. This process and its varia-

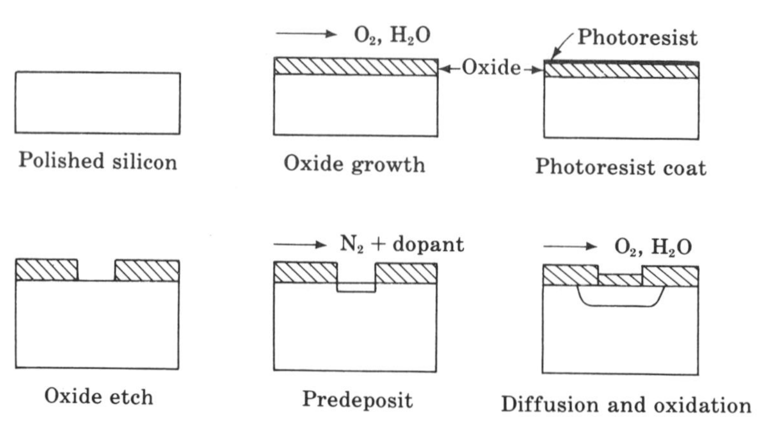

FIG. 4.1 Sequence of operations in a standard planar silicon diffusion step.

tions are described in detail in the literature and it is merely summarized here.

The starting point for the standard process is a wafer of silicon, doped p-type (usually by boron) to a level of 10^{15} impurity atoms/cm^3. The wafer is about 0.015 in. thick and about 2 in. in diameter. Both sides, or at least the side used for fabrication, are mechanically and chemically polished to a mirror finish. At this stage a selective diffusion is made, if the circuit is to contain bipolar transistors. This diffusion aids the making of low-resistance collector contacts to the transistors and is considered more fully below. The steps involved in making this diffusion are shown schematically in Fig. 4.1, for a small section of the wafer, and the steps involved in all subsequent diffusions are essentially the same. To protect the surface of the wafer from doping in undesired regions, an oxide film about 0.5 μ thick is grown by placing the wafer in a furnace in an oxidizing ambient (usually oxygen and water vapor). After the oxide is formed, the wafer is removed from the furnace and coated with a photoresist consisting of a resin which is capable of being polymerized by light. A photographic mask, containing a pattern of dark regions, corresponding to where the oxide film should be removed, is contact-printed onto the surface so that the resin becomes polymerized under the clear areas of the mask. A solvent process then dissolves away the unpolymerized resist, leaving the oxide exposed in the selected regions. Placing the wafer into an etching solution containing fluoride ions then removes the oxide from these regions. At this point, some of the wafer is covered by an oxide film which protects the silicon from dopants, while the bare regions, or "windows," are doped by exposing the wafer in a furnace to an ambient containing dopant atoms. The

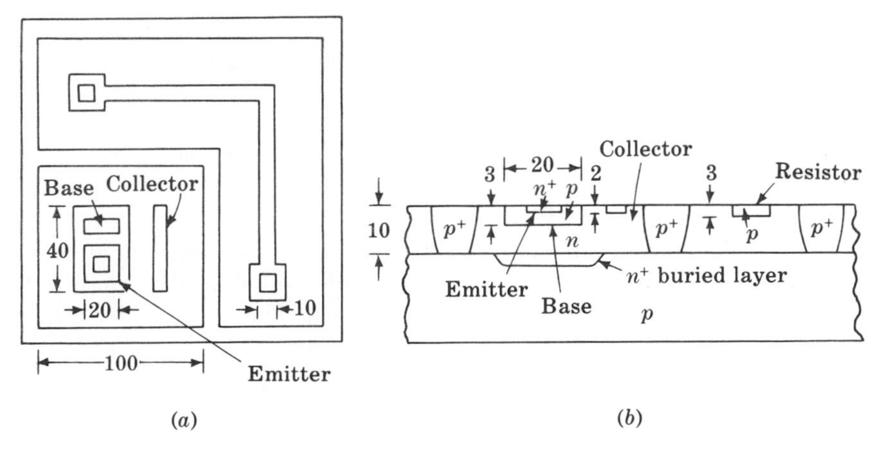

FIG. 4.2 Plan and cross-section views of a bipolar transistor and p-diffused resistor in the standard epitaxial process. Dimensions are in μ ($1\ \mu = 0.04$ mil).

doping process is normally performed in two steps. In the first, the furnace ambient contains the dopant mixed with a carrier gas (nitrogen) and after exposure the silicon in the windows has a shallow layer of heavily doped material. The wafer is then transferred to another furnace where this layer is diffused down further and the windows are covered with fresh oxide. For the first doping step in the process, the dopant is arsenic or antimony, resulting in heavily doped n regions in the p-type wafer.

The oxide is then completely removed from the wafer and a thin epitaxial layer (8–10 μ) of n-type silicon is grown on the surface by exposing the wafer to silane (SiH_4) or silicon tetrachloride ($SiCl_4$) vapor in a special reactor. The dopant (phosphorus) is introduced in this process as an additive to the vapor so that the epitaxial layer is doped uniformly. The doping level is, as in the p-type substrate, about 10^{15} atoms/cm^3.

The sequence of steps described above (oxide growth to diffusion) is then repeated at least three times for the p-type isolation walls, the p-type bases, and the n-type emitters of the bipolar transistors. After the final diffusion, windows are again cut in the oxide layer for contacts and a layer of aluminum is evaporated over the surface of the circuit. The interconnection pattern in this layer is defined by a photoetching process similar to that used for the oxide windows. At this point, the wafer is ready for separation into individual circuit die, testing, and packaging. Figure 4.2a shows a cross-sectional view of a portion of a wafer containing a bipolar transistor and a p-type resistor. This shows the relationship of the original n-type layer, buried beneath the epitaxial film, to the other three diffusions made from the top of the epitaxial film.

The basic transistor structure is also apparent, as well as the isolation walls completely surrounding the *n*-type collector with a *pn* junction which is normally reverse biased. The dimensions marked on the figure show that the horizontal and vertical scales have been distorted for the sake of clarity. Finally, a plan view of the same structure is shown in Fig. 4.2*b*.

There are many deviations from this standard process. Some are minor, involving variations in dopants, diffusion cycles, or photoetching techniques. In some cases, an inert glass is deposited on the completed circuit to protect it against contamination. The more substantial variations include replacing the *p*-type isolation walls with oxide layers, removing the silicon altogether, the formation of electroplated gold beams on the surface for mounting purposes, or the addition of thin-film resistors and capacitors on the surface. Although these variations can result in changes of device parameters and in improved high-frequency performance of the circuits, the content of this chapter applies, in general, to any version of the technology.

Resistors. As passive elements in discrete-component circuits, resistors present no particular problems. They have associated with them reactive parasitics, in the form of series inductance and shunt capacitance, which should be accounted for in very high-frequency circuits; but tolerance, size, and temperature coefficients over a wide range may be obtained without difficulty. In integrated circuits, the same degree of flexibility is not present, but there are more options available than one might first think. Some of these are examined below. As a general rule, it is noted that resistors are not used as extensively in IC design as they are in discrete circuits. This stems partly from the limitations of IC resistors, particularly the area consumed by large value units, and partly from the fact that a transistor can perform many of the functions conventionally performed by resistors. The principle underlying this change in emphasis is that the chip area used by a particular element is more closely related to the cost of a circuit than is the type of element used. In linear circuits, an obvious function which a transistor performs better than a resistor is a constant-current source for biasing a differential pair (see Sec. 4.7). Also, *pn* junctions are often more suitable as dc level shifters than a resistive divider. But the most striking illustration of this point is in digital circuits, where, in the basic logic families, resistors only perform a prime function in one case (RTL) while junction devices are the gating elements in DCTL, DTL, TTL, and ECL.

Base-diffused Resistors. The most common resistor element in a monolithic circuit is the resistor formed by diffusing a *p*-type layer into the *n*-type epitaxial background of the wafer (see Fig. 4.2). This diffusion is performed simultaneously with the base diffusions for the *npn* transistors in the circuit. However, the parameters of the diffusion are normally optimized with respect to the requirements of the transistor rather than the resistor. Thus, if the doping level is too high, the emitter injection efficiency, the current gain, and the base-emitter reverse breakdown voltage of the transistor will be low. If the doping is too low, problems arise due to surface inversion layers under the silicon oxide, and high ohmic base resistance in the device.

The geometry of a base-diffused resistor is usually very simple and a typical form is shown in Fig. 4.2. The shape is a strip which may be folded several times to fit best into the available area on the chip, and the ends are widened somewhat to accommodate the contact windows. The doping of the resistor material is characterized, for convenience, by a sheet resistance ρ_s expressed in ohms per square. The resistance between the ends of a thin conducting rectangular sheet of length l and width w is

$$R = \rho_s \frac{l}{w} \tag{4.1}$$

That is, it is proportional simply to the number of square sections of side w which separate the ends and does not depend on the absolute dimensions. If the geometry of the resistor is not simply rectangular—and it usually is not—the calculation of the equivalent number of squares is more involved. In the most general case it must be treated by conformal mapping techniques which reduce irregular shapes to forms which may be more easily calculated.[3] Results for two commonly encountered situations are given in Fig. 4.3, namely, the equivalent resistance of an

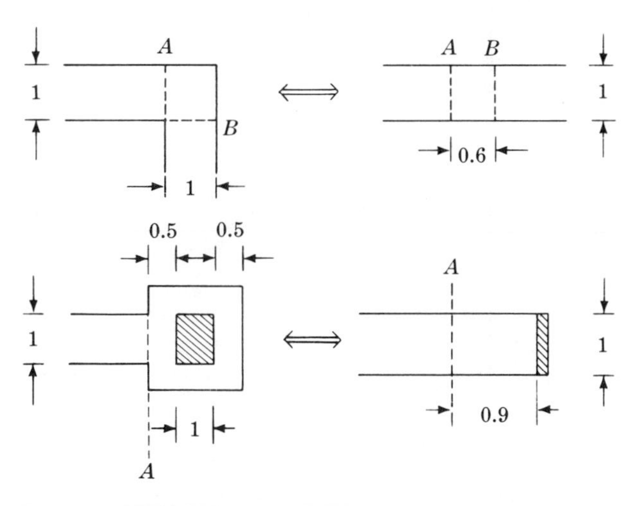

Fig. 4.3 Equivalent rectangular sections for corners and ends in diffused resistors.

end contact pad and of a right-angle bend. Some other shapes may also be reduced simply to combinations of the above.

The sheet resistance is related to the depth d, surface concentration C_s, and doping profile of the p diffusion. Detailed information on this relationship is available in the literature,[4] but a curve of average resistivity for the common case of a gaussian p-diffusion into a 1-ohm-cm n-type background is shown in Fig. 4.4. Typical parameters for a base diffusion are $d = 2.5\ \mu$, $C_s = 2 \times 10^{18}$ atoms/cm^3, giving a sheet resistance of 160 ohms per square. Thus, a 10-kΩ resistor formed from this diffusion might be 0.5 mil wide and 33 mils long. In practice the sheet-resistance range for the base diffusion is between 120 and 200 ohms per square, depending on the process and the weighting which the various factors leading to a compromise are given.

From the above, it is clear that the tolerance of base-diffused resistors depends both on the precision of the geometry and the degree of control over the final sheet resistance. With regard to the former factor, precision mask making and photolithography yield an accuracy in the number of squares in a resistor of ± 1 percent in a typical case. This accuracy depends somewhat on the geometry used since a given absolute error in the dimensions of a resistor obviously leads to greater resistance errors in long narrow structures than in those with a smaller aspect ratio. At the present state of the art, a width of 1 mil is preferred for resistors

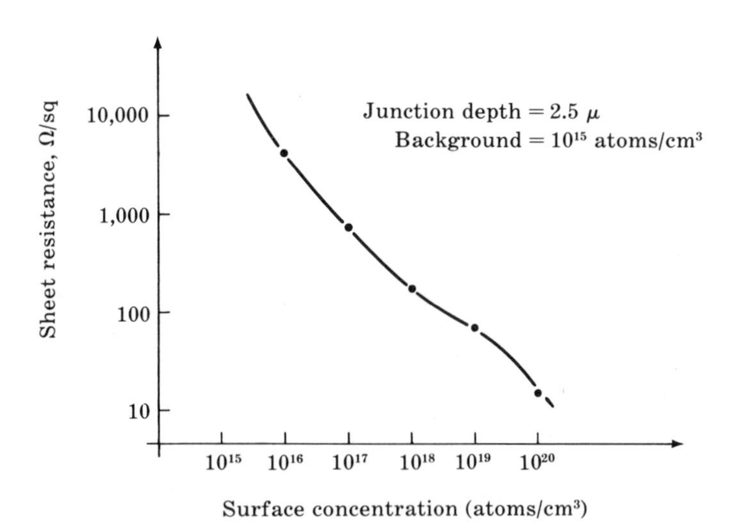

Fig. 4.4 Relation between sheet resistance and surface concentration for a diffused p layer in 1-ohm-cm n-type silicon. (*After Irvin.*[4])

where dimension control is important. Hence, accuracy is obtained at the cost of chip area. The control over sheet resistance is a greater source of errors in a conventional process. In production, this parameter is controlled in the range ± 10 percent, and thus the absolute value of resistance has a similar tolerance. It is apparent from the above, however, that all resistors in a given circuit chip, where the variations in sheet resistance are negligible, have ratios which are accurate to within the geometrical tolerance of 1 to 2 percent. Design methods for integrated circuits are therefore aimed at making the desired function more dependent on resistor ratios than on absolute values and examples of this philosophy are given in later sections.

In low-value resistors, it should be mentioned that the interface resistance of aluminum silicon contacts, which is not well controlled, is a further contributor to errors. In some processes a very thin intermediate layer of titanium is laid down between the Al and Si to improve the contact.

The sheet resistance of the emitter diffusion (n^+) is in the range 10 to 2 ohms per square and this diffusion may be used to form very small resistances, corresponding to a fraction of a square in the base-diffused case. This method is not often used, however. In low-power circuits, resistors of less than 100 ohms are rarely required, in the first place, and there are also certain basic disadvantages. To isolate an emitter resistor from the surrounding circuit, it must be placed inside a base diffusion with the resulting n^+p junction reverse biased. The reverse breakdown of this junction is low, being in the range 5 to 8 volts, so that more care must be taken in establishing the correct reverse bias than in the case of the base-diffused resistor. Also, there will not be close tracking of ratios and temperature coefficient between base- and emitter-diffused resistors because of the different doping concentrations involved. Of greater importance is the case of very large resistances, lying outside the range of these two types, and this question is taken up below.

The temperature sensitivity of diffused resistors in silicon is primarily due to mobility variations in the temperature range of normal interest, that is, -55 to $+125°C$. Silicon becomes degenerately doped at a level of about 10^{19} atoms/cm^3. Below this level, virtually all donors or acceptors are ionized and the carrier concentration is not a strong function of temperature. Mobilities in the above temperature range are limited, partly by lattice scattering and partly by scattering from ionized impurity sites, and do not vary linearly with temperature. This fact is most important in the context of active RC filters where center frequency and bandwidth are directly related to RC products. Circuit techniques which seek to minimize the temperature sensitivity of a transfer function over a wide temperature range are thus complicated by the necessity for

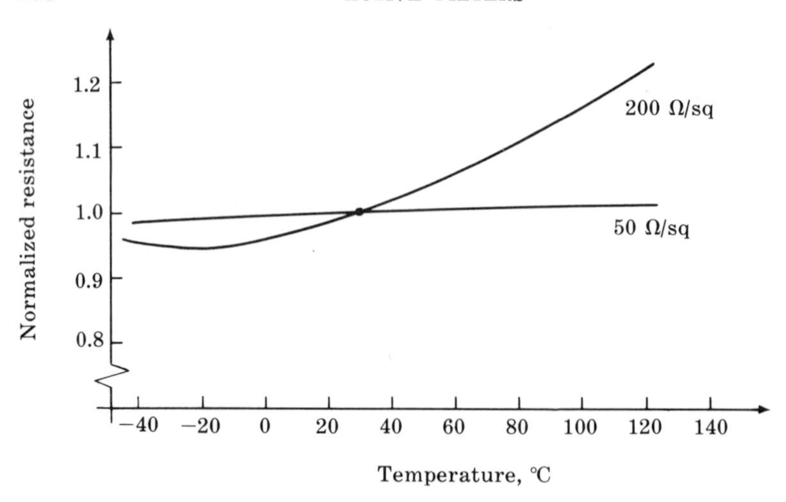

FIG. 4.5 Temperature sensitivity of typical p-diffused resistors.[1,4]

nonlinear compensation. Figure 4.5 shows some typical temperature-sensitivity data for diffused resistors of different doping levels.

When there is a requirement for higher resistance values than the base diffusion can provide, "pinch" resistors may be used. Two types of "pinch" resistor are shown in Fig. 4.6. The base pinch resistor (*a*) makes use of the high sheet resistance of the base layer of a normal transistor structure, sandwiched between the emitter and collector. The mean doping level of this layer is low, since the more heavily doped part of the base diffusion has been compensated by the emitter diffusion. Equivalent sheet-resistance values for this layer are in the range 5 to 10 kilohms per square, that is, 20 or more times that of the simple base-diffused resistor. However, the tolerance and temperature sensitivity of this resistance are worse. The sheet resistance depends not only on the original doping level but also very strongly on the positions of the two

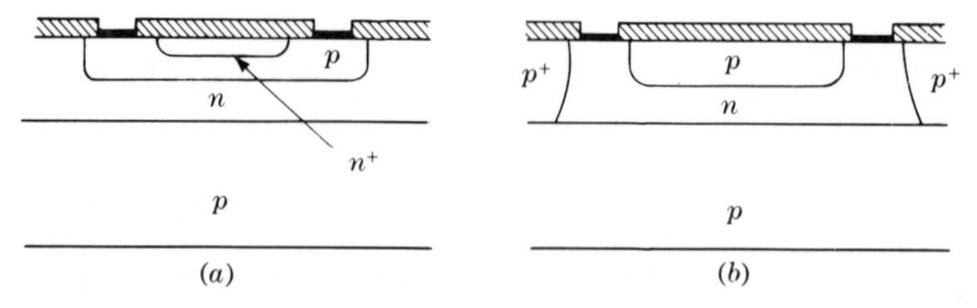

FIG. 4.6 Cross sections of two pinch resistors: (*a*) Base pinch; (*b*) collector pinch.

junctions which enclose it. As a result, the tolerance on such a resistor is usually ± 50 percent. The low doping also leads to a higher temperature coefficient, typically $+5{,}000$ ppm/°C at room temperature.[5] With the emitter-base junction reverse biased, this structure is also a field-effect device, and pinch-off effects, leading to constant current limiting, occur at higher current levels. A typical I-V characteristic is shown in Fig. 4.7. The constant-current region is potentially useful in some circuit applications, e.g., for large values of collector load resistances, but, as the curve shows, the voltage which may be placed across such a device is limited to about 6 volts by the reverse breakdown of the n^+p junction.

The collector pinch resistor, shown in Fig. 4.6*b*, has similar characteristics. The resistor region, being part of the original lightly doped *n*-type epitaxial layer, sandwiched between the substrate and base junctions, has a sheet resistance in the range 0.5 to 5 kilohms per square. Like the base pinch resistor, the temperature coefficient is high and in this case the material is *n*-type. The main advantage which this resistor offers over the base pinch resistor is the higher breakdown voltage (50 to 100 volts) of the base-collector junction. Thus, it may be used in the constant-current mode over the full range of voltage encountered in a typical integrated circuit. The same geometrical limitations described above, both in terms of junction depth and masking tolerance, apply also to the pinch resistors.

Thin-film Resistors. Thin-film resistors may be used to overcome some of these limitations of size, tolerance, and temperature sensitivity. The use of these elements in microelectronics is well established; however, the type of element and the way in which it is used is changing. The conventional method for incorporating film resistors is to deposit them on a ceramic or glass substrate and then to attach silicon chips, using

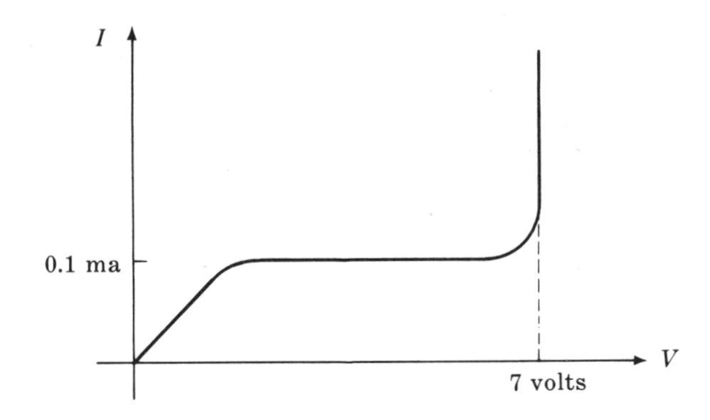

FIG. 4.7 Typical I-V characteristic of a base pinch resistor.

wire bonding, flip-chip, or beam lead techniques. These chips may contain just single transistors or small monolithic circuits. For resistors of large geometry, the deposition method may be a silk-screen method using SnO or a SiO-Cr cermet. Alternatively, thin-film materials, such as nichrome, tantalum, or cermet, may be deposited by evaporation or sputtering.[1]

Compatible thin-film techniques[6] are of greater interest here, however, because they offer a more compact method of realization. It is also probable that, because of its integral nature, the compatible thin film offers greater reliability. Basically, the method involves depositing the resistors on the monolithic circuit chip itself, rather than on a separate substrate, though the following remarks apply also to the separate-substrate case. Two materials are considered here: nichrome and tantalum. Many other materials, including cermets, Si-Cr alloys, SnO, and electroless Ni have been studied, but data on the latter group are not as comprehensive. The deposition of nichrome or tantalum involves one or two more photomasking and vacuum steps beyond the conventional six-mask process for double-diffused monolithic circuits. Consequently, there is a higher processing cost and slightly reduced yield involved, though neither of these factors precludes the use of compatible thin films where they offer a clear technical advantage.

Two typical processes are outlined here to show what is involved. For a nichrome deposition, the silicon wafer is taken after all diffusion operations have been completed and the contact windows have been opened in the silicon oxide. The wafer is placed on a substrate heater in a vacuum system and heated to about 300°C. Under a vacuum of 10^{-5} torr or better, a nichrome filament is heated by a controlled power supply to a temperature just below its melting point, causing the Ni and Cr to sublime. Using a mechanical shutter and a deposition monitor, the evaporation is carried to the desired end point. Frequently, a thicker film than is desired is deposited so that the sheet resistance of the film may be trimmed subsequently by heat treatment. If the film is heated to temperatures in the range of 300 to 500°C, in an oxidizing atmosphere, the sheet resistance gradually increases as some of the material is converted to oxide. After deposition, the film is coated with photoresist, the resistor pattern is exposed, and the film is etched directly. A solution containing ceric sulfate is one of the possible etchants. After this, the wafer is subjected to a normal aluminum evaporation and etched to interconnect both the silicon devices and the nichrome resistors. Finally, the wafer is sintered to form ohmic contacts with the aluminum layer. The sheet resistance of the nichrome is affected, as mentioned above, by the heat treatment. Also, the temperature coefficient depends on the heat treatment as well as on the rate at which the film was originally deposited. Technological problems with thin-film resistors are mainly concerned with the reproducibility of these effects and the long-term stability of the resistance. But in spite of this, the electrical properties of such elements are considerably superior to those of diffused silicon resistors.

A typical tantalum process may involve either direct etching or rejection masking. In the latter case, an aluminum layer is first deposited on the wafer, after the normal diffusions, and a pattern which is the exact negative of the resistor pattern is etched in the film. Tantalum is then laid down by sputtering from a tantalum cathode held at a potential of about -5 kV with respect to the work. The ambient here is either an atmosphere of pure argon at a pressure of about 60 μ Hg, or a mixture of argon and nitrogen. With mixed gases the sputtering process is reactive and the film consists of a mixture of Ta and Ta_3N_5, which is also a conductor. The temperature coefficient of the sheet resistance in this case varies according to the relative concentrations of the two components, as well as other sputtering parameters, and this method is inherently more flexible as a resistor-fabrication technique. After sputtering, the tantalum pattern is defined by etching the aluminum negative pattern lying beneath it, which lifts off the unwanted tantalum. Finally, a normal aluminum interconnect pattern is laid down and the wafer is heat treated as before.

Tantalum offers a feature not present in the nichrome system. The film may be oxidized anodically to Ta_2O_5 by a simple procedure and the thickness of the grown oxide film is proportional to the anodizing voltage applied. Thus, resistor arrays or individual resistors may be trimmed to the desired end point without heat treatment and the resulting oxide film forms an effective passivation layer.

The properties of the diffused and thin-film resistors described above are summarized in Table 4.1.

Capacitors. In the treatment of integrated-circuit resistors, it is assumed at the outset that virtually every resistor involved in an active *RC* filter is an integral part of the chip. This cannot be assumed in the

TABLE 4.1 TYPICAL PARAMETERS OF INTEGRATED-CIRCUIT RESISTORS

Type	*Range* Ω	*Sheet res.* Ω/*sq.*	*Temp coeff.* *ppm/°C*	*Tolerance* %	*Matching* %
Base-diffused	100–30 kΩ	100–200	500–2,000	±10	±1
Emitter-diffused	5–100	3–10	900–1,500	±15	±2
Base pinch	5 kΩ–200 kΩ	5 kΩ–10 kΩ	4,000–7,000	±50	±5
Collector pinch	10 kΩ–500 kΩ	10 kΩ–50 kΩ	4,000–7,000	±50	±10
Thin-film (Ta or NiCr)	30–100 kΩ	100–10 kΩ	0 ± 400	±2	±0.5

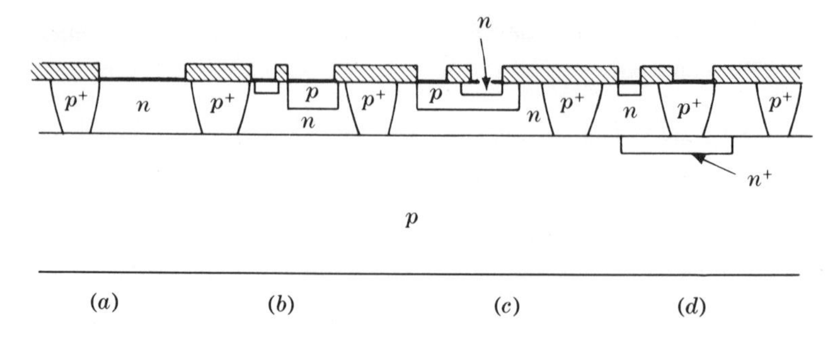

Fig. 4.8 Four junction capacitor structures: (*a*) Collector-substrate; (*b*) collector-base; (*c*) emitter-base; (*d*) isolation — n^+.

case of capacitors, so that the primary decision of a designer is whether they are to be external or integral. The considerations involved here arise out of the large areas required by capacitors of even modest values if they are to be on the chip and the problems of making these units with small parasitics and high reliability.

As for the case of resistors, discrete capacitors offer a wide range of tolerance, temperature coefficient, and parasitics, though quality and size are somewhat in opposition. Capacitors involving ceramic or electrolytically formed dielectrics have, as a rule, poor temperature sensitivity and appreciable losses at higher frequencies, so that their main usefulness is in bypass and coupling applications. It might be noted, though, that the temperature coefficient of some ceramic capacitors may be arranged to compensate for other temperature sensitivities in an active *RC* circuit. Mica and mylar are the usual dielectrics in higher-quality capacitors and give low temperature sensitivities (less than 100 ppm/°C) and low loss factors. Certain glasses and oxides have also been used to fabricate high-quality capacitors with dimensions which are compatible with microelectronics.

If a capacitor is to be fabricated directly on the chip, three main options are available: *pn* junctions, MOS structures, and thin-film techniques. For the *pn*-junction type, the collector-substrate, base-collector, and emitter-base junctions of the conventional IC structure provide usable capacitances. In addition, a fourth type may be made without additional processing.[7] This is a p^+n^+ junction formed by driving down an isolation diffusion into a buried n^+ (subcollector) layer. Under normal conditions the n^+ layer is not inverted, because of its high doping level, though an adjacent isolation diffusion not over a buried layer would penetrate through to the substrate. These four structures are shown schematically in Fig. 4.8 and are placed in order of increasing capacitance per unit area. The junction capacitors have in common the properties that they are reliable but must be correctly biased and have bias-dependent values. The principal parasitic elements are series resistance and shunt capacitance to ground, though the substrate type is further limited by the fact that it must be operated in a grounded mode. In all the structures it is clear that the series resistance results from the bulk material separating the contacts from the junctions. This problem is greatest in the *E-B* junction of Fig. 4.8*c*, which also has the lowest breakdown voltage, and in such a case an "interdigitated" or "comb" pattern of contacts should be used to minimize the resistance.

The other two capacitance structures are more conventional in the sense that they use insulators as dielectrics. Cross sections of MOS and thin-film structures are shown in Fig. 4.9. The dielectric in the MOS case is usually thermally grown SiO_2, though other dielectrics have been used. Its thickness is of the order of 1000 Å and the capacitance per unit area lies in the same range as the junction capacitances. Their advantage over the latter is a capacitance which is almost independent of voltage and which does not require an applied bias. Since one "plate" is a heavily doped semiconductor, however, similar parasitics are present. If the capacitor is operated in a floating connection, there is a parasitic capacitance to ground, by virtue of the isolation junction, and this may be as large as the principal capacitance of the structure. Again, the series resistance may be minimized by using a comb contact pattern.

Both parasitic problems are reduced greatly by the thin-film structure of Fig. 4.9*b*, in which both plates are metal. The price paid in this case is the additional process steps required. There is one problem, however, which is common to both the MOS and thin-film types. The dielectric layers, whether deposited or thermally grown, are prone to have pinholes which lead to a high probability of shorts in large-area units. Also, if the breakdown voltage is exceeded, the damage is not normally reversible. The formation of low-pinhole-density dielectric layers is therefore vital to these capacitors and requires special care in processing.

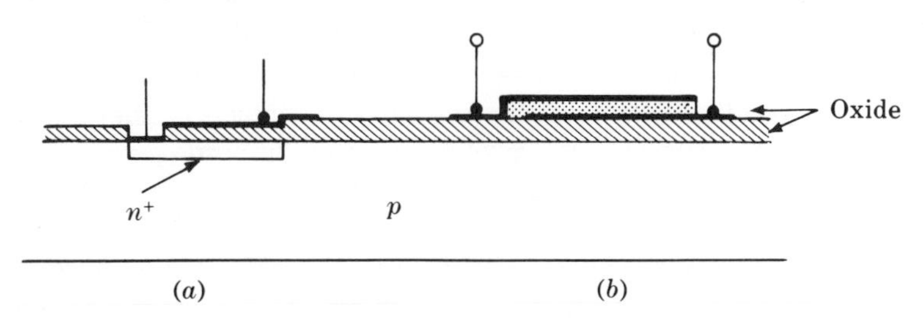

FIG. 4.9 Two surface capacitor structures: (*a*) MOS; (*b*) thin-film.

TABLE 4.2 TYPICAL INTEGRATED-CIRCUIT CAPACITANCE PARAMETERS

Type	*Cap/unit area* *pf/mil*^{2}	*Breakdown voltage* *V*	*Tolerance* *%*
Collector-substrate	0.05	100	20
Collector-base	0.09	60	20
Emitter-base	0.4	7	20
Isolation n^+	0.6	10	20
MOS	0.3	50	20
Thin-film	0.3	50	20

The properties of these various capacitor types are summarized in Table 4.2.

npn **Bipolar Transistors.** The general properties of discrete *npn* bipolar transistors will be familiar to the reader, so that we shall start by considering the ways in which these and integrated devices differ. Basically, the only structural difference between the two types is that the collector contact in a discrete unit is made to the bottom of the die, while it is made to the top surface in an integrated device (see Fig. 4.2). This is necessary in the latter case because the substrate is common to all parts of the monolithic circuit and it is also desirable to have only one interconnection system on the surface of the circuit. Parasitic capacitances are present in both types of device, of course, but their distribution is somewhat different because of the different isolation methods.

For linear operation, the doping level of the collector material is chosen as a compromise between making the reverse breakdown of the collector-base junction high and making the collector series resistance low. The resistivity normally chosen is in the range 1 to 5 ohm-cm, corresponding to a doping range of 10^{15} to 5×10^{15} atoms/cm^3. The theoretical breakdown voltage for *pn* junctions formed in this material is in the range 100 to 200 volts, but in fact it is found to be lower than this by as much as a factor of 2, owing to the existence of high-field regions at the corners of the rectangular diffusions. Very high breakdown devices are made, as discrete units, by lowering the collector doping further and by paying careful attention to the profiles of the diffused regions. The main parasitic element introduced by the use of a top contact is the ohmic resistance in series with the collector. The standard method for reducing this is the use of the buried "subcollector layer," as mentioned previously. In addition, an n^+ diffusion is made at the surface where the collector contact is formed. This is done primarily to prevent the aluminum, which is a weak *p*-type dopant in silicon, from inverting the lightly doped epitaxial layer. With the buried layer, the series ohmic collector resistance r_c is in the range 10 to 50 ohms while the same geometry device without the buried layer would have a series resistance of about 300 ohms. In linear circuits, this effect is somewhat less important than in digital circuits, where the minimization of saturation voltage is vital.

The collector of the integrated device is separated from the (grounded) substrate by reverse-biased *pn* junctions, which introduce parasitic capacitance. Though this junction is not present in discrete devices, it should be noted that the stray wiring capacitance in the latter case may easily equal or exceed the shunt capacitance in the former. The isolation capacitance is divided into two parts, for convenience: the capacitance along the planar epitaxial-substrate junction and the sidewall capacitance to the p^+ isolation diffusions. These components are calculated separately from a knowledge of the doping profiles and geometry and typical values lie in the 1-to-5-pf range. Breakdown voltages for the isolation junction are normally between 50 and 200 volts.

The base and emitter doping levels represent compromises between a number of factors including the following: a light base doping level favors high injection efficiency (see below) and a high base-emitter breakdown voltage. On the other hand, the light doping leads to large values of the parasitic ohmic base resistance between the base contact and the active region of the device. Also, if the surface concentration of the base layer falls below about 5×10^{16} atoms/cm^3, there is a risk that the surface may become inverted to *n*-type by the unneutralized charge present in the oxide layer at the surface, giving rise to a collector-emitter channel. In the case of the emitter, heavy doping favors high injection efficiency, but there is the problem that for very high doping levels ($>10^{20}$ atoms/cm^3), the resultant damage to the crystal structure in this region drastically lowers the minority carrier lifetime in the emitter and this effect tends to lower the injection efficiency. Figure 4.10 gives a typical doping profile.

Because of the high drift field in the base of a double-diffused device and the narrow base width obtainable in this process, the minority carrier transit time in the base does not play a dominant role in limiting the frequency response of the device. (Nevertheless, transistors for use in the GHz frequency range have very shallow diffusions to minimize transit-time effects.) In most circuit applications, the parasitic junction capacitances and series resistances play a more important part in determining the frequency response and these are minimized in general by using as small a geometry for the device as is allowed by the masking limits. Typical dimensions for an IC transistor were shown in Fig. 4.2, but for higher-frequency units, emitter diffusions only a few microns

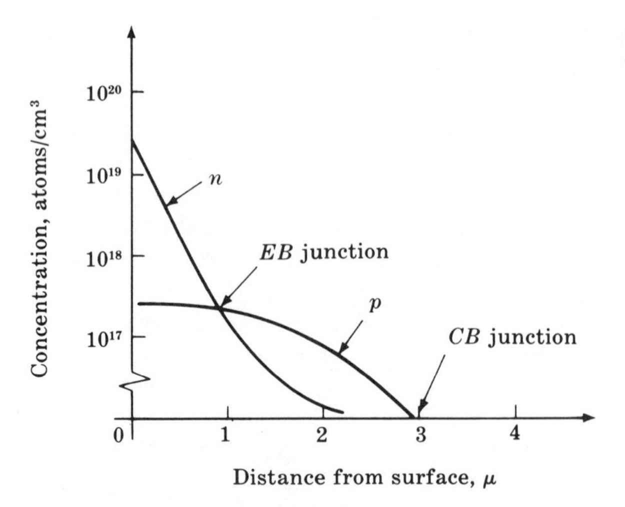

FIG. 4.10 Typical impurity-concentration profiles for a double-diffused *npn* transistor.

wide are used and the other dimensions of the device including junction depths are reduced correspondingly. The effects of the parasitic elements on circuit performance are considered more fully in Sec. 4.3.

The dc parameters directly affected by geometry and doping levels are the breakdown voltages, the maximum current ratings, and the current gain β_0. Typical values for the junction parameters are given in Table 4.2. The maximum current for a transistor is obviously proportional to its area and, in particular, to the area of the emitter. At high current levels, high-level injection effects at the emitter-base junction reduce both the emitter efficiency and the base-transport factor of the device, resulting in a degradation of current gain and frequency response. Another effect, which is closely related to these, is the base-crowding phenomenon. This arises out of the ohmic voltage drop in the base region due to the lateral flow of base current. Its effect is to enhance injection at the edge or edges of the emitter nearest to the base contact, with the result that the effective injecting area of the emitter is less than its physical area. Crowding occurs in most devices well below their maximum current rating and causes departures from the ideal diode description of the emitter-base junction. The modeling of dc and ac crowding is considered in Sec. 4.5.

In the earlier alloy junction transistors, the main source of base current was bulk recombination and this imposed the main limitation on β_0. In planar silicon devices, however, this is not the case. The base current in present silicon transistors has a number of physical origins, of which bulk recombination is one of the least important. It is worth noting, moreover, that some of the basic equations governing transistor action, which were originally derived in terms of alloy types, require a careful interpretation when applied to planar types. In the latter, the components contributing to base current include surface recombination, reverse injection at the emitter-base junction, depletion-layer recombination, junction leakage currents, and bulk recombination.[1] In terms of these components, the dc value of β_0 may be expressed generally as

$$\frac{1}{\beta_0} = \frac{I_B}{I_C} \approx \frac{N_B W_B}{N_E L_E} + \frac{\tau_t}{\tau_b} + \frac{I_{CS} + I_{rE} + I_{rS}}{I_E} \tag{4.2}$$

assuming that basewidth modulation effects may be neglected. In (4.2), the first term represents reverse injection or emitter efficiency and is given by the ratio of the doping level N_B on the base side of the emitter-base junction times the base width W_B to the emitter doping level N_E times the diffusion length in the emitter L_E. Though only approximate, this term shows the need for a large ratio of emitter doping level to base doping, as was noted previously. The second term is the ratio of the mean minority carrier transit time τ_t in the base to the bulk lifetime τ_b and is typically less than 10^{-3}. The final set of terms gives the effects of collector leakage current I_{CS}, emitter-depletion-layer recombination current I_{rE}, and surface-recombination current I_{rS}. At very low collector currents, these last terms dominate and cause the gain to drop, that is, when the base current is of the same order as the leakage currents. At moderate current levels, the injection-efficiency term is usually the most important and limits the current gain to a value in the range 40 to 200. It is possible, with special processing, however, to obtain values of β_0 in excess of 1,000 by increasing the emitter-diffusion length and reducing surface-recombination velocities. Finally, it is noted that high-power and high-voltage transistors usually have lower current gains because of large junction areas and wider base regions.

***pnp* Transistors.** One of the difficulties in making a transition from the discrete-component to the integrated-circuit medium is that of realizing *pnp* transistors. The normal practice in discrete-circuit design is to use *npn* devices as a matter of course, since a wider variety of types is available. Nevertheless, where a *pnp* device is called for, the designer has no hesitation in using it. Two common examples of their use are as collector-load current sources for *npn* stages and as complementary devices in class *B* power-output stages. The conventional integrated-circuit medium is, however, strongly oriented toward optimized *npn* transistors. So, if a *pnp* must be used in an integrated circuit, it must be made in what is basically an unsuitable medium. Of course, an integrated-circuit technology oriented toward *pnp* devices would create analogous, and perhaps more serious, problems for the *npn*. We conclude, therefore, that the designer should use *pnp* transistors in an integrated circuit somewhat more cautiously than he would in a discrete-component version.

In spite of the above remarks, one or two quite successful techniques have been developed for fabricating *pnp* transistors which are compatible with monolithic processing. Other processes have also been developed, but require additional steps.

A "substrate *pnp*" is shown in cross section in Fig. 4.11*a*. Here, the *p*-type substrate is used as the collector, the *n*-type epitaxial layer is used as the base, and a *p* diffusion is made for the emitter. The most obvious property of this device is that its collector is common to the substrate of the rest of the circuit and is normally grounded. Thus, the device is limited to common-collector (emitter-follower) operation. Though this is a limitation, such devices may be used successfully in complementary class *B* output stages for applications such as operational amplifiers. The emitter diffusion is carried out in the same operation as the resistor and *npn* base diffusion, so that an extra process step is not added. However, the thickness of the epitaxial layer must be carefully controlled, since this directly affects the base width. Even with good control, the base width of the *pnp* is normally larger than that of the *npn* devices. Another point, which is relevant to the use of these transistors in power stages, is that the base is very lightly doped (being the original epitaxial material) and the maximum current density for high-gain operation is much lower than in the *npn*. Substrate *pnp* transistors, therefore, are usually large-area devices with a poor frequency response.

The triple-diffused *pnp* structure of Fig. 4.11*b* is isolated from other devices on the chip. But this isolation is achieved at the expense of at least two extra masking and diffusion steps. This structure is the least compatible with conventional processing, since the desired parameters of the three *pnp* diffusions are different from those of the two *npn* dopings. The collector of the *pnp* must be much more lightly doped than the normal base doping for an *npn* to preserve a reasonable collector breakdown voltage and the *p*-type emitter diffusion must be far heavier than either of the two diffusions which precede it. Finally, there is the problem of controlling simultaneously the base widths of the two complementary devices. The triple-diffusion technique has been used in commercial devices, but its use is not common.

Dielectric isolation methods, which do not involve reverse-biased *pn* junctions, provide alternative solutions to the complementary-device problem. One such scheme is the beam lead type,[8] in which the silicon joining different parts of a circuit is completely removed, leaving the beam lead as the only physical connection. A complementary pair of transistors, in which the *pnp* is a substrate device, is shown in Fig. 4.11*c*. Another isolation scheme involves a physical bond formed with polycrystalline silicon which is insulated from the single-crystal regions by thin oxide films. The fabrication process used here is rather complicated, but a cross section of a complementary pair in this medium is shown in

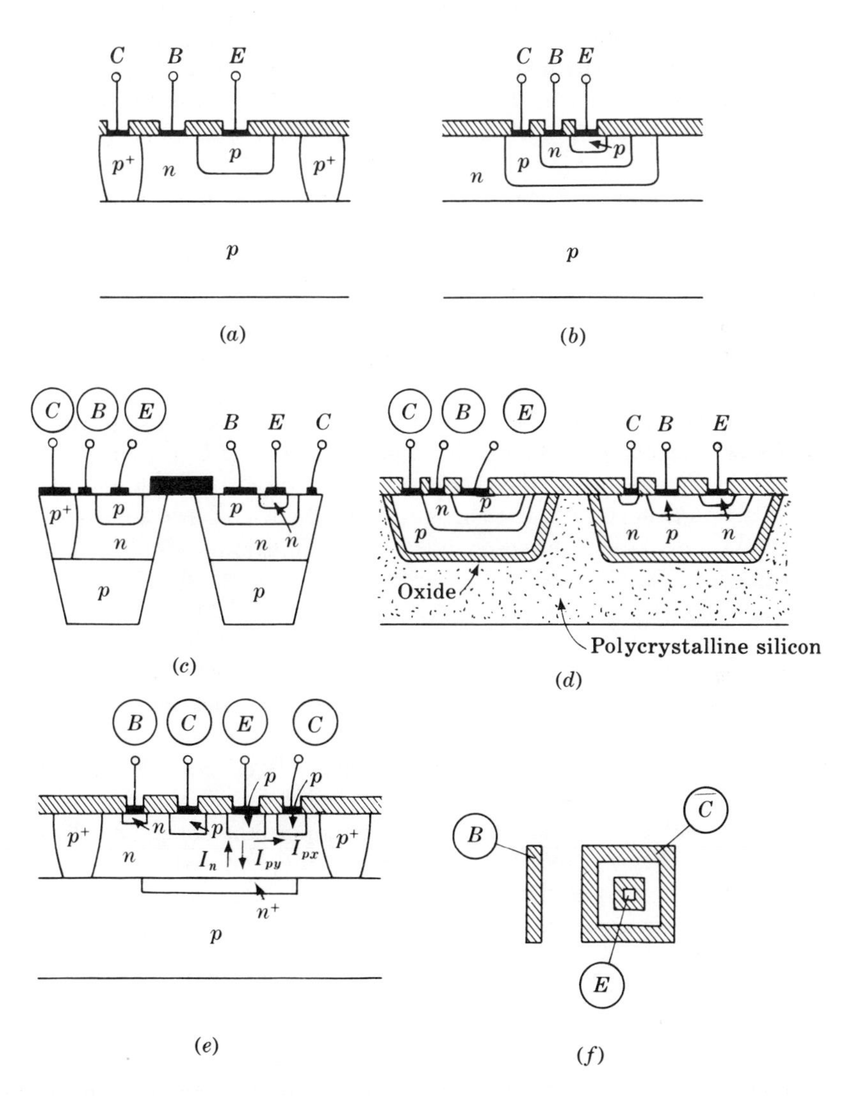

FIG. 4.11 Compatible *pnp* geometries: (*a*) Substrate *pnp*; (*b*) triple diffused; (*c*) beam lead isolated; (*d*) dielectric isolated; (*e*) lateral (cross section); (*f*) lateral (plan view).

Fig. 4.11*d*. Because both methods involve considerable extensions of the basic monolithic process, with their consequent cost and yield problems, they are not treated further here. However, both find important uses in special applications.

The lateral *pnp*[2] is attractive from the viewpoint that it is truly compatible with standard processing. Cross-section and plan views of a typical device are shown in Fig. 4.11*e* and *f*. As implied by its name, the basic conduction process in this device is parallel to the surface of the chip, in contrast with the vertical devices. The emitter and collector diffusions are the same as the normal base and resistor diffusions and the buried n^+ layer is also made in the same process as the subcollector of the *npn*'s. Thus, no additional processing is required. Also, since the base width is determined primarily by surface geometry, a certain amount of independence exists between the process control of base width in the *npn* and the *pnp*. This is not a great advantage, however, because the practical limitations of photolithography restrict the base width of the latter to 5 to 10 μ, compared with 0.3 to 2 μ for the *npn*. Consequently, the current gain and high-frequency performance of the lateral *pnp* are inferior.

Let us now examine the device in more detail, referring to Fig. 4.11*e* and assuming that it is biased as a normal small-signal amplifier. Injection takes place across the whole surface of the emitter-base junction, but only that portion of the current injected parallel to the surface gives rise to lateral gain. The components of the emitter current are separated as follows:

$$I_E = I_{px} + I_{py} + I_n \tag{4.3}$$

where I_{px} is the component of injected hole current parallel to the surface, I_{py} is the normal component, and I_n is the reverse injection electron component. To a first approximation, I_{py} is subject only to bulk recombination, while I_{px} is subject to both surface and bulk effects. The substrate-base junction is reverse biased for normal operation, so that the substrate acts as a collector in competition with the surface collector. In other words, this device contains a parasitic vertical *pnp* structure. If the current gain is computed in a similar way as for the vertical *npn* [see (4.2)], with $I_n/I_E \approx N_B W_B / N_E L_E$, the following expression is found:

$$\frac{1}{\beta_0} \approx \frac{I_B}{I_C} \approx \frac{N_B W_B}{N_E L_E} + \frac{\tau_t}{\tau_b} + \frac{I_{py} + I_{cS} + I_{rE} + I_{rS}}{I_E} \tag{4.4}$$

The value of β_0 is lower than for the vertical device for reasons which are explicit in (4.4). First, the vertical injection term I_{py}, which is not present in (4.2), may be large compared with the other components of base current. It is minimized in practice through the use of a buried layer in the base (see Fig. 4.11*e*), which sets up a drift field to inhibit vertical injection. Parasitic action is also reduced by maximizing the ratio of the "sidewall" area of the emitter to the area parallel to the surface. Because of the geometry of the device, both the bulk- and surface-recombination components, represented by τ_t/τ_b and I_{rS}/I_E in (4.4), are larger than in a typical vertical device. These contributions are minimized by reducing the collector-emitter spacing and by treating the surface to reduce the density of recombination centers. The typical range of β_0 in a lateral *pnp* is 5 to 15, though values up to 50 have been obtained with specially optimized processing.

Another characteristic of the lateral *pnp* which degrades its performance somewhat is the base-width modulation effect. In a vertical *npn*, the collector is more lightly doped than the base, so that under normal bias most of the *CB* depletion layer lies in the collector where variations in its width have just a small effect on the base width. In the lateral *pnp*, the situation is the opposite. Hence, base-width modulation effects are much stronger, though they are offset to a certain extent by the larger base width of the lateral geometry. For a typical vertical device, the output resistance due to base-width modulation is about 50 kΩ for low-current operation, while it is about 10 kΩ for a similarly biased lateral.

To conclude this section we look briefly at the high-frequency performance of the lateral *pnp*. The device analysis is complicated by the fact that it does not have a planar geometry and there is a range of path lengths for carriers in transit from the emitter to the collector. We should expect, therefore, that the driving-point and transfer functions of the device should have the character of a distributed parameter system. This is found in practice, but for an optimized geometry the distributed effects become weaker and dominant characteristics may be approximated by the normal lumped-element charge-control model. It was stated above that vertical *npn* devices are not normally limited by base transit-time effects. But lateral devices, because of their large base width, are. According to the charge-control model, the base transit time τ_t, ignoring parasitic effects, is given in the field-free case by:

$$\tau_t = \frac{W_B^2}{2D_p} \tag{4.5}$$

where W_B is the base width and D_p is the diffusion constant for holes in the base. If we take a typical value of $W_B = 8\ \mu$ and $D_p = 12.5$ cm^2 sec^{-1}, the transit time is found to be 25 nsec, which implies an intrinsic α cutoff frequency of 6.2 MHz. Measurements on typical devices give a value of the order of 10 MHz. Note also that the low value of D_p, compared with $D_n = 35$ cm^2 sec^{-1} for electrons, contributes to this low cutoff frequency.

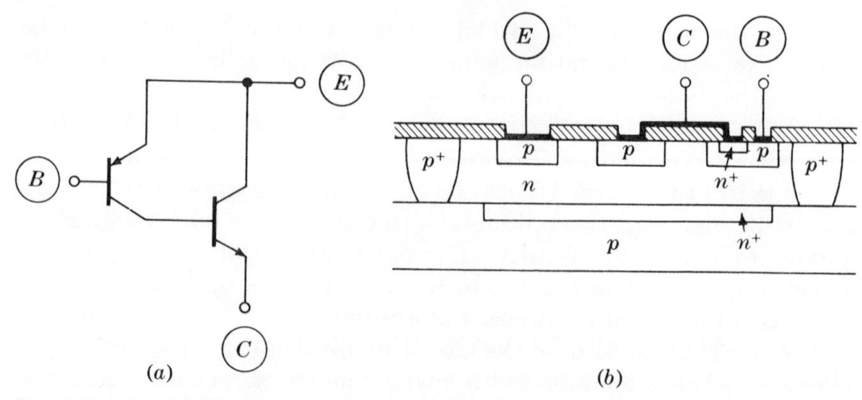

FIG. 4.12 (*a*) A high-gain composite *pnp-npn*; (*b*) composite geometry.

In the majority of its applications, the lateral *pnp* is used only for low-frequency or dc signals, since there are no simple ways to improve its frequency response. However, the low-frequency β_0 of the lateral may be enhanced by coupling it with a high-gain *npn*. Figure 4.12 shows an example of such a composite device[2] whose effective current gain is approximately the product of the two device gains.

Field-effect Transistors. The usefulness of FETs in *RC* active filter realization has not been clearly established and the literature contains few examples of their use. For the sake of completeness, however, we consider briefly the possibilities of making these devices in the conventional IC process. A *p*-channel JFET can be made from the basic *npn* diffusions with a simple change in geometry, letting the base region become the channel and the emitter the top gate, which is connected to the original epitaxial material for the lower gate. This approach has problems, however. As noted in the case of the base pinch resistor, Figs. 4.6 and 4.7, the gate-to-channel breakdown voltage will be about 7 volts, which may in fact be less than the pinch-off voltage. Since *npn* transistors would normally be made at the same time as the FET, reducing the channel (and base) width to obtain low pinch-off voltages would reduce the base width to below its optimum value and would lead in turn to excessive base-width modulation.

These problems may be avoided by using a separate and lighter doping for the channel, giving the structure of Fig. 4.13.[9] This gives a higher gate breakdown voltage and allows, to a certain extent, separate control of base and channel widths. Typical parameters for this device are:

Gate-source breakdown voltage:	50 volts
Pinch-off voltage V_p:	3 volts
Transconductance g_m:	6 mmhos

In principle, MOSFETs may also be made on the IC chip, but these devices are not simply compatible. Additional oxide growth and photomasking steps are necessary for the gate, and the optimum epitaxial layer for an enhancement-type device is somewhat different from that required for the standard IC process.

4.3 Circuit Models for Integrated-circuit Devices

Effective circuit models of the various elements in an electronic circuit form an important link between the physical considerations above and the circuit treatments which follow. Of these models, the bipolar transistor model clearly deserves the major part of our attention, though passive elements will also be considered later in this section. A multiplicity of transistor models have been proposed and may have had a wide usage, e.g., the simple resistive T model for low frequencies, the y-parameter and h-parameter models for high frequencies, leading up to very complex distributed models for computer analysis. However, the hybrid-pi model is used here for three reasons: its elements have a one-to-one correspondence with the physical processes in a bipolar transistor, it is valid over a wide range of frequencies and operating conditions, and its usage is widespread.

The derivation of the hybrid-pi model from the charge-control theory of a bipolar transistor is treated in many texts on transistor electronics and is not repeated here.[10,11,12] Figure 4.14 shows an elementary version of the model in which only the fundamental processes of charge-controlled conduction in a hypothetical one-dimensional transistor are considered. The transistor is in a grounded-emitter configuration and is biased for normal conduction, i.e., emitter-base junction forward biased and collector-base reverse biased, and signal voltages are small compared with kT/q (26 mV). The capacitance C_B is not a physical capacitance in the device, but models the relationship of the small-signal base voltage and

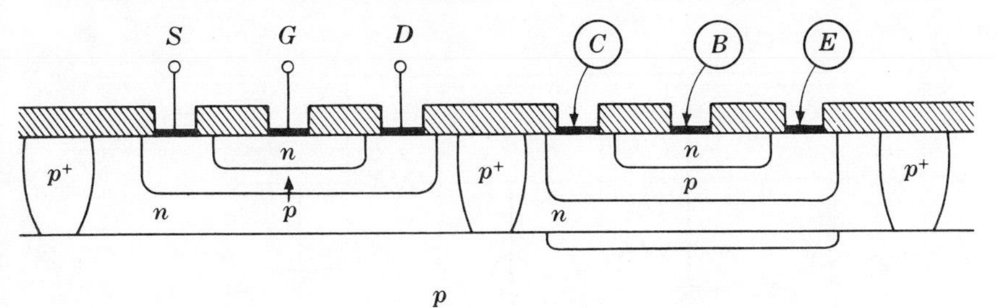

FIG. 4.13 Compatible JFET geometry (with *npn* for comparison).

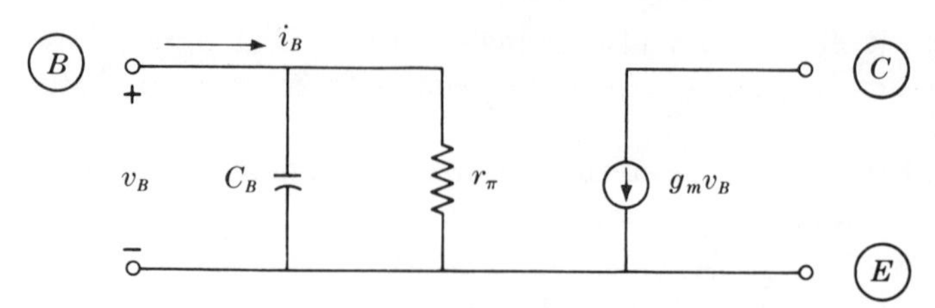

FIG. 4.14 Elementary hybrid-pi model of a bipolar transistor.

current, v_B and i_B, to the charge stored in the base by the excess minority carriers. The resistor r_π, which is also not a physical element in the device, models the processes which cause an in-phase small-signal base current to flow when the transistor is conducting. The origin of this current is described by (4.2). According to the charge-control theory, it is proportional to the excess carrier charge though it is not due solely to bulk recombination. Finally, the control of collector current by the base charge is modeled by the g_m current generator, it being implied in the simple model that collector voltage does not influence collector current.

When the additional effects present in a real device are considered, it is necessary to add more elements to the model and this is done in one step here to produce the circuit of Fig. 4.15. This is not the final model, however, since some of the elements shown may be neglected in most cases. But it is necessary to justify the final simplification and to determine its range of validity.

The parasitic capacitors added to the basic model are almost entirely physical. The depletion layer capacitances of the EB and the CB junctions are shown as C_{jE} and C_{jC}, respectively, while the parasitic capacitances associated with conductors connecting the device to other circuit elements and the isolation capacitance are represented as C_x, C_y, and C_z.

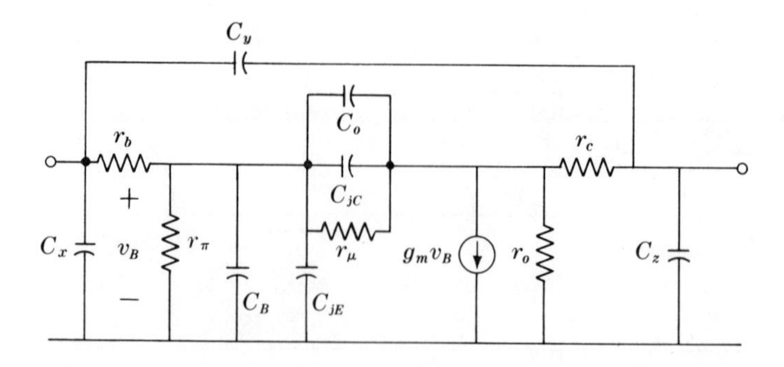

FIG. 4.15 Full hybrid-pi model.

In the monolithic case, the largest of these is C_z, which is comprised mainly of the capacitance across the reverse-biased collector-substrate junction. Included in these capacitances are also components which model the influence on the base control charge of base-width modulation, but these are usually negligible compared with the direct physical components.

Of the four additional resistance elements, r_0 and r_μ model the low-frequency effects of base-width modulation. As the collector-base voltage changes, the width of the depletion layer across this junction changes, thus altering the effective width of the quasi-neutral base region. The consequence of this change is that, for a given charge concentration in the base, more collector and less base current will flow as the base becomes narrower. This effect is modeled by the combination of the two resistances. Like r_π, these two elements are not physical resistances in the base. On the other hand, r_b and r_c are physical resistances. The former models the effective ohmic resistance of the base material between the base contact and the region at the emitter junction where injection takes place. This element is one of the most troublesome in the model, in terms both of knowing its value and of its degrading effect upon the gain-bandwidth and noise performance of the device. If we refer back to the cross section of the transistor in Fig. 4.2, we see that the ohmic base resistance consists of a part in the "open" region of the base diffusion, in series with a section lying beneath a portion of the emitter junction. In addition, it is distributed along the capacitance of the CB junction. This complexity results in considerable measurement difficulties. For example, if a low-frequency, low-level measurement is made of the input resistance of the transistor and the r_π contribution subtracted, a value corresponding to the full "open" component plus about one-quarter of the resistance lying under the emitter will be found. A measurement performed at a high current level gives a lower value because the "base-crowding" (see Sec. 4.2) effect forces the heaviest emitter injection to occur at the emitter edge nearest the base contact. At very high frequencies, an even lower value would be obtained, due to an ac "crowding" effect as well as to the shunting effect of the distributed CB capacitance. Finally, a value obtained by measuring the equivalent noise voltage of the base resistance will, in general, be different again. Manufacturers tend to quote a value obtained from the most convenient measurement technique, which is usually the high-frequency value, but this may be misleading in some applications. So the most satisfactory solution to this problem is to use, where possible, a value obtained by a measurement technique which corresponds most closely to the mode of operation of the circuit being studied. The typical values of r_b used in this chapter are intended for medium-frequency, low-level injection operation.

The other resistive element r_c, which models the ohmic resistance between the collector contact and the CB junction, also has a complex structure. But its effects are not nearly as important as those of r_b—at least at moderate frequencies—and a value obtained from a dc measurement in saturation is usually sufficient.

The final modification of the basic charge-control model is the addition of an excess phase operator e^{-st_0} to the g_m generator. Though not very important in most instances, this operator partially compensates for a simplifying assumption made in charge-control theory, namely, that the charge distribution in the base of the device is always in equilibrium with the terminal currents. This is not true, though it is a good approximation at low frequencies. In fact, a change in the injected current at the emitter takes a finite time to produce a corresponding change in the collected current. In terms of the above operator, the mean delay is t_0 and its value is found by solving the equation for charge transport across the base of the device and then approximating the resulting transcendental solution by a delay one-pole response,[12] following the method given in Sec. 4.4.

Much of the above is summarized in Table 4.3. This gives both fundamental and hybrid-pi model element values for integrated-circuit transistors for both the model of Fig. 4.15 and the simplified version developed below. Listed in this table are also some alternative symbols for the elements we have been considering, and which may be more familiar to some readers. In some cases, the temperature and bias-point dependence of the elements follow directly from their defining equations and the rest are quoted here without derivation.

Before simplifying the hybrid-pi model to a more useful version, we should consider its range of validity. There are two aspects of this question. One is whether the element values, as given in terms of physical parameters, are valid, and the other is whether this network, regardless of its element values, adequately models the transistor. The question may be answered as follows: For low current levels, normal bias, and small signals, the model is valid up to about $0.25f_T$ or 100 MHz, whichever is lower. For high current levels, the model will still be valid if the current-dependent elements are given modified values which are either measured experimentally or computed from the physics of the device. The elements most strongly affected are g_m and r_π. Similar rules apply at high signal levels. For VHF operation, a y- or h-parameter model derived from experimental measurements is more appropriate, while for saturated operation a different model again should be used. In the context of active RC filters, however, a transistor is normally operated in a mode where the hybrid-pi, as given, is valid, though the question of more complex models is taken up in Sec. 4.5.

Model Simplification. We come now to the problem of reducing the model of Fig. 4.15 to one which is simple enough to use in hand calculations, but which still gives reasonably accurate information on the gain, bandwidth, and nondominant frequency effects in a transistor circuit. This reduction is done most conveniently by considering a single transistor in a typical common-emitter amplifier configuration, as shown in Fig. 4.16*a*. This, of course, is not the most general case, but the conclusions drawn from this example are correct for most cases. For analysis purposes, the circuit is modeled as in Fig. 4.16*b* where C_B and C_{jE} of Fig. 4.15 have been lumped into C_π and $C_{jC} + C_0 = C_\mu$.

The first element in the model to be neglected in typical cases is r_μ. As Table 4.3 shows, this is typically greater than 1 MΩ. Its effect at the input may be seen by simple Miller-effect reasoning which says that the

TABLE 4.3 TYPICAL PARAMETERS OF MONOLITHIC *npn* TRANSISTORS

Parameter	*Symbol*	*Typical value*	*Bias dependence*	*Temperature dependence*
Breakdown voltages	BV_{CBO}	60 V	....	Low
	BV_{CEO}	30 V	....	Low
	LV_{CEO}	20 V	....	Low
	BV_{BEO}	6 V	....	Low
Leakage current	I_{CO}	5 na	Const. to 0.9 BV_{CBO}	+10%/°C
Low-frequency current gain	β_0, h_{fe}	80	Falls at high and low I_c. Falls for $V_{CE} < 1$ V.	+0.6%/°C
Unity-gain frequency	f_T, ω_T	600 MHz	Similar to β_0	Low
Base-emitter voltage	$V_{BE,\text{on}}$	0.65 V at 1 ma	$\sim \ln I_c/I_0$	−2.5 mV/°C
Small-signal hybrid-pi parameters at $I_c = 1$ ma $V_{CE} = 6$ V	r_b, r_x	75 Ω	Falls with incr. I_c	+0.2%/°C
	r_π, $r_{b'e}$ $\left(= \dfrac{\beta_0 kT}{qI_c}\right)$	2 kΩ	I_c^{-1}	+0.9%/°C
	r_μ, $r_{b'c}$	5 MΩ	$\sim I_c^{-1}$, $\sim V_{CE}^{-1/2}$	
	r_0, r_{ce}	50 kΩ	$\sim I_c^{-1}$, $\sim V_{CE}^{-1/2}$	−0.6%/°C
	r_c, r_{sat}	50 Ω	....	+0.2%/°C
	g_m $(= qI_c/kT)$	39 mmhos	I_c	−0.3%/°C
	C_π, $C_{b'e}$ $(= g_m/\omega_T)$	15 pf	$\sim I_c$	~−0.3%/°C
	C_μ, C_{be}	0.5 pf	$\sim V_{CE}^{-1/2}$	+0.01%/°C
	C_x	1.0 pf		
	C_y	0.3 pf		
	C_z, C_s	3 pf	....	+0.01%/°C
	t_0	0.2 nsec		

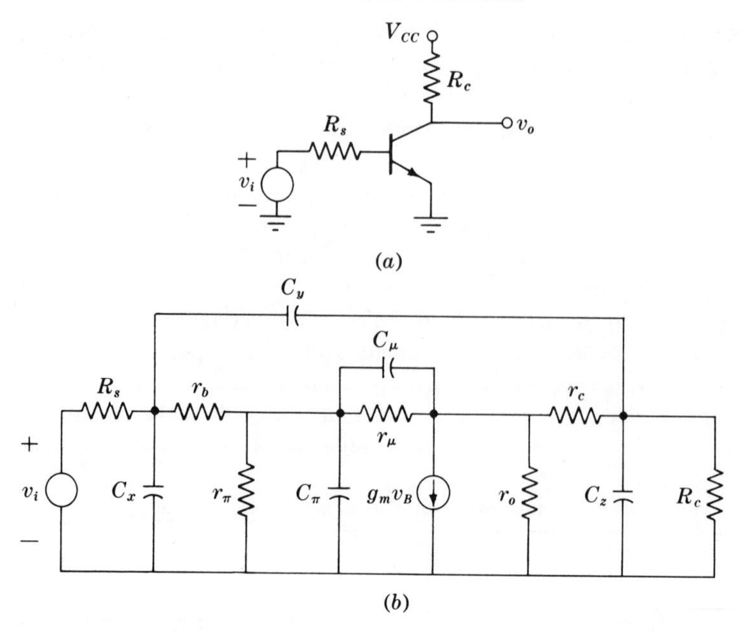

FIG. 4.16 (a) Simple common-emitter stage; (b) its full-circuit model.

equivalent resistance in shunt with r_π at the input at low frequencies is $r_\mu/(1 - A_v)$, where A_v is the voltage gain of the stage. If we take, for example, $r_\mu = 5$ MΩ and $A_v = -50$, the shunt resistance at the input due to r_μ is approximately 50 kΩ. This may be neglected in the wideband case, where r_π is of the order of 2 kΩ, though in certain low-current, high-impedance applications, neglecting r_μ cannot be justified. The case of r_0 is simple, since it may be placed in parallel with R_c if the latter is high (>10 kΩ) or neglected altogether in the low-load case. Next, the collector resistance r_c may, by a similar argument, be neglected in most cases of moderate-to-high values of R_c, or included in series with the load for frequency-response calculations at low impedance levels.

The case of the capacitors could be treated formally by examining the contribution of each to the natural frequencies of the circuit transfer functions. But this is a most cumbersome procedure in view of the large number of circuit elements. Thus, the following reasoning may be used. The output capacitor C_z may not be small enough to neglect, but it may be regarded as part of the load impedance Z_L, and a method is developed later for treating the case of an RC load. We cannot, a priori, omit this element. The role played by C_x and C_y depends on the source impedance. However, as C_z may be lumped in with the load, C_x may similarly be lumped with the source circuit. In many treatments of the model, C_y

is omitted, but this is difficult to justify in modern transistors where both C_{jC} and C_y are small and of the same order. Therefore it is retained in this treatment. Finally, we justify the omission of the delay operator e^{-st_0} on the grounds that its effects are swamped by other nondominant phase shifts due to the parasitic capacitances in the circuit.

We now propose two simplified versions of the hybrid-pi which are widely useful. The first, shown in Fig. 4.17*a*, is purely a low-frequency model which is valid over a wide range of circuit impedance levels. This would be useful where frequency information is not required, but where accurate transfer and driving-point calculations are needed. The second model, intended for wideband computations at moderate impedance levels, is shown in Fig. 4.17*b*. This is the most familiar form of the hybrid-pi and most of the treatments in the remainder of this chapter are based on this model. Its weakness at the high and low extremes of circuit impedance levels should be borne in mind, however, and there will be occasions when a circuit designer must resort to computer-aided analyses based on a more complete model for accurate information (see Sec. 4.5). So we have the familiar situation of a trade off between simplicity and accuracy. Thus, the choice of a model should be made in full knowledge both of the accuracy required in a calculation and the degree of confidence which may be placed in a given model.

In the case of *pnp* transistors, the conventional vertical devices are modeled in the same way as the *npn*, with little difference between the element values. But the lateral device, having a quite different geometry, might be expected not to conform to this pattern. One problem here is that the distributed effects described in Sec. 4.2 are more noticeable. For the purposes of first-order circuit calculations, however, the simple hybrid-pi models may be used, with C_π chosen to correspond to an f_T of about 10 MHz, r_π for a β_0 of about 10, and r_0 about 20 percent of its value for the *npn*. For more detailed models, particularly in the

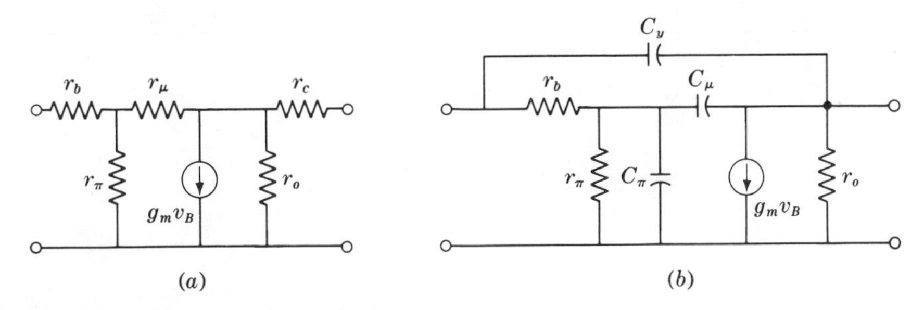

FIG. 4.17 Two simplified versions of the hybrid-pi. (*a*) For low-frequency, high-impedance circuits; (*b*) wideband medium-impedance model.

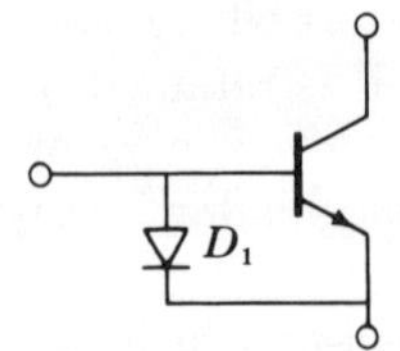

FIG. 4.18 Modified device model of the lateral *pnp*. D_1 is a wide-base diode.

case where parasitic vertical action is important, an equivalent circuit of the type shown in Fig. 4.18 has been proposed. The diode is treated as a wide-base unit and models the vertical action, while the transistor is taken as a conventional unit. This model more accurately predicts the distributed-parameter driving-point and transfer functions.

JFETs and MOSFETs are usually treated together for modeling purposes, though the physics of the two devices is somewhat different. Charge-control models are valid and useful in both cases. The control charge is that on the gate, as opposed to the majority charge in the base of a bipolar transistor, and the charge in transit is majority charge in the channel, as opposed to the minority carrier charge in the base. When used as amplifiers or current sources, FETs are normally operated above pinch off (saturation) and a model for this region is given in Fig. 4.19.

The capacitors in this model are all associated with physical capacitances in the device, either across a reverse-biased *pn* junction or through a layer of insulating material. The g_m generator and the output resistance r_d are both incremental quantities and the latter models the dependence of the length of the pinched-off region of the channel upon the source-to-drain voltage. The parasitic source resistance r_s is the ohmic resistance of the region between the source contact and the point in the channel where pinch off begins (see Fig. 4.13). Its effect is to degrade the transconductance of the device, but this may be neglected in the

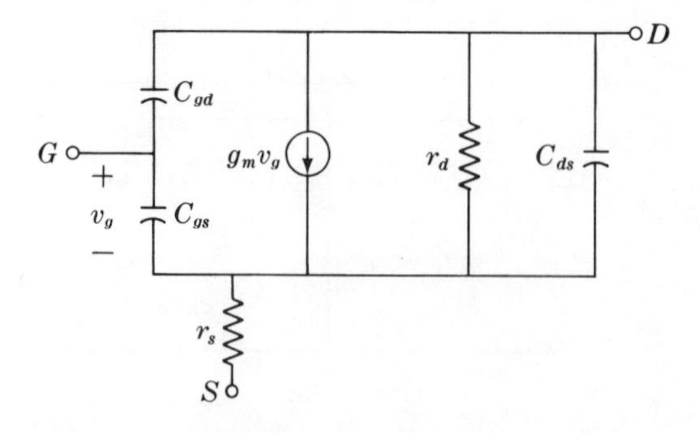

FIG. 4.19 Small-signal circuit model for JFETs and MOSFETs above pinch off.

TABLE 4.4 TYPICAL SILICON FET PARAMETERS

Parameter	Symbol	*p-channel MOSFET*	*n-channel JFET*
Breakdown voltages	BV_{DS}	30 V	30 V
	BV_{GS}	25 V	30 V
Threshold voltage	$V_{GS,\mathrm{th}}$	−5 V	....
Pinch-off voltage	V_p	....	6 V
ON resistance	$r_{ds,\mathrm{on}}$	1 kΩ	500 Ω
Small-signal parameters	C_{gd}	0.5 pf	1.2 pf
	C_{gs}	2 pf	3 pf
	C_{ds}	0.5 pf	1 pf
	g_m	0.5 mmho	3.5 mmhos
	r_d	40 kΩ	50 kΩ

usual case where the channel is very narrow. In view of the geometry of the device, the representation of C_{gs} and C_{gd} as capacitors lumped between the gate and the two ends of the channel is obviously an approximation. In fact, the gate capacitance is distributed along the channel. Once again, the question arises of what is the range of validity of the simple representation. To answer this question, we define a frequency figure of merit:

$$f_1 = \frac{g_m}{2\pi}\,(C_{gs} + C_{gd}) \tag{4.6}$$

At this frequency, the reactance of the total gate capacitance equals the transconductance of the device. It has a value of the order 100 MHz, typically. We may therefore place a conservative upper limit on the validity of Fig. 4.19 at about 20 percent of f_1, that is, 20 MHz in this case.

Table 4.4 gives some typical parameter values for JFETs and MOSFETs above pinch off.

4.4 First-order Circuit Analysis

Though an extensive treatment of electronic-circuit design is not the purpose of this chapter, the establishment of sound device models translates many aspects of the analysis and design processes into routine circuit analysis. In view of the number of elements in the circuit model of just one transistor, the full analysis of a circuit containing two or more transistors is obviously a complex process and a designer would understandably be reluctant to attempt such a job. The analysis problem must therefore be divided into two operations, the first of which involves simplified hand calculations to give approximate answers and the second involving computer-aided analysis with models of higher complexity and accuracy. Though the two operations should complement each other, the first is almost always the more important, since it gives to the designer a basis upon which to judge the merits of a design and an insight into what factors within the circuit enhance or degrade its performance. For example, if a circuit is to meet certain requirements on input impedance and gain, the methods given in this section provide, very quickly, a reliable indication of the number, type, and bias conditions of the devices required.

We are concerned here with the specific class of design problem associated with active *RC* filters and in this context the following approach is an effective one. The establishment of a circuit configuration and the evaluation of its first-order performance is most efficiently carried out by hand, using whatever experience, intuition, and imagination the designer can apply to the problem. Then, the computer should be used to study design optimization, second-order effects, temperature sensitivities, component-tolerance effects, and the effects on the performance of the circuit parasitics.

The techniques which provide rapid first-order information on circuit performance arise out of the application of some standard results in circuit theory.

The low-frequency gain and driving-point impedance problem is not necessarily the simplest one, but it is a logical starting point. The key to a simplified approach is that the active devices which we treat here are considered to be unilateral. Thus, the gain of a cascade of amplifying stages is expressed as the simple product of individual stage gains. An example is now introduced to illustrate this point and the same example is carried through this chapter as various aspects of the analysis problem are considered. The circuit, shown in Fig. 4.20, is a three-stage voltage

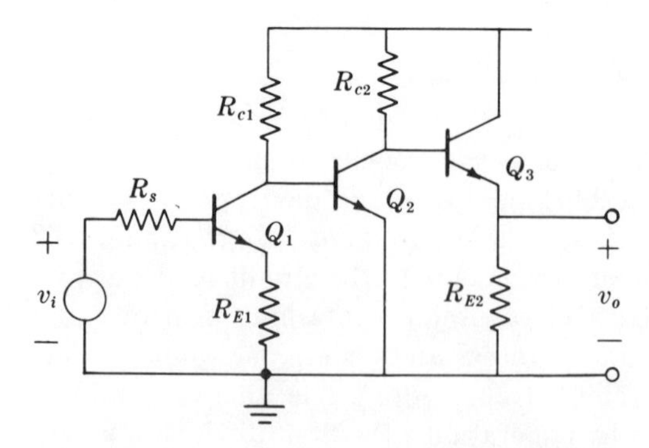

FIG. 4.20 Three-stage voltage amplifier example.

amplifier with an emitter-follower output stage. Series feedback is incorporated in the first stage to raise the input resistance and to broadband it slightly with respect to the second stage, though the main motivation for using it in this particular case is the presentation of a useful simplification of the model for a series feedback stage. Thus, in this simple circuit, we have three commonly used transistor configurations: the series feedback, the grounded emitter, and the common collector, or emitter follower.

The first step in modeling the circuit is to replace each device with its appropriate hybrid-pi model and this is done in Fig. 4.21, using the standard model of Fig. 4.17*b*. Further model reductions are possible, however, before beginning the analysis process, and we start with the first stage.

It can be shown by a simple calculation, which is left to the reader, that a transistor stage with a series feedback resistor R_E in the emitter may be modeled by a grounded-emitter hybrid-pi with modified parameter values. These modified values are denoted by primed symbols and are related to the original device parameters as follows:

$$r'_\pi = r_\pi(1 + g_m R_E) \qquad r'_0 = r_0(1 + g_m R_E)$$

$$g'_m = \frac{g_m}{1 + g_m R_E} \qquad C'_\pi = \frac{C_\pi}{1 + g_m R_E} \tag{4.7}$$

Other elements in the model are left unchanged. In a wideband model, this conversion is strictly true only if a small capacitance C_E is placed in shunt with R_E with a value such that $R_E C_E = 1/\omega_T$. However, the capacitance which satisfies this relation in a typical case is so small that it would be provided by stray capacitance in the circuit and is not usually shown in a schematic.

Next, the emitter-follower stage may be represented more simply as a unity-gain voltage amplifier with the following input and output impedances at low frequencies:

$$Z_{\text{in}} \approx (r_b + r'_\pi)\|(C_\mu + C'_\pi)$$

$$R_{\text{out}} \approx \frac{1}{g_m} + \frac{R_b}{\beta_0} \tag{4.8}$$

where R_b is the effective source resistance of the stage. At high frequencies and particularly in the presence of a complex load impedance, this model is poor, however, because of reactive components in the output impedance of the stage. Therefore, if these effects in an emitter follower are found to be important, this model reduction should not be used. In the case being considered here, however, the reduction is valid and a simplified model of the complete circuit is shown in Fig. 4.22. In the figure $C_{L2} \approx C_{\mu 3} + C'_{\pi 3}$.

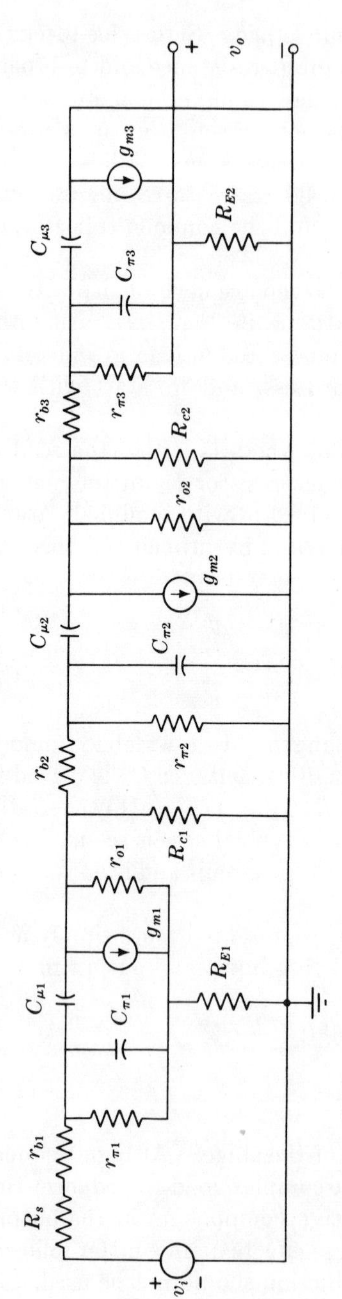

FIG. 4.21 Circuit model for amplifier of Fig. 4.20.

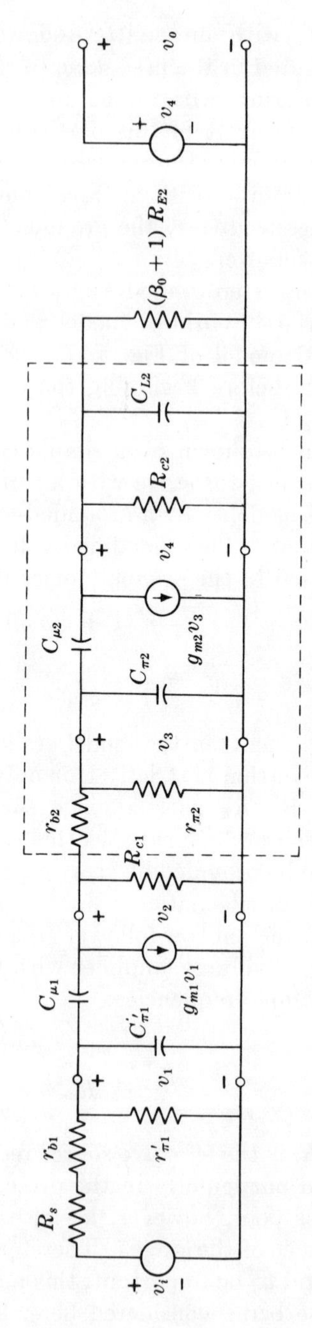

FIG. 4.22 Reduced-circuit model for the amplifier of Fig. 4.20.

For a low-frequency gain calculation, all capacitors in the model are ignored and the voltage-gain expression now may be written down by inspection:

$$A_v = \frac{v_o}{v_i} = \frac{g'_{m1} r'_{\pi 1} R_{c1}(r_{b2} + r_{\pi 2}) g_{m2} r_{\pi 2} R_{c2}(r_{b3} + r'_{\pi 3}) R_{E2}}{(R_s + r_{b1} + r'_{\pi 1})(R_{c1} + r_{b2} + r_{\pi 2})(r_{b2} + r_{\pi 2})(R_{c2} + r_{b3} + r'_{\pi 3})(R_{E2} + 1/g_{m3})} \tag{4.9}$$

Using (4.7), the following typical parameter values are now substituted in (4.9):

$$R_s = R_{E2} = 1\text{ k}\Omega \qquad R_{E1} = 100\text{ ohms} \qquad R_{c1} = 3\text{ k}\Omega \qquad R_{c2} = 2\text{ k}\Omega$$

$$r_b = 50\text{ ohms} \qquad r_\pi = 1.5\text{ k}\Omega \qquad g_m = 40\text{ mmhos}$$

and the resultant value of A_v is 505. A tolerance of 10 percent is placed on this answer, because of the modeling approximations. Finally, the low-frequency driving-point impedances are calculated from:

$$R_{\text{in}} \approx (r_b + r'_{\pi 1}) = 5.3\text{ k}\Omega \qquad \text{and} \qquad R_{\text{out}} \approx 1/g_{m3} + R_{c2}/\beta_0 \approx 58\text{ ohms}$$

Though a more accurate calculation could have been made by incorporating the r_μ elements into the model, the fact that the true gain of the amplifier is not very stable or precise itself, because of its dependence on active device parameters, makes this refinement rather pointless. On the other hand, in cases where a precisely defined gain is required, a feedback configuration would be used, in which the gain calculation as well as the precision of the true performance would be correspondingly enhanced.

If coupling and bypass capacitors are used in a wideband amplifier block, low-frequency-response limitations are introduced. We do not treat this case here, however, because it is relatively straightforward and rarely arises in integrated circuits where the use of large capacitors is avoided.

The high-frequency analysis of a circuit is helped by another set of simplifications which are introduced first on a formal basis.[12] If the circuit to be analyzed is realizable, stable, and lumped, its characteristic polynomial $P(s)$ will be of the Hurwitz form:

$$P(s) = a_0 + a_1 s + a_2 s^2 + \cdots + a_n s^n \tag{4.10}$$

where the a_i's are positive and real. The roots of $P(s) = 0$ are the poles of the transfer and driving-point functions of the circuit and are, of course, of prime interest in characterizing the frequency response of the circuit. Less attention is paid here to the transmission and driving-point zeros since these are usually easier to compute. Also, they are usually, though not always, less important in determining the frequency response.

Though the factors of $P(s)$ may not be simple in form, particularly in the case of feedback circuits, the polynomial can nevertheless be expressed as a product of its factors:

$$P(s) = a_0\left(1 - \frac{s}{s_1}\right)\left(1 - \frac{s}{s_2}\right) \cdots \left(1 - \frac{s}{s_n}\right) \tag{4.11}$$

where the s_i correspond to the poles and have negative real parts. The characteristic equation is now said to have a dominant root if the above s_i are written in order of increasing magnitude, and $|s_1| \ll |s_2| \ll |s_3|$ etc. That is, the frequency response of the circuit is dominated by a single pole s_1 which is negative and real by assumption and which is much closer to the origin than all the others. If this is the case, a great simplification in our first-order analysis is possible. Without evaluating the s_i, it may also be shown that the dominant-root condition is satisfied if, in (4.10),

$$\frac{a_1^2}{a_0 a_2} \gg 4 \tag{4.12}$$

The relevance of this property to the approximate analysis problem may now be seen, if (4.11) is multiplied out and compared with (4.10). By equating coefficients of s in the two expressions, it follows that

$$\frac{a_1}{a_0} = \sum_{i=1}^{n} \frac{1}{s_i} \approx -\frac{1}{s_1} \tag{4.13}$$

since it was assumed that the magnitude of s_1 is much less than the magnitudes of the other roots. Thus, if a dominant-pole description of a circuit transfer function is valid, the problem of finding the approximate location of this dominant pole reduces to the simpler problem of finding a_1/a_0 in the characteristic polynomial.

For the next step, we need another property of linear, lumped time-invariant RC networks: the "zero-value" or "open-circuit" time constant. If we assume that the circuit contains no inductors and we examine any capacitor C_i in the network, a charging resistance R_{0i} for this capacitor may be computed by setting all other capacitors in the network to zero. The time constant $T_{0i} = C_i R_{0i}$ is defined as the "zero-value" time constant associated with C_i. In a passive network, the calculation of the R_{0i}'s is trivial and may be done by inspection in most cases. If there are controlled sources present, the calculation may be a little more involved, depending on their location. These time constants have an important bearing on the approximate analysis problem, since it may be shown that they are related to the coefficients in the Hurwitz polynomial of (4.10) by:

$$\frac{a_1}{a_0} = \sum_{i=1,n} R_{0i}C_i \tag{4.14}$$

Thus, if the network response described by this characteristic polynomial has a dominant pole s_1, then, from (4.13) and (4.14):

$$s_1 \approx -\Big(\sum_i R_{0i}C_i\Big)^{-1} \tag{4.15}$$

which shows that the position of the dominant pole, and hence the approximate frequency response of the network, may be determined simply by summing the zero-value time constants. This process is far simpler than analyzing the network formally by state-space or nodal-admittance-matrix techniques (though it is also less accurate).

To illustrate this technique, consider the second stage of our three-stage amplifier example, as shown between the dotted lines of Fig. 4.22. First, the analysis is carried out using the familiar nodal-admittance technique. The y matrix for this stage is

$$[y] = \begin{bmatrix} \frac{1}{r_{b2}} & 0 & 0 \\ -\frac{1}{r_{b2}} & \frac{1}{r_{\pi 2}} + \frac{1}{r_{b2}} + s(C_{\pi 2} + C_{\mu 2}) & -sC_{\mu 2} \\ 0 & g_{m2} - sC_{\mu 2} & \frac{1}{R_{L2}} + s(C_{L2} + C_{\mu 2}) \end{bmatrix} \tag{4.16}$$

where R_{L2} is the parallel combination of R_{c2} and $(\beta_0 + 1)R_{E2}$.

The voltage gain of this stage is found, from (4.16), to be

$$\frac{v_3}{v_4} = -\frac{g_{m2}R_{L2}}{1 + r_{b2}/r_{\pi 2}} \cdot \frac{1 - SC_{\mu 2}/g_{m2}}{1 + s(C_{\pi 2}R_{\pi 0} + C_{\mu 2}R_{\mu 0} + C_2R_{L0}) + s^2R_{L2}R_{\pi 0}(C_{\pi 2}C_{L2} + C_{\pi 2}C_{\mu 2} + C_{\mu 2}C_{L2})} \tag{4.17}$$

The positive transmission zero at $g_m/C_{\mu 2}$ is characteristic of the common-emitter stage, but its existence is usually ignored because its magnitude is greater than that of either of the negative real poles in (4.17). For example, if $g_m = 40$ mmhos and $C_{\mu 2} = 3$ pf, the zero is at 1.3×10^{10} sec^{-1}, making it important only as frequencies approach 1 GHz. The form of the coefficient of s in (4.17) follows directly from (4.14) and the zero-value charging resistances are

$$\begin{aligned} R_{\pi 0} &= r_{\pi 2} \| r_{b2} \\ R_{\mu 0} &= R_{L2} + R_{\pi 0}(1 + g_m R_{L2}) \\ R_{L0} &= R_{L2} \end{aligned} \tag{4.18}$$

It might be noted, however, that the coefficients of higher powers of s do not bear a simple relation to the zero-value time constants.

Numerical values are now substituted in (4.17), using the parameters of the low-frequency calculations above, and the gain is found to be

$$\frac{v_4}{v_3}(s) = -74 \frac{1 - (0.75 \times 10^{-10})s}{1 + (52.5 \times 10^{-9})s + (18.5 \times 10^{-18})s^2} \tag{4.19}$$

If the dominant-root test, (4.12), is applied, it is found that

$$a_1^2/a_0a_2 = 149 \gg 4$$

showing that a strongly dominant pole exists. This is confirmed by solving for the two poles of (4.19):

$$s_1 = -2.1 \times 10^7 \qquad s_2 = -2.8 \times 10^9 \text{ sec}^{-1}$$

and in this case the value of s_1 found here is virtually identical with the estimate provided by (4.15). The approximate bandwidth of the voltage gain of this stage is therefore $(1/2\pi)s_1 \approx 3$ MHz.

The conclusion to be drawn from this example is obvious: an estimate of the bandwidth of the stage may be found with sufficient accuracy for first-order calculations without analyzing the stage. In fact, this process may be carried further to estimate the bandwidth of the complete amplifier modeled by Fig. 4.22. The sum of all zero-value time constants for the circuit is 6.1×10^{-8} sec^{-1}, which implies an approximate bandwidth of 0.42 MHz. In Sec. 4.5 the results of a computer analysis of this circuit are given; here the actual bandwidth is found to be 0.425 MHz, showing that the estimate is in error by less than 2 percent.

Finally, we consider a difficulty with this technique which the reader may have already noticed. The zero-value time-constant sum gives the ratio a_1/a_0, but the dominant-root test requires also the value of a_2, which can only be determined by full analysis. Therefore, it appears that one can only apply the approximation if the exact answer is already known! This is strictly true, but it has been found by experience that almost all transistor circuits which do not contain feedback loops around more than one stage satisfy the dominant-pole condition. Thus, if a circuit is not pathological, it is reasonably safe to use the approximation without testing its validity. Even if the pole with the smallest magnitude is not dominant but is merely real, the estimate of bandwidth provided by (4.15) is still correct to within a factor of 2. For this reason, it cannot be applied to complete active-filter circuits, which normally have dominant complex-conjugate poles in their transfer functions, but it may be applied effectively to the wideband gain blocks which comprise the complete filter.

4.5 Computer-aided Analysis

Common to virtually all programs for computer-aided analysis and design of linear circuits is a basic program which converts circuit data, usually in topological form, into transfer functions and other response data. In the more complex programs, such as those which produce integrated-circuit layouts on graphic outputs or perform design optimization, the basic analysis program becomes merely a subroutine which is called a number of times during the solution of a problem. We concentrate here on this basic analysis function, however, because the other aspects of circuit computation depend strongly on the use to which they are put. The factors which must be considered in developing an analysis program include the compromise between speed and accuracy, the memory capacity of the machine it is to be used on, the type of output required, and, to a considerable extent, the preferences of the programmer. In view of this, it is easy to explain the large number of available programs which perform essentially the same functions using quite different methods.[13]

Topological data is the usual type of input to a circuit-analysis program and, for linear circuits, consists of a specification of each lumped element in the circuit giving the type of element, its value, and the identification numbers of the nodes it appears between. Most programs allow the user to specify the type of transfer or driving-point function he wants and the form of the output, i.e., as a polynomial in s, with or without poles and zeros computed, or as gain-phase information. Some programs have "built-in" models of transistors, which feature eliminates some of the required input data, but at the expense of flexibility. The larger programs also have separate sections for dc, frequency-domain, and time-domain computations, but it is desirable, for the sake of speed and economy, to be able to load only those parts of the program required for a particular problem.

The following sequence of operations is typical of a program with moderate speed and accuracy:

1. The topological input data is read and checked for consistency and completeness. The element values are then arranged in an array corresponding to a nodal admittance matrix.
2. The correct minors of the matrix are chosen according to the desired transfer function.
3. These minors are then converted to polynomials in s by a gaussian elimination algorithm with pivoting to maximize accuracy.
4. For frequency-domain output, the polynomials are evaluated in magnitude and phase at a sequence of selected frequencies.

5. For an s-plane output, the polynomials are factored by a routine which makes successive quadratic approximations to the polynomial near a root until the root is found with the desired degree of accuracy. This root is factored out of the polynomial and the process is repeated.
6. For graphical output, the appropriate routine is then called, or the results are presented in tabular form.

Other subroutines which are valuable in the determination of performance sensitivities to temperature, manufacturing tolerance, and aging are those which automatically generate new sets of topological data corresponding to random variations in circuit parameters or the variation of these parameters with temperature. The latter case is far from trivial, because the temperature sensitivity of most circuit elements is nonlinear and frequently depends on dc bias points which are themselves functions of temperature.

Models of circuit elements which have been developed for first-order analysis are a result of a closely drawn compromise between simplicity and accuracy. But in modeling for computer-aided analysis, most of the constraints on simplicity are removed. In principle, therefore, one is free to develop models for specific problems, starting with the geometry and physics of the device. This freedom means, also, that there is no widely accepted scheme of modeling for the computer. Some representative examples are given here by way of illustration, but it is not implied that these are standard.

The full hybrid-pi model of Fig. 4.15 is an example of a higher-order model which is suited to most computer programs with the exception of one element. The delay operator of the g_m generator, e^{-st_0}, is not in a convenient form, but it may be modeled for computer analysis by the circuit of Fig. 4.23, which makes a one-pole approximation to the delay. The hypothetical R and C elements are chosen so that $t_0 = RC$ and the approximation is valid up to a frequency $\omega \approx 0.1/RC$. This approximation to a delay or excess phase operator also has uses in many other cases where a transfer function has a small excess phase term.

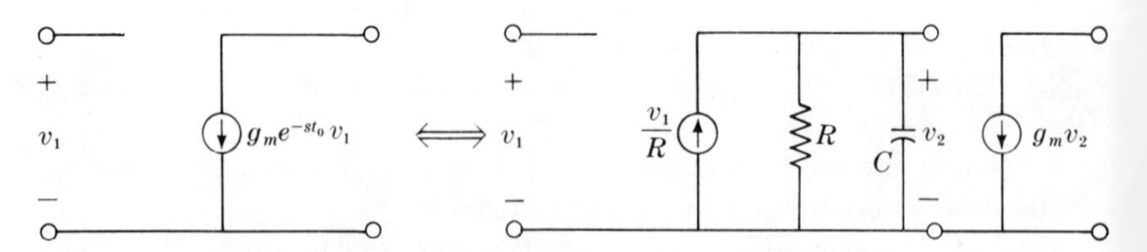

Fig. 4.23 Lumped-circuit approximation for the delay operator in a g_m generator.

It was noted previously that inaccuracies arise in the hybrid-pi model because of distributed effects, particularly in the base. This suggests the use of lumped RC and ladder models and two examples are given here. The first is shown in Fig. 4.24a. It is a more detailed model of the npn transistor in the lateral direction and approximates both ac crowding

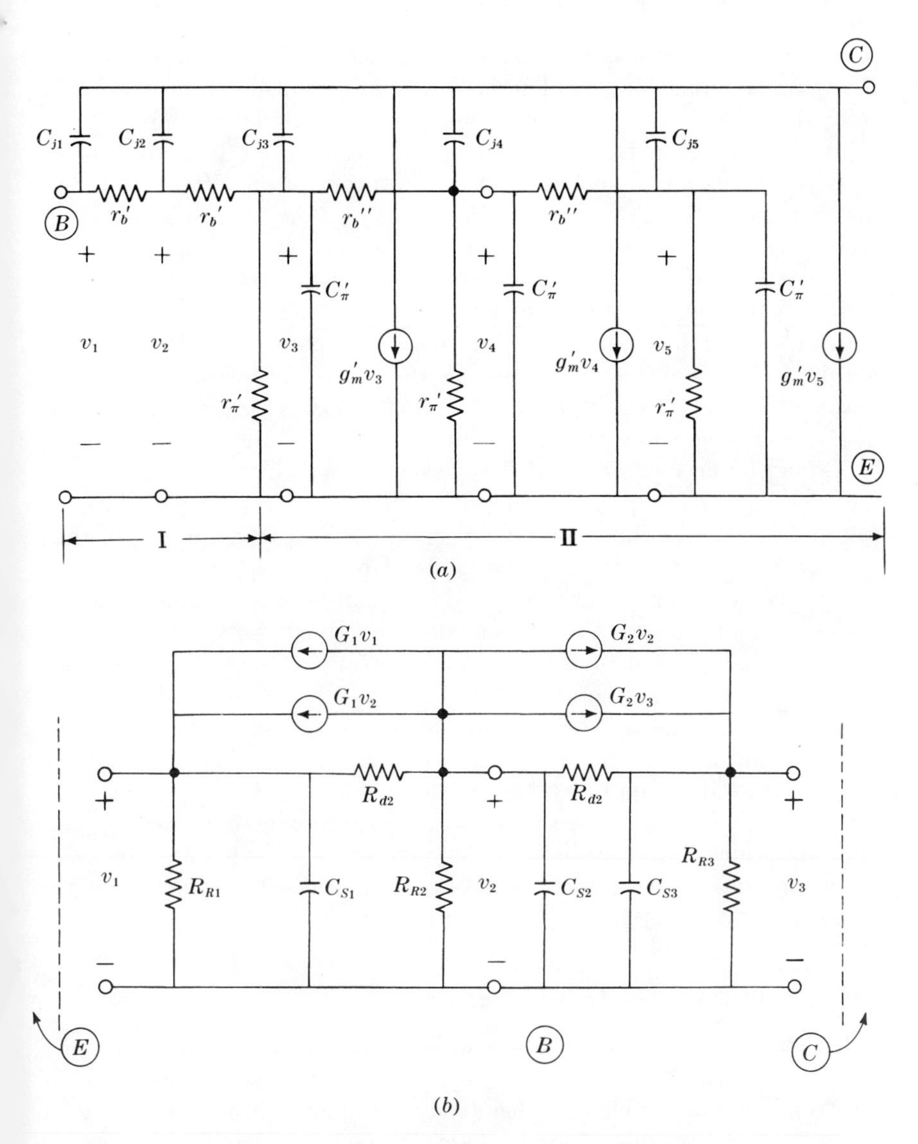

FIG. 4.24 Two expanded models for regions in a planar transistor. (a) Section in base parallel to the surface; (b) active region of the base, normal to the surface.

and distributed capacitance. Region I corresponds to the "open" part of the base and II to the portion under the emitter for a single base contact structure. The complexity of such a model is arbitrary and the accuracy could be increased, for example, by adding resistance elements along the collector. What limits the allowed complexity is obviously the capabilities of the program and the ability to determine the values of the elements used.

A circuit analog for the base transport process in one dimension normal to the emitter and collector planes is shown in Fig. 4.24*b*. This is derived from the Linvill "lumped-model" scheme.[11,12] Again, the complexity is arbitrary, though a two-lump model is chosen here. The voltages v_1, v_2, v_3 are analogs of minority carrier concentration of the *EB* junction, the midpoint, and the *CB* junction, respectively. The R_R elements model recombination processes, the C_S elements model charge storage, the R_d elements are diffusion models, and the controlled-current sources model drift conduction. In this version, the drift elements closer to the emitter oppose the normal current flow, while those nearer the collector aid it. This is usually the case in double-diffused structures. The reader is referred to the literature on lumped models for the computation of element values from the physics and geometry of the base region.

Note also that computer-aided analysis allows the use of more complex models of passive integrated-circuit elements. For example, diffused resistors have a distributed shunt capacitance arising from the *pn* junction surrounding them, which may be modeled by an *RC* ladder. Alternatively, the excess phase shift in such an element may be accounted for by the equivalent nondominant-pole method described above.

To conclude this section, the results of a full computer-aided analysis of the circuit example of Fig. 4.21 are given and compared with the approximate solutions developed in previous sections. Table 4.5 shows the data and leads to the conclusion that the results of simple analysis may be trusted within fairly broad tolerances. Furthermore, the accurate results may now be used to construct an approximate function of the amplifier which is intermediate in accuracy between the full and the first-order solution. This approximation may then be used for a characterization of the circuit as a block in a larger array.

4.6 Circuit Design of Active Filters

It should be apparent at this stage that a gap exists between the formal treatment of active *RC* filter synthesis and the design principles of electronic circuits. Though there may be many reasons for this gap, not all of them technical, one contributing factor is clear. When one is concerned with a circuit containing a large number of electronic com-

TABLE 4.5 VOLTAGE GAIN OF SAMPLE AMPLIFIER DESIGN (See Fig. 4.20)

	First-order calculation (see Sec. 4.4)	*Computer solution*	
Midband voltage gain	505 ± 50	480	
	Pole	Poles	Zeros
Poles and zeros (units of 10^6 sec^{-1})	−2.6	−2.67	
		−93.78	
		−564.6	
		−2,024	−1,983
		−3,111	−3,561
		−7,055	−7,835
		−28,419	−30,142
			+1,211
			+4,328
3-db bandwidth	420 ± 100 kHz	425 kHz	
Phase at −3 db	45°	47°	

ponents, the first-order characteristics of the array are not dependent on the detailed nature of many of the components. For example, if a filter calls for the use of an operational amplifier or a gyrator, it may not matter how such a block is made, provided that it has the desired terminal characteristics. Thus, the filter synthesizer may feel liberated from many considerations which detract from his principal concern, and filter syntheses are normally cast in terms of interconnecting "black-box" gain blocks and passive *RC* elements. Before looking at the dangers of such a dichotomy, let us emphasize that this common practice has a very positive aspect. It allows a filter synthesizer to propose virtually any functional block, real or hypothetical,† in an arbitrary embedding and to examine its general properties. Such liberties are an important part of the creative process and produce important results. However, what will be pointed out in this section is that, at best, the results of such a process are *proposed* solutions to a problem and not actual solutions. The transition between the former and the latter is not simple and is our principal concern here.

The basic problem is that functional blocks never have the ideal terminal characteristics which the synthesis calls for. Furthermore, resistors and capacitors are also complicated by parasitic effects and temperature sensitivities as we have seen already. What we must look for, therefore, are efficient methods by which an ideal schematic of a circuit may be translated into electronic-circuit terms and evaluated.

† The "nullor" is a good example of a hypothetical functional block.

With one or two exceptions, all active *RC* filter techniques which give rise to complex-conjugate transfer-function poles involve feedback. The exceptions are some two-terminal devices such as tunnel diodes and thermistors which exhibit inductive effects. These effects are rarely used, either because they are unstable or because they are associated with strong parasitic effects. On the other hand, such two-port functional blocks as gyrators and negative immittance converters are not exceptions to the above statement. The approach to be used here, therefore, is to represent an active *RC* scheme explicitly as a feedback circuit. By doing this, it is then possible to apply the well-established and powerful analytical methods available for feedback structures. These include both the type of direct circuit analysis described in earlier sections of this chapter and root-locus techniques. In many active *RC* circuits the feedback loops are clearly identifiable but, for the purposes of illustration, we consider here two types of circuit in which they are implicit: the NIC selective amplifier and the gyrator selective amplifier. The operating principles of these circuits are described earlier in the book. The filter examples taken here are narrowband two-pole types, since this is one of the more important categories and is one which shows the effects of circuit limitations very clearly.

Figure 4.25*a* shows a voltage NIC selective amplifier in its conventional representation.[14] The VNIC, on a zero-order basis, is described by the two equations:

$$I_1 = -I_2 \qquad \text{and} \qquad V_1 = -kV_2 \tag{4.20}$$

where $k > 0$ is real and is defined as the conversion constant of the VNIC. Using (4.20), an alternative representation of Fig. 4.25*a* may be drawn as in Fig. 4.25*b*. It has been implied up to this point that the common node in the circuit is grounded, but there is no special reason for this to be so and we may now shift the ground point to the output node and redraw the circuit in Fig. 4.25*c*. No basic change has been made, but the feedback path through the shunt *RC* arm is now clear. It is also clear that, at low frequencies, the feedback is positive. More important still is the fact that the NIC-derived circuit is essentially identical to another selective amplifier configuration derived from the Wien bridge, shown in Fig. 4.25*d*. The realization problem now is not one of constructing a strict analog of the ideal VNIC, but rather that of realizing a voltage gain block with a gain of $k + 1$. A possible way of doing this is shown in Fig. 4.26 where a series-shunt feedback gain block is used. With this, the essential function of the NIC filter is obtained with a relatively simple circuit.

There are now two questions to be considered. Is the realization of Fig. 4.26 the simplest and most suitable for integration, and how do the

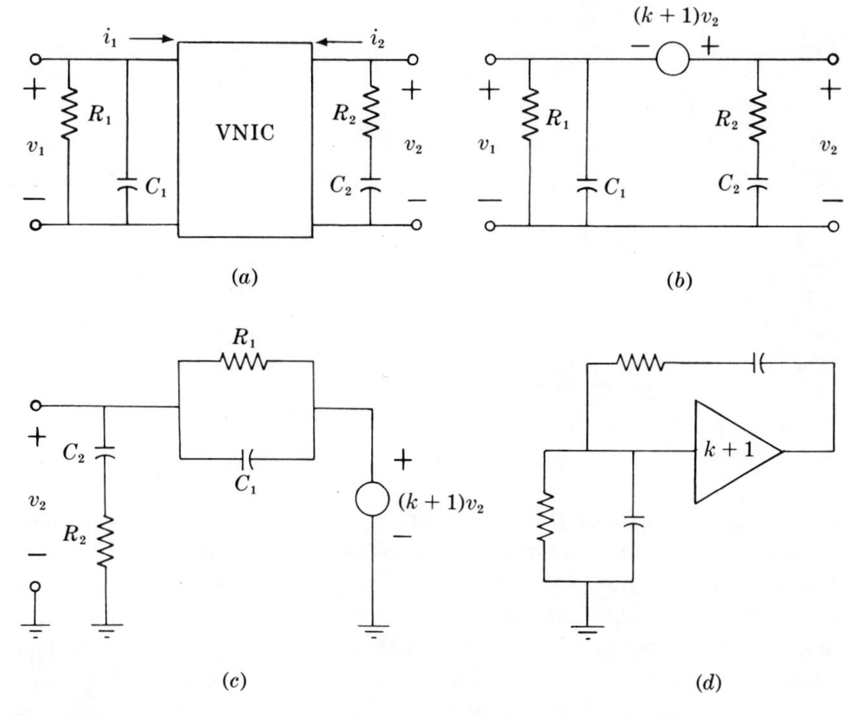

FIG. 4.25 Alternative and equivalent representations of an NIC-derived selective amplifier.

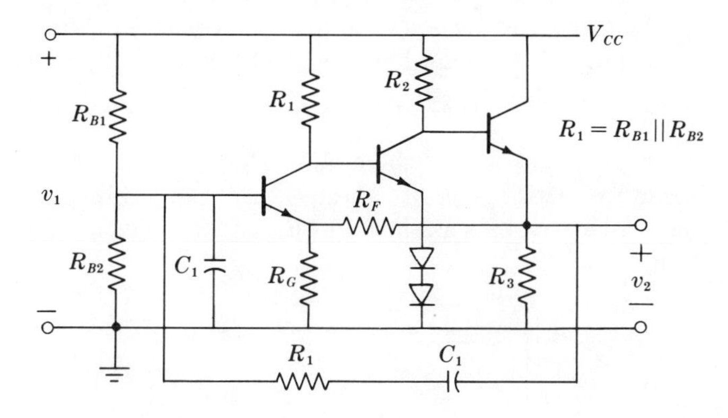

FIG. 4.26 Practical version of a VNIC-type selective amplifier.

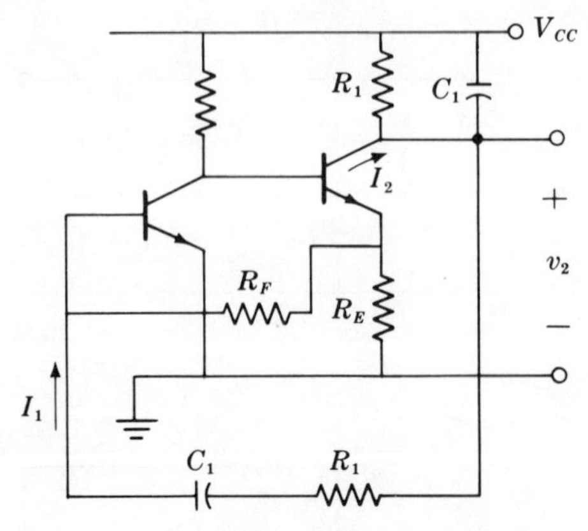

FIG. 4.27 Practical version of INIC selective amplifier.

practical limitations of the devices affect its performance? These questions cannot be answered independently, since some compromise may be made between simplicity and performance. However, it may be seen that we do not have the simplest version. If the same reduction is applied to a current NIC (INIC), it turns out to have a simpler realization. This version is shown in Fig. 4.27. Here the gain block is a shunt-series current feedback pair, which is easier to bias and, in fact, uses one of the two frequency-determining resistors as a collector load for the second transistor. We next consider which of the two circuits is more suitable from the viewpoint of performance.

The characteristic equation for both versions of the selective amplifier, assuming that the gain block is ideal for the moment, is

$$1 + sC_1R_1(3 - A) + s^2C_1^2R_1^2 = 0 \tag{4.21}$$

where $A = 1 + k$ and where $R_2 = R_1$, $C_1 = C_2$. The root locus for this structure, with respect to the feedback path through R_1C_1—again on a zero-order basis—is as shown in Fig. 4.28. The parameter for the locus

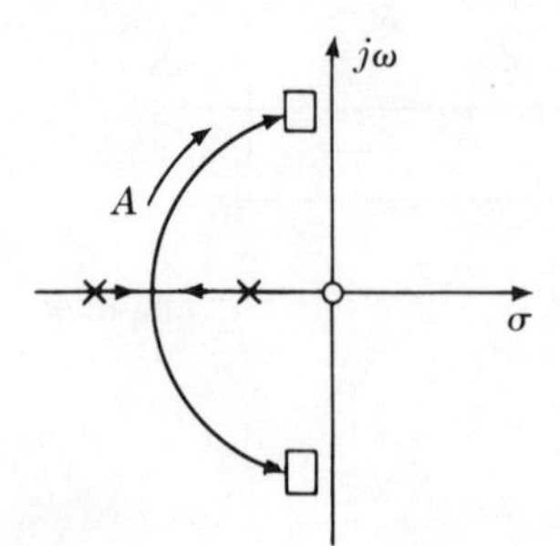

FIG. 4.28 Characteristic root locus of NIC and Wien-type selective amplifiers.

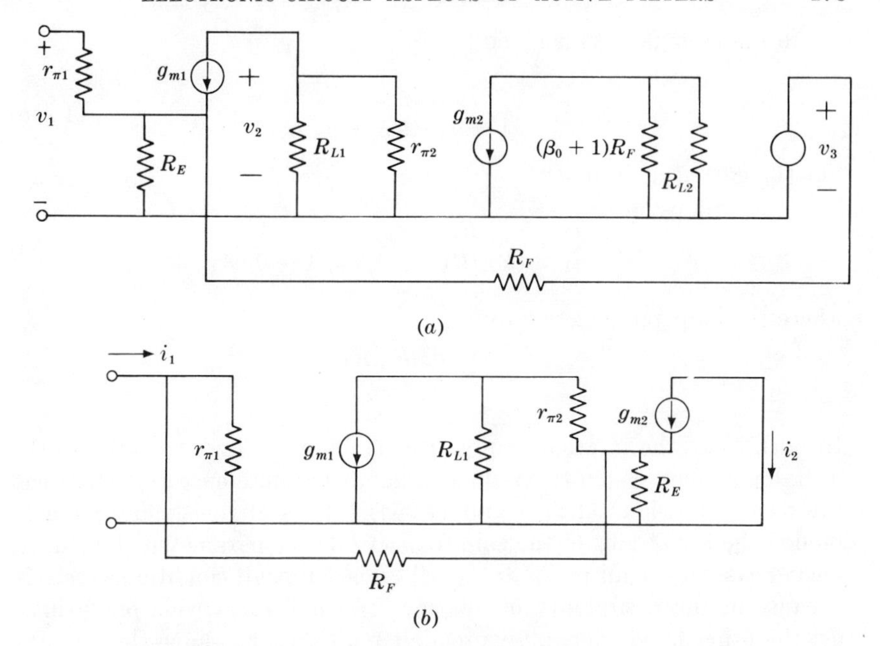

FIG. 4.29 Simplified circuit models for (a) Fig. 4.26 and (b) Fig. 4.27.

is A and, as is clear from (4.21), the closed-loop poles marked as boxes lie on the imaginary axis for $A = 3$. For the narrowband case, it follows from (4.21) that the Q of the bandpass characteristic is approximately

$$Q = \frac{1}{3 - A} \tag{4.22}$$

and the sensitivity of Q to A is

$$S_A{}^Q = QA \tag{4.23}$$

Though this sensitivity is very high, it does not immediately mean that the Wien/NIC approach is a bad one for narrowband selective amplifiers, since we do not yet know how well defined A is. But it is clear that a gain block must be used which has a very stable gain.

With this objective in mind, we compare the two amplifier circuits. In both cases, the full-circuit models are complex, requiring some simplification for a first-order analysis, and this carried out according to the procedures of Sec. 4.3. Figure 4.29*a* and *b* show the simplified models whose gains at low frequencies are approximately as follows.

For the voltage-gain block:

$$A_v = \frac{v_3}{v_1} \approx \frac{R_E + R_F}{R_E} \frac{1}{1 + 1/A_1} \tag{4.24a}$$

where the loop gain A_1 is given by:

$$A_1 \approx \frac{g_{m1}R_{L1}g_{m2}R_{L2}}{1 + R_F/R_E + g_{m1}R_F} \tag{4.24b}$$

For the current gain block:

$$A_I = \frac{i_2}{i_1} \approx \frac{R_F + r_{\pi 1}}{(R_F \| R_E) + 1/g_{m2}} \frac{1}{1 + 1/A_1} \tag{4.25a}$$

where the loop gain is given by:

$$A_1 \approx \frac{g_{m1}R_{L1}R_F}{R_F + r_{\pi 1}} \tag{4.25b}$$

In both cases the internal loop gain may be made high to stabilize the overall gain, but even if this is done, a significant difference exists between the two expressions (4.24*a*) and (4.25*a*). It is obviously desirable to reduce the sensitivity of the gain to active device parameters, but in the second case this requires $R_F \gg r_{\pi 1}$. This is a difficult condition to satisfy because of the desirability of keeping R_F small for a wide bandwidth. On the other hand, the voltage mode circuit does not have this requirement and is therefore preferable. As (4.25*a*) shows, its gain is determined, to a first order, by the ratio of the two resistors R_E and R_F, with a small modification due to the finite loop gain. Thus, the consideration of sensitivity may now be separated into two parts. The sensitivity of the Q to the ratio R_F/R_E is of the order $3Q$, as given by (4.23). But it was stated earlier in this chapter that resistor ratios may be held to a small tolerance in integrated circuits. Furthermore, two resistors formed in the same diffusion step track each other closely with temperature. If we propose that these two resistors may be held to ± 1 percent of their nominal ratio, in spite of process variations and a range of operating temperatures, and a Q stability of ± 20 percent is required, what value of Q could be obtained? Let $R_F/R_E = \alpha \approx 2$; then, from (4.24) and (4.25*a*)

$$S_\alpha{}^Q = Q \frac{\alpha}{1 + 1/A_1} \approx \alpha Q \tag{4.26}$$

so that, for $S_\alpha{}^Q = 20$, the maximum Q obtainable is 10. The second consideration is the effect on Q of variations in the internal loop gain A_1. This may be determined from:

$$S_{A,1}{}^Q = \frac{QA}{A_1 + 1} \tag{4.27}$$

In the circuit of Fig. 4.27, a loop gain of about 100 may be obtained with typical parameter values. Thus, for $Q = 10$, $S_{A,1}{}^Q = 1/3$ and a variation of ± 60 percent in the loop gain may be tolerated for a Q variation of ± 20 percent.

The conclusion from the above is that the NIC or Wien-type selective amplifier is not particularly suitable for high-Q applications, though we could expect satisfactory bandwidth stability for Q values of the order of 10. More important, however, is the fact that this approach to the problem shows more clearly the nature and magnitude of the circuit contribution to Q sensitivity than does the simple expression of (4.22). Other circuit configurations have lower transfer-function sensitivities, but before looking at these we should consider the performance limitations of this circuit at higher frequencies.

Following the dominant-pole development of Sec. 4.4, the transfer function of the gain block is approximated as:

$$A_v(s) = \frac{A_0}{1 + sT_1} \tag{4.28}$$

where it is assumed that the magnitudes and phase response of A_v may be represented, below the -3-db frequency, by a single pole at $-1/T_1$. The value of this approximate pole would be found by analysis. To determine the effects of this on the center frequency and bandwidth of the selective amplifier, (4.28) for $A_v(s)$ is substituted in (4.21) for A to yield the modified characteristic equation:

$$s^3TC_1^2R_1^2 + s^2(C_1^2R_1^2 - C_1R_1T_1) + s[T_1 + C_1R_1(3 - A_0)] + 1 = 0 \tag{4.29}$$

Since this is now a cubic, a simple algebraic expression for the roots cannot be written down, so that it is solved numerically. Let the center frequency of the system be normalized to 1 and the Q be 20 for the case $T_1 = 0$. Then let $T_1 = 0.05$, implying a dominant amplifier pole at 20 times the center frequency of the selective passband. After substituting these values and $3 - A_0 = 0.05$ into (4.22), the closed-loop poles for the transfer function are found to be:

$$s = -0.0253 \pm j1.055, \ -18.957$$

with respect to the normalized center frequency. The effect of the gain block pole, therefore, is to shift the center frequency up by 5.5 percent and the Q up by 5 percent to 21. Though these figures may not be very significant in themselves, they will be used as a basis for comparison with other schemes below.

The second scheme to be considered from an electronic-circuit viewpoint is the gyrator-derived selective amplifier.[15] A formal schematic is

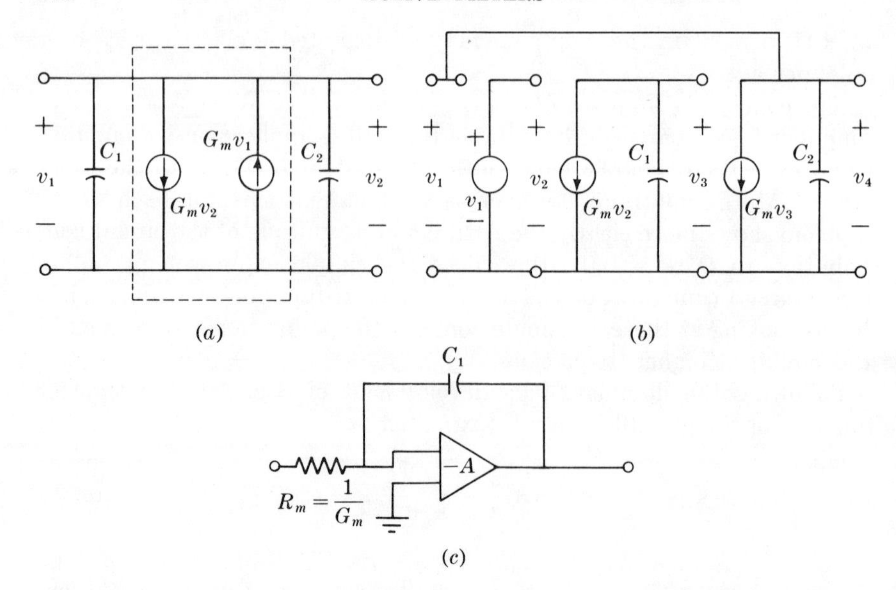

FIG. 4.30 (*a*) Gyrator-derived selective amplifier; (*b*) feedback representation of (*a*); (*c*) conventional operational integrator.

shown in Fig. 4.30*a*. Again, we wish to show that the gyrator circuit is, in fact, a feedback amplifier. This is done simply by redrawing the circuit as in Fig. 4.30*b*. The feedback path, from node 4 to node 1, is now clear and the only change made to the circuit, which is not essential, is the addition of a unity-gain inverting block, so that both current generators will have the same polarity. One notable difference between this circuit and the NIC version is that here the feedback is negative at low frequencies. Many electronic-circuit realizations of gyrators have been proposed and it is not easy to judge the relative virtues of these proposals. Therefore, it is desirable to treat the realization problem in the light of Fig. 4.30*b* rather than Fig. 4.30*a*. If a block comprised of a controlled current source and a shunt capacitor is considered on its own, it is seen that the function performed by this block is that of an integrator. One way to realize this circuit, therefore, would be to use conventional integrators, as shown in Fig. 4.30*c*. For this state, if A is large $R_m = 1/G_m$, then the transfer function of the integrator is identical, on a first-order basis, to that of the capacitance–current source combination.

A realization consisting of two operational amplifiers connected as integrators and a third inverting amplifier with a low wideband gain is, in fact, the well-known "analog-computer" realization of a selective amplifier characteristic.[16] Thus, the formal gyrator approach to selective amplifiers and the operational amplifier approach are fundamentally

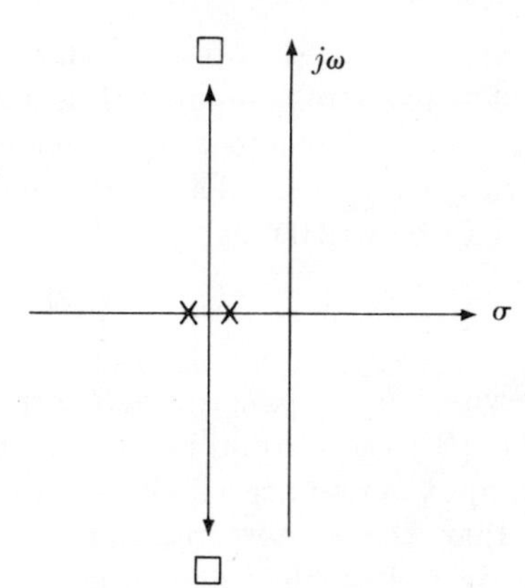

FIG. 4.31 Feedback root locus for circuit of Fig. 4.30*b*.

equivalent and consequently both have the idealized root loci shown in Fig. 4.31.

To examine the performance limitations imposed by a practical realization, we take as an example a circuit in which the controlled-current-source approach is used. A schematic is shown in Fig. 4.32. To minimize the magnitude of the pole associated with each of the capacitors, *pnp* current sources Q_1 and Q_5 are used to supply the collector current to the two controlled-source transistors Q_2 and Q_4. Emitter followers

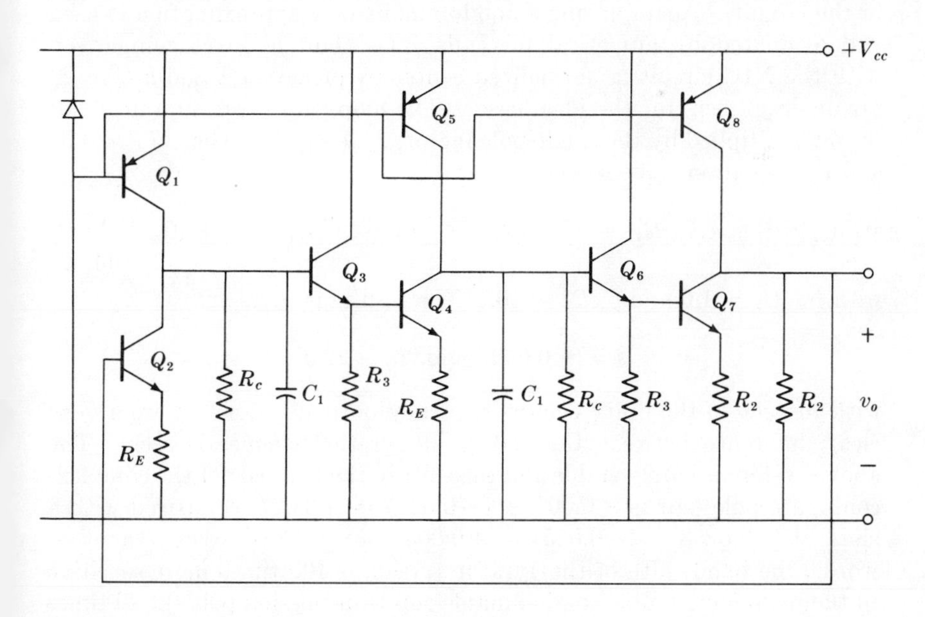

FIG. 4.32 Practical version of the circuit of Fig. 4.30*b*.

are also included to minimize the resistance shunting of these capacitors. The inverting amplifier Q_7 is designed for a voltage gain of approximately unity. If charge-storage effects are neglected, initially, and the two integrator stages have the same parameters, the loop gain of this circuit may be written as

$$A_1(s) = -\left(\frac{G_m R_1}{1 + sC_1R_1}\right)^2 \tag{4.30}$$

where $G_m \approx g_m/(1 + g_m R_E) \approx 1/R_E$, and R_1 is the shunt combination of R_c, the output resistances of the *npn* and *pnp* transistors, and the input resistance of the emitter follower at the node in question. So that the center frequency and bandwidth of the selective amplifier are well defined and insensitive to transistor parameters, R_c should be small in comparison with the other shunt elements. If the equation $A_1(s) - 1 = 0$ is solved for the poles of the transfer function, it is found that the center frequency of the passband is approximately $\omega_0 = G_m/C_1$, with a bandwidth $\omega_b = 2/C_1R_1$, giving a Q of $G_mR_1/2$. In contrast to the positive-feedback NIC version, the sensitivity of Q to the parameters R_1 and G_m is low:

$$S_{R,1}{}^{Q} = S_{G,m}{}^{Q} = 1 \tag{4.31}$$

The center-frequency sensitivity is also low, as it was in the NIC case.

Next, we consider the effects of charge storage in the active devices in the circuit. Again, a single nondominant-pole approximation is used and evaluated by numerical methods. To allow a direct comparison with the NIC circuit, a normalized center frequency of 1 and a Q of 20 are again chosen, for the ideal case. The loop-gain expression in (4.30) is now multiplied by the single pole factor, $1/(1 + sT_1)$, where $T_1 = 0.05$ and the equation to be solved is

$$s^3T_1C_1{}^2R_1{}^2 + s^2(C_1{}^2R_1{}^2 + 2T_1C_1R_1) + s(T_1 + 2C_1R_1) + 1 + (G_mR_1)^2 = 0 \tag{4.32}$$

with $G_m/C_1 = 1.0$ and $G_mR_1 = 40$. The resultant roots are

$$s = -0.002 \pm j0.96, \ -22.0$$

with respect to the unity center-frequency normalization. Here a very clear difference between the NIC and gyrator schemes is seen. The above system is barely stable and the shift in the real part of the complex-conjugate pole pair is $+0.002 - (-0.025) = +0.027$, compared with a shift of $-0.0253 - (-0.025) = +0.003$ for the NIC case. In other words, the bandwidth of the gyrator version is 100 times more sensitive to the introduction of a nondominant loop transmission pole (at 20 times the center frequency) than is the NIC version. It might be noted, however, that the center-frequency sensitivity to T_1 is lower in the gyrator version.

The above comparison is not completely fair unless one also asks which version is likely to have the larger value of T_1 in a given case. This question cannot be answered simply, however, but it may be stated that in both versions the functional blocks are operated at about the same voltage gain in the passband of the selective amplifier. Therefore, to the extent that gain and frequency response are inversely related, the charge-storage effects in both cases should be of the same order. Furthermore, since more active devices are used in the gyrator version, we would expect the nondominant frequency effects to be greater. These conclusions are confirmed in practice. The gyrator approach does provide a wider range of stable Q values than the NIC, but the former is very difficult to use at center frequencies above about 100 kHz. NIC selective amplifiers, on the other hand, may be used successfully for low-to-medium Q values in the MHz range of frequencies.

To complete this section, a third selective amplifier scheme is considered which combines some of the better features of the two previous approaches, namely, the low loop gain of the Wien/NIC and the negative-feedback feature of the gyrator/integrator version. In fact, it is also derived from a formal network theory function—the positive immittance inverter (PII).[17] We proceed directly to a feedback representation of the PII-derived circuit as shown in Fig. 4.33. It is seen to consist of two grounded unity-gain voltage amplifiers A_1 and A_2 with a transconductance block shown as the controlled-current source $G_m v_1$. The fact that the voltage amplifiers have a gain of unity rather than 2 or 3 as in the Wien case is very significant because it is much simpler to realize a precise gain of unity in a feedback amplifier than it is to obtain a precise

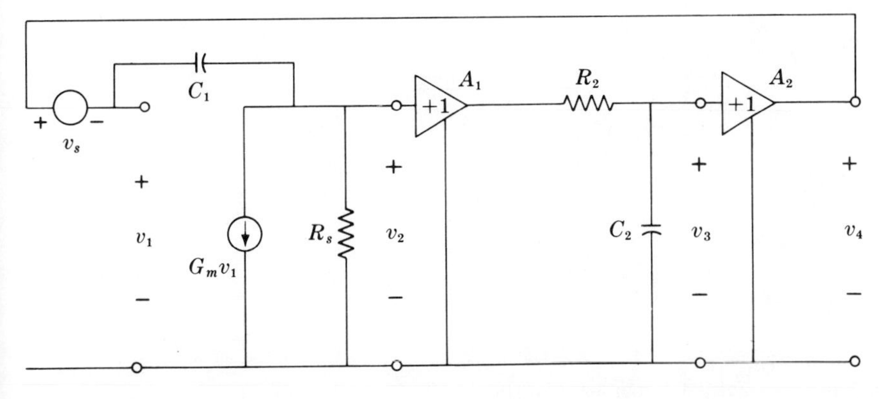

FIG. 4.33 Feedback representation of the PII-derived selective amplifier.

nonunity gain. This is a consequence of the fact that a finite resistor ratio is not involved in a unity-gain stage. (Consider, for example, the emitter follower, which has approximately unity gain but no feedback resistors.)

A loop-gain expression for the circuit of Fig. 4.33 may be found by breaking the loop between the output of A_2 and the control node of the G_m block. That is, for $v_s = 0$,

$$A_1(s) = \frac{v_4}{v_1} = \frac{\mu_1\mu_2 G_m R_s(1 - sC_1/G_m)}{(1 + sC_1R_s)(1 + sC_2R_2)} \tag{4.33}$$

where μ_1 and μ_2 are the actual voltage gains of A_1 and A_2, respectively. The root locus corresponding to (4.33), with $\mu_1\mu_2$ as a parameter, is plotted in Fig. 4.34. The boxes denote the closed-loop positions for $\mu_1\mu_2 \approx 1$ and the positions are given by:

$$s = -\tfrac{1}{2}\left(\frac{1}{R_sC_1} + \frac{1 - \mu_1\mu_2}{R_2C_2}\right) \pm j\left(\frac{1 + \mu_1\mu_2 G_m R_s}{R_sR_2C_1C_2}\right)^{1/2} \tag{4.34}$$

A comparison between this locus and the Wien locus of Fig. 4.28 shows that the essential difference is the transfer of the zero from the origin to the positive real axis. The unity-gain and negative-feedback features noted above are a consequence of this difference.

From (4.34) it is seen that the sensitivity of the center frequency to the circuit parameters is of the same order as in the previous two cases. The bandwidth situation is somewhat different, however. After expressing the Q of the passband of this amplifier as half the ratio of the imaginary

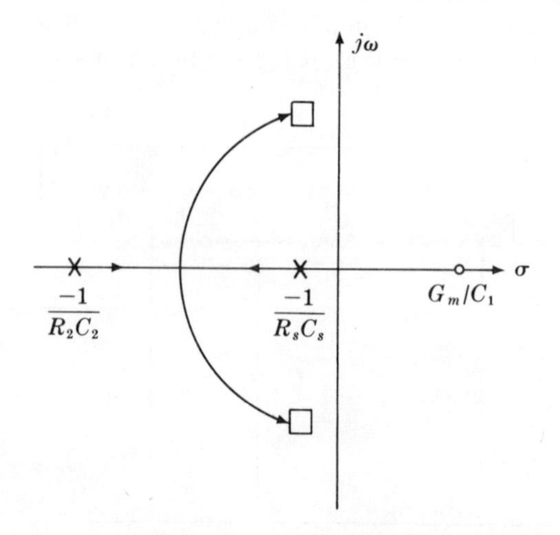

FIG. 4.34 Feedback root locus for the circuit of Fig. 4.33.

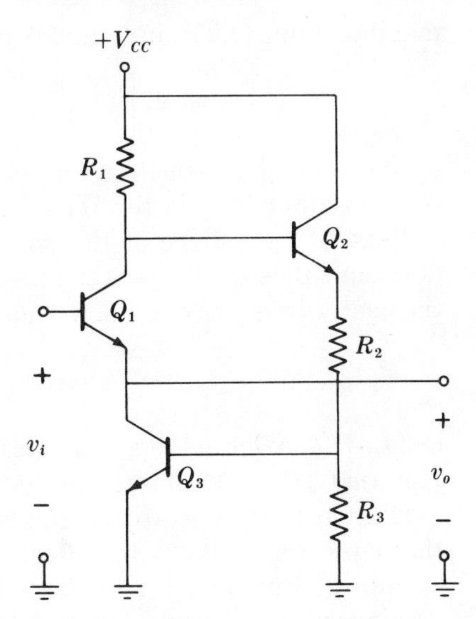

FIG. 4.35 Schematic of a unity-gain amplifier with internal feedback.

part to the real part of (4.34) it may easily be shown that the Q sensitivities to the resistor and capacitance values are less than unity. But its sensitivity to the gain of the unity-gain stages $\mu_1\mu_2$ is high and is given approximately by:

$$S^{Q}_{\mu_1\mu_2} \approx 1 + \frac{1}{2(1 - \mu_1\mu_2)} \tag{4.35}$$

As in the Wien case, however, this should be examined further to determine the sensitivity of $\mu_1\mu_2$ to more fundamental circuit parameters.

Figure 4.35 shows a schematic of a unity-gain block in which Q_1 acts as an emitter follower with an additional negative-feedback loop through Q_2 and Q_3. This loop corrects for the error voltage $v_1 - v_2$ and brings the gain closer to unity than it is for the emitter follower. If simple circuit models are used for the devices, the voltage gain is found approximately to be:

$$\mu = \frac{v_2}{v_1} \approx \frac{A_l}{1 + A_l} \tag{4.36a}$$

where

$$A_l \approx g_{m1}R_E + \frac{g_{m1}g_{m3}R_1R_3}{R_2 + R_3} \tag{4.36b}$$

and g_{m1} and g_{m3} are the transconductances of Q_1 and Q_3, respectively. In a typical case, μ is only 0.01 percent less than unity.

The sensitivity expression (4.35) is now rewritten in terms of A_l, rather than μ, since the power involves circuit parameters directly. After some manipulation, (4.35) and (4.36a) yield:

$$S^{Q}_{A_l} \approx A_l \left[\frac{2}{(1 + A_l)^2} + \frac{1}{1 + 2A_l} \right] \approx \frac{2}{A_l} + \frac{1}{2} \tag{4.37}$$

for $A_l \gg 1$. This sensitivity is very low, as it was in the gyrator case, and is much lower than in the Wien case.

Next, the tolerance of the positive-zero selective amplifier to a nondominant pole of 20 times the center frequency is tested by normalizing the center frequency to unity and substituting:

$$\mu_1\mu_2(s) = \frac{\mu_1\mu_2}{1 + 0.05s} \tag{4.38}$$

in (4.34). After solving numerically with $R_sC_1 = 40$, $R_2C_2 = 0.4$, and $1 + \mu_1\mu_2G_mR_s = 16$, it is found that the degree of narrowbanding is almost as great as in the operational amplifier case (i.e., a shift in the real part of the poles from -0.025 to -0.008) and much greater than in the Wien example. But there is a further difference between this last case and the previous two. As was noted earlier, the gain stages in the first two examples are operating at high gains in the passband of the selective amplifier. But in this case, the voltage gains of all three stages, the two unity-gain blocks and the G_m block, are very low in the passband of the amplifier. Therefore, for equivalent bias points and device parameters in all three cases, the positive-zero version is likely to suffer least from the effects of nondominant poles.

Finally, a complete circuit example is given in Fig. 4.36 for a positive-zero selective amplifier in which the three gain blocks and the RC embedding may easily be identified.[18]

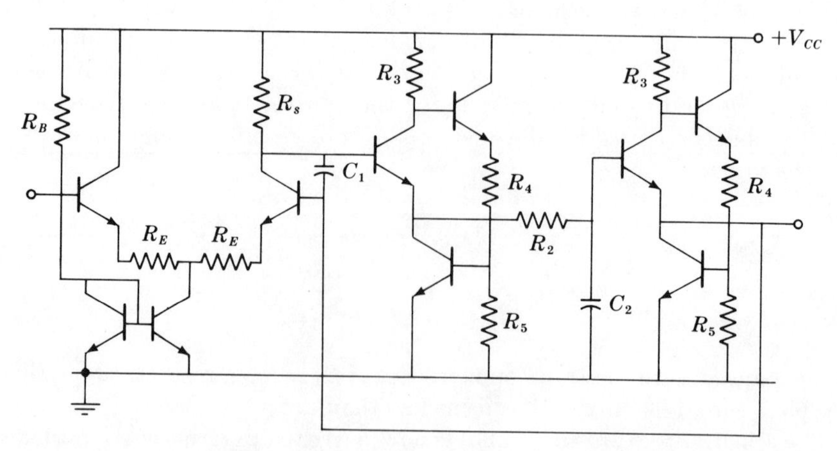

FIG. 4.36 Practical version of the PII-derived (positive-zero) selective amplifier.

4.7 Integrated-circuit Building Blocks

The circuit examples given earlier in this chapter are generally conventional, in the sense that they could be equally well realized in discrete or integrated circuits. There are, however, some circuit techniques which make direct use of the matching and tracking properties of integrated-circuit elements. Some of these are described in this section. Frequently, these new techniques have evolved simply because it was not possible to translate a discrete-component method into integrated circuits. The necessity for eliminating large coupling and bypass capacitors is a good example of this. In other cases, there are circuit functions which may be performed better in the integrated-circuit medium and an example of this is the differential amplifier.

The first group of circuit techniques to be described contains circuit configurations whose characteristic feature is the matching of emitter-base characteristics between bipolar transistors on the same chip. The "diode-biased" transistor[19] is shown in its simplest form in Fig. 4.37*a*. Initially, it is assumed that Q_1 and Q_2 have emitter-base junctions with identical characteristics, and the effects of small mismatches are considered below. Though Q_1 may appear simply to be a diode, since its base and collector are shorted, it is in fact biased in the active region of the transistor characteristics. For transistors with a saturation voltage less than 0.6 volt, which is normally the case at low current levels, this connection does not allow the collector junction to inject. Thus, from Fig. 4.37*a* it may be seen that the emitter current of Q_1 is $I_1 - I_{B2}$, where I_{B2} is the base current in Q_2. Thus, V_1 is the forward emitter-base voltage corresponding to a current of $I_1 - I_{B2}$ and is applied to the emitter junction of Q_2. From the above assumption, and neglecting voltage drops across the ohmic base resistance, we see that the emitter current in Q_2 must also be $I_1 - I_{B2}$ and its collector current consequently is $I_1 - 2I_{B2} = I_1(1 - 2/\beta_0) = I_2$. Thus, this connection biases Q_2 to a collector current which is equal to the reference current I_1 within an error of $2/\beta_0$, that is, about 2 percent. This relationship applies so long as the two devices are at the same temperature and have matched characteristics.

Another interesting property of this circuit is that, if $V_{cc} \gg V_{BE,\text{on}}$ and the collector of Q_2 is returned through a resistor R_2 to V_{CC}, its collector-emitter voltage is approximately $V_{CC}(1 - R_2/R_1)$; that is, it is dependent on a resistor ratio and not on transistor parameters.

As it stands, the circuit of Fig. 4.37*a* is a commonly used current source, with the collector of Q_2 connected, for example, to the emitters of a differential transistor pair. Its output resistance is, on a small-signal basis, the r_0 of Q_2, in shunt with the collector-base plus collector-

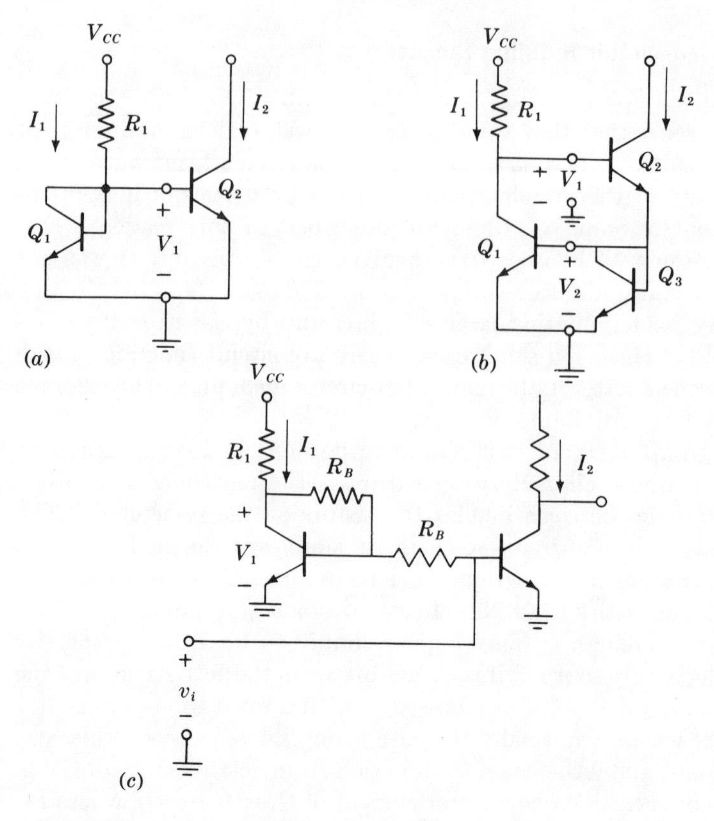

FIG. 4.37 (*a*) Diode-biased current source for integrated circuits; (*b*) modified version of (*a*) with higher output resistance; (*c*) diode-biased grounded-emitter gain stage.

substrate capacitances. This technique may be extended either by using Q_1 to bias several current-source transistors for different parts of a circuit or by fabricating Q_2 with a different emitter area to that of Q_1. If the emitter area of Q_2 is n times that of Q_1, the collector current in the former will also be approximately n times I_1. Because of crowding effects, however, I_2 may be made an integral multiple of I_1 more accurately by placing n separate emitter diffusions in the base of Q_2, each one being identical to the emitter of Q_1. Though the performance of the basic circuit is satisfactory for most applications, another modification is possible[9] which increases the output resistance of the currents ource. This modification is shown in the schematic of Fig. 4.37*b*. Here, a negative-feedback loop is introduced, through Q_3, which senses a change in I_2 due to a change in the output voltage of the current sources and causes the same change to take place in I_1. The bias voltage V_1 then changes to compensate for these effects. Using simple circuit models of the devices, it may be shown that the output resistance of this current source, r_0', is approximately:

$$r_0' = r_0\left(1 + \frac{g_m R_1}{1 + 2R_1/r_\pi}\right) \tag{4.39}$$

since the three transistors are each conducting about the same collector current. This result shows that an increase in output resistance by a factor of 20 or more is easily obtained and this is confirmed by reported results.

The basic circuit of Fig. 4.37*a* is not useful as a small-signal gain element, since the base of Q_2 is shunted by the low impedance presented by Q_1 (of the order of $1/g_m$). To reduce this shunting, a resistance may be added between the two devices. If only one were added, though, a current imbalance would result, because of base current flowing through the resistance. Thus, the balanced configuration of Fig. 4.37*c* is used. Here, V_1 is higher than the base-emitter voltage of Q_1 by $I_{B1}R_B$, but the base voltage applied to Q_2 is $V_{BE} + I_{B1}R_B = I_{B1}R_B \approx V_{BE}$, so that the equal-current condition is maintained. A similar cancellation of voltage drops across the base resistances of the two devices also takes place. If an input signal v_i is applied as shown, the shunt resistance at this point to ground is approximately R_B, which may be made sufficiently high for this biasing technique to be generally useful for ground emitter stages.

Similar techniques may be used for biasing *pnp* current sources as, for example, collector-current sources for *npn* stages. Because of the lower current gain of lateral *pnp*'s (Sec. 4.2), this method of biasing produces larger current errors than with high-gain *npn*'s. Also, the reference junction in this case must be formed by another lateral device.

The second circuit configuration which makes direct use of the junction-matching properties in integrated circuits is the familiar differential pair, shown in Fig. 4.38*a*. As the basic building block of differential amplifiers, this circuit has received a great deal of attention in the literature, but its usefulness is more general than this. Because of its ease of biasing and the availability of outputs of complementary polarity, it tends to replace the grounded emitter stage as the basic small-signal gain block in an integrated circuit. Obviously a penalty is paid in terms of greater circuit area per function and it should not be used where a grounded emitter stage would be equally suitable.

There are many potential sources of imbalance but only a few of these are easily isolated and characterized.[20,21] The most familiar source is "V_{BE} mismatch," which is identified by the situation in which the internal base-emitter voltages of the two transistors are equal, but the collector currents are not. This, however, is not always the dominant source of

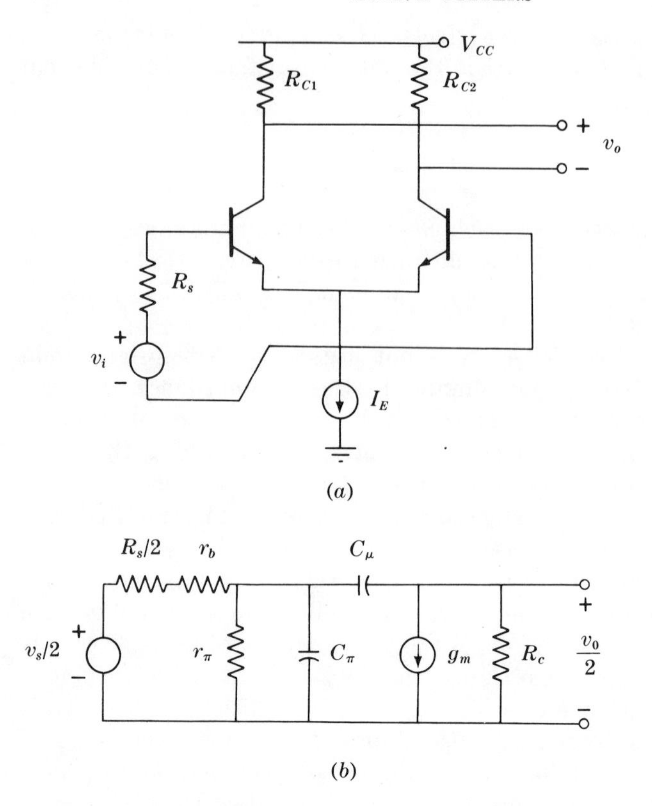

Fig. 4.38 (a) General schematic of a differential pair; (b) typical "half-mode" model of (a).

imbalance, though it usually is when stages are operated at very low collector currents (for example, 10 μa). Another source, which is important at higher current levels, is the voltage drop caused by base current in the internal and external base resistors. Imbalances both in these resistors and the β_0''s of the transistors can lead to significant mismatch. Next, the two collector load resistors will, in general, be unbalanced as a result of mask tolerances and random variations in sheet resistance. Also, where the pair is operated over a wide temperature range, voltage drops caused by collector saturation currents flowing in the base circuit can contribute significantly to imbalance.

In addition to the above, which could be regarded as fundamental to the physics of the devices, there are other effects which are less predictable. These include thermal gradients within the circuit, interface resistance between the devices and their interconnections, and contact emfs at these interfaces. Methods exist for compensating the pair for these effects by providing adjustments of critical circuit elements, but such methods are necessarily empirical and can be tedious.

What is more relevant to us here, however, is the question of how well matched the pair may be without resorting to a compensation scheme. The answer is dictated partly by the accuracy with which the processing masks are made and partly by random variations in the diffusions themselves. The imbalance is conventionally expressed as an equivalent input offset voltage or current, that is, the value of V_i which must be applied to set $V_0 = 0$. Typically it is 2 to 3 mV, with an upper limit for most circuits of 5 mV. The corresponding offset currents depend on the bias current and β_0 of the devices. The input offset changes with temperature and the rate of change with temperature is called the equivalent input drift. Typical values of the voltage drift are ± 5 to 10 μV/°C and the drift of a given unit varies considerably and may change sign over wide temperature range.

The small-signal differential gain of the stage v_0/v_s may be analyzed directly by replacing the transistors in the circuit with their small-signal models. But it is found that the small-signal properties of the pair, with differential input and output, are the same as those of a single grounded emitter stage operated at an emitter current of $I_E/2$. Thus, it is standard practice in analyzing cascades of such circuits to replace each differential pair by a single grounded emitter model, which is sometimes referred to as the "differential mode half-circuit."[22] Other combinations of input and output may be used, however. If the output is differential and the input is applied in a single-ended mode, that is, one base is at ac ground and the signal is applied to the other, the gain remains almost the same. The difference in this case is that the common-emitter point is not a virtual ac ground, as it was in the previous case. But if the output impedance of the emitter current source is very high, there is no signal shunting at this point. There can be changes in the positions of some of the nondominant poles and zeros of the transfer function, but these effects are more properly treated by computer-aided analysis. In the third case the input is applied either balanced or single ended, but the output is taken between only one collector and ground. Here, the voltage gain is almost exactly halved, in comparison with the first two cases. Again, the nondominant frequency effects are slightly different.

The conclusion from the above three cases is that the half-circuit may be used validly for first-order calculations in any of the cases, while higher-order effects may be investigated with the computer. Figure 4.38*b* shows a typical half-mode model. Note, however, how the input and output voltages are defined in the model. The use of a factor of $\frac{1}{2}$ in these quantities preserves the correct driving-point impedances.

Level Shifting. In the light of the treatment of integrated-circuit transistors given in Sec. 4.2, it is generally preferable to use *npn* devices as gain elements. In a cascade of such devices, therefore, dc level shifting is frequently required. Described here are some of the techniques which may be used in integrated circuits instead of the standard capacitance-coupling approach of discrete-component circuits. One property which alleviates the level-shifting problem somewhat is the fact that transistors may, in fact, be direct coupled in the grounded emitter mode, making the V_{CE} of one stage approximately equal to the V_{BE} of the next. The circuit example given earlier (Fig. 4.20) used this approach. But the transistor under these conditions is close to saturation and its small-signal gain properties are not optimum. The main source of degradation is the small bias across the collector-base junction, which leads to high values of C_μ. In high-gain cascades, this capacitance usually dominates the frequency response, and the overall gain-bandwidth performance of the cascade is lower than it would be for high values of V_{CE}.

Figure 4.39 shows some of the common level-shifting circuits. The first one, Fig. 4.39*a*, is the zener-diode type, which appears to be more useful than it is in fact. The diode is normally fabricated as an emitter-base junction and is operated in reverse breakdown. Depending on the base doping level, the breakdown voltage lies in a typical range of 5 to 8 volts and the dynamic resistance of the device may be made low—in the range of 10 to 50 ohms. The attractive feature of this technique is that a relatively large voltage shift may be obtained with one simple device and with little attenuation of the signal. The two factors which detract from its usefulness, however, are the tolerance problem, since it is difficult to control the breakdown voltage to closer than about ± 0.2 volt, and the noise problem. Junctions operated in the zener or avalanche breakdown modes generate large components of noise current. (In fact, they are frequently used as noise generators.) Therefore, if they are to be used in signal paths, they may only be used in places where the signal level is high. On the other hand, they are very useful as dc voltage references, where their noise contributions are unimportant.

An example of the next scheme, involving forward-biased diodes, is shown in Fig. 4.39*b*. The diodes are normally emitter-base junctions, so that their forward drop will track that of other transistor junctions in the circuit. The two problems mentioned above, noise and voltage tolerance, are greatly reduced by this scheme but the price paid is the lower dc voltage shift (0.6 to 0.7 volt) per junction. At constant current, the temperature coefficient of the forward diode voltage is about -2.5 mV/°C in silicon, so that the shifted voltage V_2 in the figure will tend to increase with temperature, if the bias point of Q_1 is held constant. As a consequence, the bias stability of subsequent stages may be degraded, and

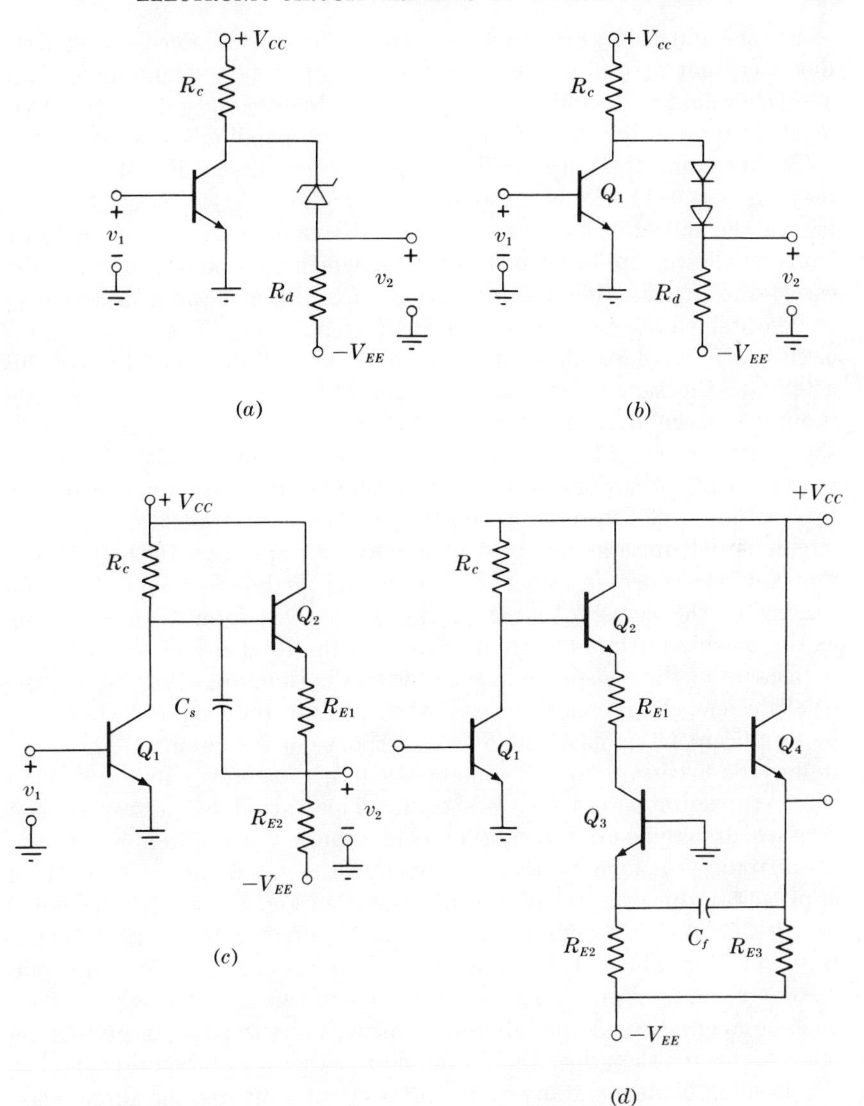

FIG. 4.39 Integrated dc level-shift circuits. (a) Zener diode; (b) forward-biased diodes; (c) resistive divider; (d) active divider.

balanced configurations should be used in conjunction with this level shifter. To minimize the signal attenuation in the shifter and to ensure that the diode string does not add dominant frequency effects to the overall transfer function, it is desirable to operate the diodes at a current level comparable to that of the transistors. This imposes certain limits on the resistance and voltage levels in the circuit, because R_c must conduct the diode current as well as the collector current of Q_1. Thus, for a given V_{CC}, R_C would be approximately half the value it could take without the level shifter and the gain of this stage is correspondingly lowered.

To overcome this current limitation, the circuits of Fig. 4.39*c* and *d* may be used. The former uses an emitter follower Q_2 to reduce the loading on the collector of Q_1 as well as a resistive divider. On its own, the emitter follower produces a drop of V_{BE} which may be sufficient in some cases, and the resistive divider enables an additional and arbitrary drop to be obtained at the expense of signal attenuation. The more complex circuit of Fig. 4.39*d* has similar properties but does not appreciably attenuate the signal—at least at midband frequencies. By inspection, it may be seen that the dc level shift between the collector of Q_1 and the emitter of Q_4 is $2V_{BE} + I_{C2}R_{E1}$, but because of the high impedance presented at the lower end of R_{E1}, the signal attenuation between these two points is of the order of unity. In terms of realization, this last circuit is not quite as involved as it may first appear. Both Q_2 and Q_1 are common-collector structures, requiring little area, and the diode biasing of the current source Q_3 may be derived from the same source as the biasing for other current sources in the total circuit.

Because of the resistance in series with the signal path in the last two level shifters, shunt capacitances in the circuit introduce poles which may be significant in terms of the overall response of the circuit. Therefore, it may be necessary to add a compensating capacitor. In Fig. 4.39*c* a shunt capacitor across R_{E1} is shown. This introduces a transmission zero which may be used to cancel a pole, though the interaction of C_s and R_c introduces a higher-frequency pole at the same time. A feedback approach to this compensation is illustrated in Fig. 4.39*d*, with the capacitance C_f. A fairly detailed analysis is required to study the effect of this capacitor, which is chosen to be of the same order as the load capacitance driven by R_{E1}. This analysis is not given here, though the feedback-compensation method does give rather more freedom in placing the poles and zeros than does the simple shunt capacitor of the previous case.

Summing Nodes. Many active-filter circuits involve the summing or subtraction of voltages or currents at a node and we should consider briefly some of the circuit techniques for doing this. Of the above four possible operations, only two come naturally in electronic circuits: current summing and voltage subtraction. The latter may be performed simply at the input of a differential amplifier. The essential requirement for the former is a low impedance node or a unilateral device such that the net current into the node is converted, by a well-defined transfer function, into a voltage or current at another node. The simplest device which satisfies these requirements is the common-base transistor, shown in

Fig. 4.40 with resistor inputs. The current gain and transimpedance of the stage are both well-defined functions and may be derived directly from the small-signal models of the stage. The input impedance at the emitter node is given approximately by:

$$Z_{\text{in}}(s) \approx \frac{r_b + r_\pi + sC_\pi r_b r_\pi}{1 + \beta_0 + sC_\pi r_\pi} \tag{4.40}$$

At low frequencies, the input resistance, from (4.40), is low, being about 30 ohms for an emitter current of 1 ma. The inductive component of the input impedance appears at higher frequencies but, with typical parameter values, its magnitude is only 5 ohms at 10 MHz. Biasing the common-base stage is not quite as simple as biasing for common-emitter operation. Either a negative supply rail is needed or, if this is not available, the base must be returned to a positive level which has a very low impedance to ground.

Though the above examples are, by no means, an exhaustive list of circuit techniques which are relevant to active RC filters, they include some of the more widely used techniques in integrated circuits. There is, as well, another group of basic circuit building blocks which may involve the use of high-gain or operational amplifiers. These include current summing nodes, unity-gain blocks, low-gain wideband amplifiers, and integrators. Examples have been given already of ways to perform some of these functions without using operational amplifiers, but there are many instances in which the use of the latter is preferable. Therefore, some of the basic properties of operational amplifiers are considered in the next section.

Operational Amplifiers. Before the advent of the integrated operational amplifier, discrete-component operational amplifiers were associated almost invariably with analog computation and closely related

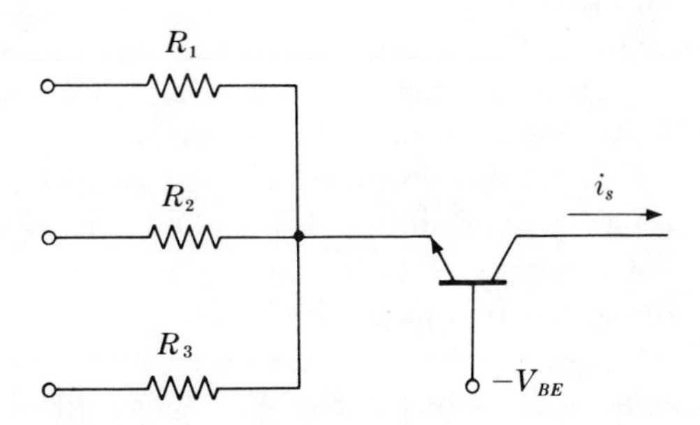

FIG. 4.40 Common-base transistor summing node.

functions. Now that the integrated operational amplifier is one of the most familiar and widely used linear-integrated circuits, one may readily conclude that there were many latent applications for such circuits which only became apparent after inexpensive, high-performance circuits became readily available. Another conclusion may be drawn, though it is true to a lesser extent: operational amplifiers are sometimes used in applications where another circuit—not so readily available—would be better, or they are occasionally used in applications to which they are poorly suited. However, rather than citing cases of good or poor use we will consider here the generic properties of operational amplifiers so that the reader may have a basis on which to make such a judgment in a given case.

It should be said at the outset that the grounds for an operational amplifier being suitable or unsuitable in a particular case depend far more strongly on its frequency characteristics than its gain and impedance levels. The latter topics will be looked at first, however.

Briefly, the generic properties of operational amplifiers are these: the input is differential and is normally applied to a differential pair; the input offset voltage and current, as well as the total bias current, are kept as small as possible. The input circuit is not usually self-biased. Common-mode range and rejection are maximized, both by the careful matching of circuit elements and by the use of common-mode feedback. The output is taken either from a class A emitter-follower stage or a class B output which usually contains one or two *pnp* transistors. The aim in the design of the output stage is to provide a maximum voltage swing as close as possible to the power-supply voltages applied across the circuit, and to minimize the output impedance. Normally, two power supplies (positive and negative) are used so that the input and output dc levels are close to ground. Steps are taken to minimize the coupling between the supply voltages and the output voltages. This coupling is expressed as a power-supply rejection factor, in μV/volt, referred to the input.

Two typical circuit schematics of commercial operational amplifiers are given in Figs. 4.41 and 4.42. The purpose in giving these is not to make comparisons between various types, but to point out methods by which the above properties are realized. For simplicity, the former is referred to as circuit A and the latter as B.

Circuit A is fairly conventional in terms of the circuit techniques described in this chapter.[23] The input differential pair Q_1 and Q_2 are supplied by the diode-biased current source Q_3. In this case, the simple scheme described in Fig. 4.37*a* is modified by the addition of two resistors and an extra diode, which stabilizes the overall circuit against temperature effects. The differential output of the first pair drives the second pair,

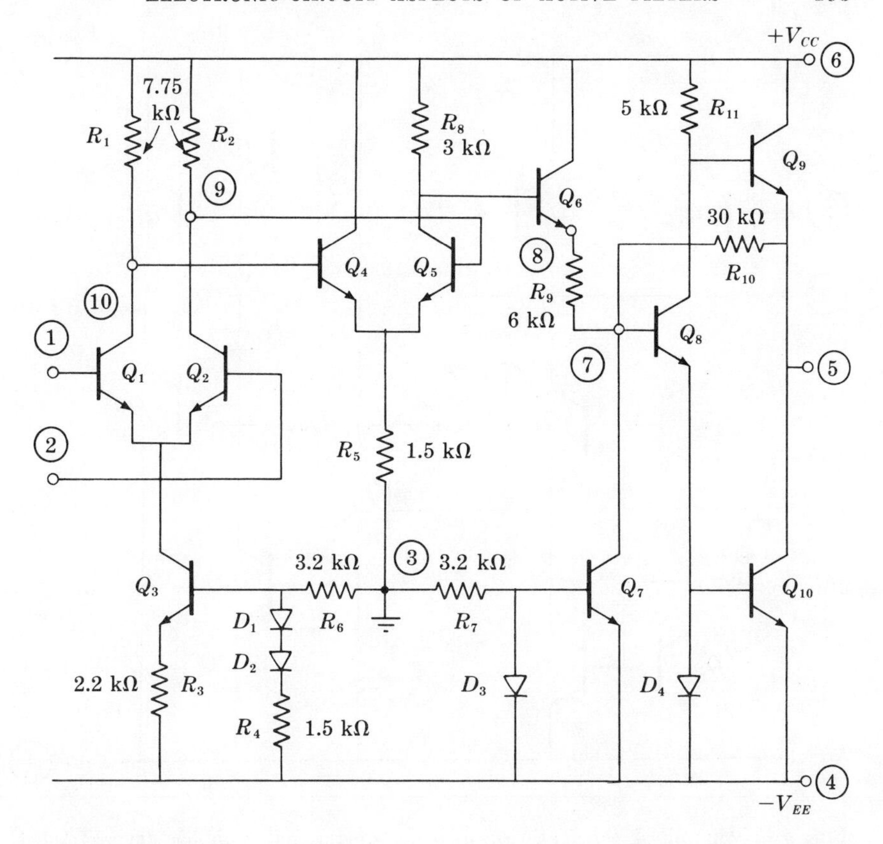

FIG. 4.41 Schematic of monolithic *npn* operational amplifier (type MC1530, Motorola Inc.).

Q_4 and Q_5. This pair, however, has a resistive emitter source, rather than a transistor. A single-ended output is taken from the second pair to drive the output stage through a level shifter Q_6. This level shifter is similar in essence to the version shown in Fig. 4.39*d*, with the addition of the feedback resistor R_{10}, whose purpose is to stabilize the output voltage with respect to ground. The driver transistor Q_8 operates as a phase splitter to provide a push-pull drive to the pair of output transistors Q_9 and Q_{10}. This output stage configuration is frequently used in operational amplifiers, since it provides a large output swing without recourse to a complementary configuration involving *pnp* transistors. Q_{10} is diode biased by D_4, but in this case the area of the diode is only one-third of the emitter area of the transistor. This means that Q_{10} is biased to three times the current in Q_8, resulting in a saving on total current drain and power dissipation.

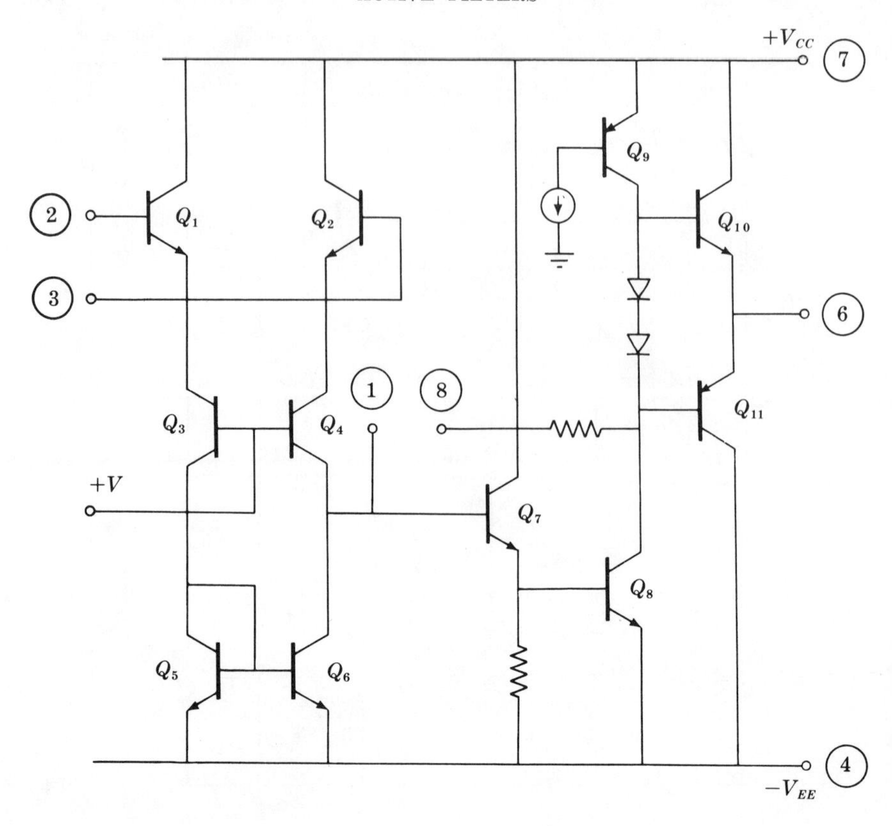

FIG. 4.42 Simplified schematic of *npn-pnp* operational amplifier (type LM101, National Semiconductor).

The overall voltage gain of this circuit is approximately 5,000, with an input resistance of 5 kΩ and an output resistance of less than 100 ohms.

Circuit B uses quite a different approach,[24] through the extensive use of lateral *pnp* devices. Only two gain stages are used, compared with three in circuit A, and there are fewer resistors in the circuit. Note, however, that much of the biasing and short-circuit protection circuitry has been omitted from the schematic of Fig. 4.42, in the interest of simplicity. The input pair Q_1 and Q_2 are operated in the common-collector mode, with their emitters driving the emitters of a pair of lateral *pnp*'s, Q_3 and Q_4. Instead of collector load resistors for the latter pair, *npn* current sources are used, resulting in a very high voltage gain (>500) for the first stage. The compound second stage, consisting of the emitter follower Q_7, the common-emitter stage Q_8, and its *pnp* collector current source Q_9, is also a high-gain configuration. Furthermore, because of the use of the *pnp*'s, a separate level-shifting stage is not required. The output stage Q_{10} and Q_{11} is an example of a complementary class B

configuration, in which the *pnp* is actually a compound device of the type described in Sec. 4.2.

The open-loop gain of amplifier *B* is about 160,000, with an input resistance of 800 kΩ.

Next, we come to a consideration of frequency response. The open-loop bandwidth of operational amplifiers lies typically in the range 100 kHz to 1 MHz, without compensation. Since they involve two or more high-gain stages, a large number of poles are associated with this gain and a thorough characterization of the transfer function would not be possible without extensive computer-aided analysis and experimental measurement. This task is rarely attempted. For most design purposes, which involve the embedding of the amplifier in a feedback network, a characterization of the amplifier in terms of its dominant transmission pole is adequate, though it is desirable to know the position of the next most dominant pole as well. Nodes are provided in most operational amplifiers to which an external *RC* network may be connected to modify the transfer function so that it suits a particular feedback application. In circuit *A*, a capacitor may be connected between nodes 9 and 10 to produce a dominant transmission pole, while a capacitor connected between nodes 7 and 8 produces a transmission zero as well as an additional nondominant pole. Similarly, in circuit *B*, a capacitor between 1 and 8 produces a dominant pole, though in this case a smaller capacitor is required to produce a given pole position, because of the very high impedance level at node 1.

Specific information on compensation networks for commercial integrated operational amplifiers is provided by manufacturers and usually covers a wide range of applications. Therefore, we do not treat the details here. However, we should consider a point raised earlier, concerning whether operational amplifiers might be unsuitable for some applications, where, on first sight they appear to be appropriate. The main consideration here arises in cases where an operational amplifier is used, with heavy negative feedback, to produce a wideband stable transfer function. An example of this is the Wien- or VNIC-type selective amplifier of Sec. 4.5, where a precise gain of the order of 3 is required. With the very large midband gain of an operational amplifier and the consequently large loop gain of the feedback circuit, it is necessary to place the dominant pole of the forward-path gain very near to the origin of the s plane. In a typical case, where the loop gain might be 10^4 and where the second pole of the amplifier might be at 10^7 sec^{-1}, the dominant pole should, for simple theory, be placed at about 10^3 sec^{-1} to ensure stability. The positioning of the dominant pole is accomplished with the compensation network.

If the low-gain amplifier thus constructed were used at very low frequencies, the sensitivity of the transfer function to parameters of the operational amplifier would be desirably low. On the other hand, if the amplifier were used in an active *RC* circuit with a center frequency of 1 MHz, for example, the loop gain would be only $10^4 10^3/(2\pi \times 10^6) \approx 1.6$. Thus, at this frequency, the operational amplifier is hardly performing a useful function and a better approach would be to use a wideband amplifier with a smaller midband gain as the forward path in the feedback structure.

Though the above example is a case where the operational amplifier is of dubious value, it is obviously true that there are very many cases where it provides an effective means of realizing a function. Nevertheless, the circuit designer should approach this technique with a degree of caution.

4.8 Summary

This chapter has attempted to cover a wide area in a relatively small space, so that it leaves some unanswered questions and requires that the reader provide much of the background information himself. Some important points have been brought out, however, and bear reiteration. The first is that we are mainly interested in those approaches to active *RC* filter design which, to use the language of patents, can be reduced to practice. A familiarity with the principles and limitations of electronic circuits should be applied to this problem in two ways, namely, the initial selection of designs which are amenable to optimum realization and, second, the actual circuit design itself. Two levels of design and analysis were developed in this chapter, and these should be used to complement each other. At the first level, the emphasis is on first-order estimates of gain, bandwidth, and impedance levels. Out of this comes not only the initial design of a filter structure, but also a physical understanding of the nature and source of the performance limitations. The second level, involving computer-aided analysis, is more straightforward in many ways, since the emphasis is on adequate circuit models, rather than on how to perform approximate calculations.

Finally, it should be recalled from the latter parts of this chapter that the representations of network structures which have grown out of network theory are not usually amenable to electronic-circuit manipulations. It appears necessary, therefore, to maintain two sets of terminology to suit the former and the latter respectively.

References

1. Warner, R. M., Jr., and J. N. Fordemwalt (eds.): "Integrated Circuits, Design Principles and Fabrication," McGraw-Hill Book Company, New York, 1965.

2. Lin, H. C.: "Integrated Electronics," Holden-Day, Inc., Publisher, San Francisco, 1967.
3. Integrated Silicon Device Technology, *Research Triangle Institute Rept. ASD-TDR*-63-316, vol. I, chap. 4, June, 1963.
4. Irvin, J. C.: Resistivity of Bulk Silicon and of Diffused Layers in Silicon, *Bell Syst. Tech. J.*, vol. 41, pp. 387–410, March, 1962.
5. Camenzind, H. R., and A. B. Grebene: Signetics Corp., private communication.
6. Bennett, C. J., and P. G. Luke: Tantalum Thin Film Resistors and Their Use in Integrated Circuits, *IEE Eastbourne Conf. Record (Integrated Circuits)*, pp. 39–45, 1967.
7. Gay, M. J., and J. S. Brothers: Capacitors for Monolithic Integrated Circuits, *Plessey Company, Tech. Pub. PS*1214*A*, January, 1967.
8. Lepselter, M. P.: Beam Lead Technology, *Bell Syst. Tech. J.*, vol. 45, pp. 233–254, February, 1966.
9. Wilson, G. R.: A Monolithic Junction FET-NPN Operational Amplifier, *ISSCC Dig. Tech. Papers*, pp. 20–21, February, 1968.
10. Gray, P. E., et al.: "SEEC Notes Vol. 2, Physical Electronics and Circuit Models of Transistors," John Wiley & Sons, Inc., New York, 1964.
11. Gibbons, J. F.: "Semiconductor Electronics," McGraw-Hill Book Company, New York, 1967.
12. Pederson, D. O.: "Electronic Circuits" (preliminary edition), McGraw-Hill Book Company, New York, 1965.
13. Calahan, D. A.: "Computer Aided Network Design," McGraw-Hill Book Company, New York, 1968.
14. Linvill, J. G.: Reactive Filters, *Proc. IRE*, vol. 42, p. 555, March, 1954.
15. Rao, T. N., and R. W. Newcomb: Direct Coupled Gyrator Suitable for Integrated Circuits and Time Variation, *IEE Electron. Letters*, vol. 2, p. 250, July, 1966.
16. Green, B. J.: The Design of Active Filters for Hybrid Integrated Realization, *IEE Eastbourne Conf. Record (Integrated Circuits)*, pp. 76–90, 1967.
17. Lampard, D. G., and G. A. Rigby: The Application of a Positive Immittance Inverter to the Design of Integrated Frequency Selective Amplifiers, *Proc. IEEE*, vol. 55, pp. 1101–1102, June, 1967.
18. Rigby, G. A., and D. G. Lampard: Integrated Selective Amplifiers for Radio Frequencies, *ISSCC Dig. Tech. Papers*, pp. 24–25, February, 1968, and *IEEE J. Solid State Circuits*, vol. SC-3, December, 1968.
19. Widlar, R. J.: Some Circuit Design Techniques for Linear Integrated Circuits, *IEEE Trans. Circuit Theory*, vol. CT-12, pp. 586–590, December, 1965.
20. Hoffait, A. H., and R. D. Thornton: Limitations of Transistor dc Amplifiers, *Proc. IEEE*, vol. 52, no. 2, p. 179, February, 1964.
21. Baldwin, G. L., and G. A. Rigby: New Techniques for Drift Compensation in Integrated Differential Amplifiers, *ISSCC Dig. Tech. Papers*, February, 1968, pp. 64–65, and *IEEE J. Solid State Circuits*, vol. SC-3, December, 1968.
22. Middlebrook, R. D.: "Differential Amplifiers," John Wiley & Sons, Inc., New York, 1964.
23. Wisseman, L., and J. L. Robertson: High Performance Integrated Operational Amplifiers, *Motorola Inc. Application Note AN*-204.
24. Widlar, R. J.: A New Monolithic Operational Amplifier Design, *National Semiconductor Tech. Paper TP*-2, June, 1967.

CHAPTER 5

DIGITAL FILTERS

John V. Wait
University of Arizona
Tucson

5.1 General Scope

The availability of very fast, low-cost digital integrated circuits has created new interest in using digital (discrete-time) filtering techniques in signal-processing-systems design. Digital filters will often provide an outstanding degree of accuracy and stability; moreover, in low-frequency applications, they can yield important reductions in size, weight, and cost, as well as improved reliability.

This chapter covers methods for the design of fixed (time-invariant), linear discrete-time operators to match frequency or time-domain specifications. We want to emphasize that we are dealing with discrete-time operators, and thus the performance of such filters will only approximate the behavior of continuous-time filters. Methods for achieving a satisfactory approximation will be covered.

It will normally be assumed that the filter is operating on a discrete-time sequence of *numbers* (of course, these may be derived from a suitably band-limited continuous-time signal, e.g., by analog-digital conversion); these numbers are then manipulated by a digital processor to produce an output sequence of *numbers*, which may subsequently be used to reconstruct a continuous-time signal via suitable data-reconstruction schemes.

Section 5.3 introduces some basic concepts and viewpoints used in the analysis of continuous- and discrete-time systems. Section 5.4 discusses the sampling theorem and the linking of the two types of systems by means of the s-domain to z-domain transformation.

Section 5.5 describes a variety of practical data-reconstruction filters. Section 5.6 presents a summary of the associated data-reconstruction errors. From this it is seen that the choice of sampling interval must be made from a consideration of output-reconstruction accuracy as well as input-signal bandwidth.

Various means for representing digital filters are presented in Sec. 5.7, and the flow-diagram structure of several important realization forms is presented. Sections 5.8 and 5.9 present design methods for nonrecursive (finite-memory) and recursive filters; both frequency- and time-domain viewpoints are used. Some typical design examples follow in Sec. 5.10.

In the treatment of quantization and round-off errors (Sec. 5.11) the discussion will be limited to some of the more generally accepted viewpoints that should permit avoiding obvious problems, but not necessarily the more subtle troubles that are in general difficult to anticipate without tedious analysis.

Note that in some applications it is important that the digital-filter realization match the desired transfer characteristic without introducing appreciable delay (e.g., in real-time digital simulation, and other closed-loop application); in other cases, a reasonable delay can be tolerated provided the gain and relative phase curves meet the desired specification. Design methods for handling both situations are covered.

This chapter is intended primarily to present the background necessary to understand the nature of digital filters, and to describe some generally useful design methods. The recent literature on the subject is extensive; in particular, the reader is referred to the works of Kuo and Kaiser,[1] Rader and Gold,[2] and Monroe[3] for more detailed discussions. Certainly the work of Hurewicz[4] must be cited as an early "modern" treatment of the subject; the author also originally received much inspiration for the state-space methods of Sec. 5.8 from the excellent treatment given by Schwarz and Friedland.[5]

5.2 Summary of Definitions and Symbols

Although the later sections will amplify the following definitions, it is felt that a presentation of the more important definitions and a summary of commonly used symbols (Table 5.1) may help to orient the reader's thoughts toward the conventions that will be followed hereafter.

Fourier-series Representations of Periodic Functions. Given the periodic function $f(t)$, satisfying the Dirichlet conditions (Ref. 5, p. 137), with period T, that is,

$$f(t) = f(t + T) \qquad -\infty < t < \infty$$

then it can be represented by a Fourier (exponential or trigonometric) series

$$f(t) = \sum_{n=-\infty}^{\infty} F_n e^{jn\omega_0 t} \tag{5.1}$$

where

$$\omega_0 = \frac{2\pi}{T}$$

and

$$F_n = \frac{1}{T}\int_{t_0}^{t_0+T} f(t)e^{-jn\omega_0 t} \tag{5.2}$$

TABLE 5.1 COMMONLY USED SYMBOLS AND CONVENTIONS

Symbol	Meaning
A, B, C	Boldface capitals will normally denote constant vectors and matrices
I	The identity matrix
u	Input variable
y	Output variable
x	Internal state-variable vector
t	Continuous-time independent variable
T	Sample interval for discrete-time systems; that is, $t = nT$
f	Frequency in Hz (cps)
f_s	Sample frequency in Hz; $f_s = 1/T$
ω	Frequency in rad/sec
ω_s	Sampling frequency in rad/sec; $\omega_s = 2\pi/T$
s	Laplace-transform complex variable
z	Z-transform complex variable

which converges uniformly to $f(t)$ in the open interval

$$t_0 < t < t_0 + T$$

everywhere except at points of discontinuity, where it converges to

$$\tfrac{1}{2}[f(t_{1+}) + f(t_{1-})]$$

Normally t_0 is chosen to be either 0 or $-T/2$, but it may be chosen to be any convenient value.

Alternatively, we may express $f(t)$ in one of two other equivalent forms:

$$f(t) = a_0 + \sum_{n=1}^{\infty} (a_n \cos n\omega_0 t + b_n \sin n\omega_0 t) \qquad (-\infty < t < \infty) \tag{5.3}$$

with a_0, a_n, and b_n given by

$$a_0 = \frac{1}{T}\int_{-T/2}^{T/2} f(t)\, dt \tag{5.4a}$$

$$a_n = \frac{2}{T}\int_{-T/2}^{T/2} f(t) \cos n\omega_0 t\, dt \tag{5.4b}$$

$$b_n = \frac{2}{T}\int_{-T/2}^{T/2} f(t) \sin n\omega_0 t\, dt \tag{5.4c}$$

or

$$f(t) = a_0 + \sum_{n=1}^{\infty} A_n \cos (n\omega_0 t + \phi_n) \tag{5.5}$$

where a_0 is given above and

$$A_n = \sqrt{a_n^2 + b_n^2} = 2|F_n| \tag{5.6a}$$

$$\phi_n = -\tan^{-1}\frac{b_n}{a_n} = \arg F_n \tag{5.6b}$$

Fourier-transform Relationships.[5,9] Given the function $f(t)$ satisfying the Dirichlet conditions (Ref. 5, p. 160), we will designate†

$$F(\omega) = \int_{-\infty}^{\infty} f(t)e^{-j\omega t}\,dt \tag{5.7}$$

as the Fourier transform of $f(t)$. In this context, it is often found to be useful to admit also those functions $f(t)$ for which the transform equation (5.7) exists in the Cauchy principal value sense, where

$$\lim_{T\to\infty} \int_{-T}^{T} |f(t)|\,dt < \infty \tag{5.8}$$

The corresponding inverse transform relationship is

$$f(t) = \frac{1}{2\pi}\int_{-\infty}^{\infty} F(\omega)e^{j\omega t}\,d\omega \tag{5.9}$$

Laplace-transform Relationships. Given the function $f(t)$ satisfying

$$\int_{0_+}^{\infty} |f(t)|e^{-\sigma t}\,dt < \infty \qquad \sigma < \infty \tag{5.10}$$

we will designate

$$F(s) = \int_{0_+}^{\infty} f(t)e^{-st}\,dt \tag{5.11}$$

as the Laplace transform of $f(t)$ with the corresponding inverse relationship

$$f(t) = \begin{cases} \dfrac{1}{2\pi j}\displaystyle\int_{\sigma-j\infty}^{\sigma+j\infty} F(s)e^{st}\,ds & t > 0 \\ 0 & t \le 0 \end{cases} \tag{5.12}$$

***Z*-transform Relationships.** Given the discrete-time function (number sequence) $f(t = nT) = f(n)$ with properties (Ref. 5, p. 238) (1) $f(n) < \infty$ for all finite n; (2) there exist positive constants N, r, and K such that

$$|f(n)| \le Kr^n \qquad \text{for } n \ge N$$

then we designate the summation

$$F(z) = \sum_{n=0}^{\infty} f(n)z^{-n} = \mathcal{Z}\{f(n)\} \tag{5.13}$$

as the (one-sided) Z transform of $f(n) = f(nT)$. The corresponding inverse relationship is

$$f(n) = \frac{1}{2\pi j}\oint_C F(z)\,z^{n-1}\,dz \tag{5.14}$$

† We elect to use the notation $F(\omega)$; some writers use $F(j\omega)$. Usually the explicit functions are written in an identical fashion; e.g., the transform of $e^{-at}U(t)$ is usually written as

$$\frac{1}{j\omega + a}$$

regardless of the notation used.

where C is a circle of radius sufficiently large to enclose all singularities of $F(z)z^{n-1}$. References 6, 7, 8, and 10 provide extensive tables of Z transforms; some important Z-transform theorems and transform pairs are given in Appendix B. In this context, consider the continuous-time function

$$Tf^*(t) = T \sum_{n=0}^{\infty} f(t)\delta(t - nT) \tag{5.15}$$

which *is* a valid approximation to $f(t)$ in the sense that the area under both functions is approximately the same in the interval $nT \leq t \leq (n+1)T$ if T is sufficiently small. This has justifiably led some authors (notably Wilts[11] and Kuo and Kaiser[1]) to define and use the Z transform with the factor T included; that is, it is defined as

$$T \sum_{n=0}^{\infty} f(nT)z^{-n}$$

We prefer to define the Z transform as in (5.13) without the T, since the more commonly available Z-transform tables appearing in sampled-data control system texts (e.g., Refs. 6 and 7) use this definition.

5.3 Discrete and Continuous Systems

It is important at the outset to make careful distinctions between the mathematical models used to describe digital filters and the analogous models used to describe continuous-time analog signal filters. It is true that most often we are concerned with processing information derived from real-world analog signal sources (transducers, etc.) for which an *analog signal model* (Fig. 5.1*a*) is best suited, viz., where we assume that the signals are single-valued functions of a continuous independent time variable, and that the function may take on any value in a continuous range. The function may contain discontinuities, but there is no discretization of either the independent or the dependent variable.

On the other hand, digital processing systems manipulate *number sequences* (data) for which a different model is required. We often assume the *discrete-time signal model* (Fig. 5.1*b*) involving the use of a single-valued function of a discrete independent variable; in so doing we overlook that fact that the real-world implementation of a digital signal-processing system involves the use of quantized values of the dependent variable as well.

In other instances, the effect of finite digital word length (number of binary digits in dependent-variable representation) must be considered. Here we must use a more precise *quantized signal* (digital data sequence) *model* (Fig. 5.1*c*) where we assume that the dependent variable (function

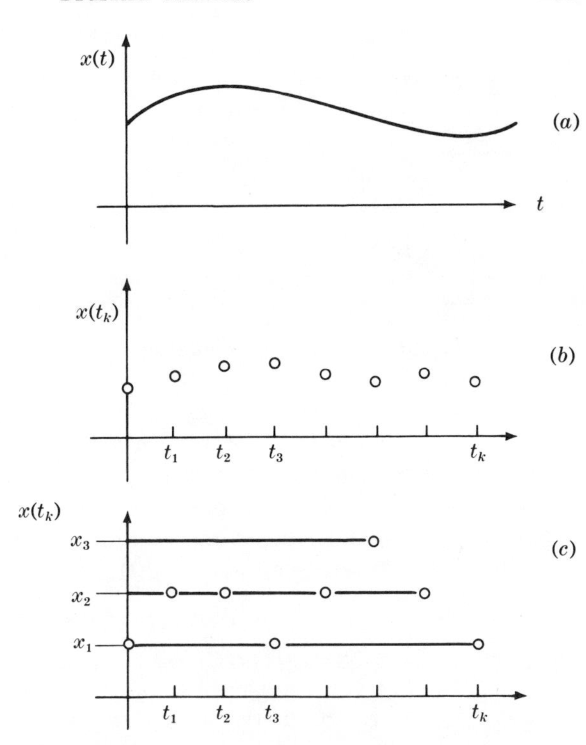

FIG. 5.1 Types of signals. (*a*) Continuous-time signal; (*b*) discrete-time signal; (*c*) quantized signal.

value) is also quantized (function defined only over a countable set of values).

Similarly, we may speak of continuous-time, discrete-time, and quantized systems, depending upon the type of variables used in the system description. In our discussions we will for the most part consider discrete-time systems where the signals involved may be modeled as in Fig. 5.1*b*. Where finite-word-length effects (round-off errors) are to be considered, the quantized signal model of Fig. 5.1*c* is more applicable (see, e.g., Sec. 5.11).

One of the major areas of confusion arises in *mixed* or *hybrid analog-digital* signal-processing systems (Fig. 5.2) where analog signals are sampled and the samples converted to digital data sequences (A/D conversion), then processed by a digital filtering system (digital computer, DDA, etc.), and then the resulting processed data sequence is converted back to an analog signal by means of a data-reconstruction system (D/A converter and hold circuit). We must always remember that the discrete-time and continuous-time models apply to different systems;

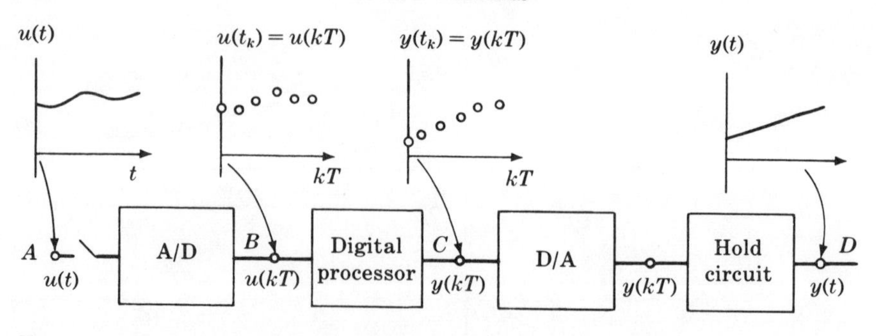

FIG. 5.2 Mixed or hybrid analog-digital signal-processing system.

special care must be taken when we are discussing *approximations* to continuous-time operations by discrete-time digital filters.

Fixed Systems. A continuous-time system is considered to be *fixed* if and only if its input-output relationships are time-invariant such that if an input $u(t)$ produces an output $y(t)$ governed by

$$y(t) = H\{u(t)\}\dagger \tag{5.16}$$

then

$$H\{u(t - \tau)\} = y(t - \tau) \tag{5.17}$$

for any $u(t)$ and τ. Similarly, a fixed discrete-time system governed by

$$y(t_k) = H\{u(t_k)\} \tag{5.18}$$

is fixed if and only if

$$H\{u(t_{k-n})\} = y(t_{k-n}) \tag{5.19}$$

for any $u(t_k)$ and n. Here the notation t_k means that the discrete set of times is indexed by the integer k. Normally in this chapter we assume that the times are equally spaced, i.e., that the discrete-time functions take on values only at time $t = kT$, where T is a time increment associated with a particular data-processing system.

Linear Continuous-time System: Time-domain Description. We assume that in this chapter we are considering the class of linear fixed continuous-time causal systems[5] described by the vector differential equations

$$\frac{d\mathbf{x}(t)}{dt} = \dot{\mathbf{x}}(t) = \mathbf{A}\mathbf{x}(t) + \mathbf{B}u(t) \tag{5.20a}$$

and

$$y(t) = \mathbf{C}\mathbf{x}(t) + Du(t) \tag{5.20b}$$

where $y(t)$ is a scalar output time function
$u(t)$ is a scalar input time function

† The symbol $H\{u\}$ means "H operating on u."

$\mathbf{x}(t)$ is a $k \times 1$ state vector time function
$\mathbf{A}$ is a $k \times k$ constant matrix
$\mathbf{B}$ is a $k \times 1$ constant matrix
$\mathbf{C}$ is a $1 \times k$ constant matrix
D is a constant

The general structure of this class of systems is shown in Fig. 5.3*a*; although we will discuss only the scalar-input, scalar-output case, most of the results can be extended to the multiple-input, multiple-output case.

Note also that if D is nonzero this indicates that the system is idealized, and not truly physically realizable, since it will have infinite bandwidth. Further, we may wish to extend the definition of (5.20) to include such idealized cases as a differentiator, viz.,

$$y = \dot{u}$$

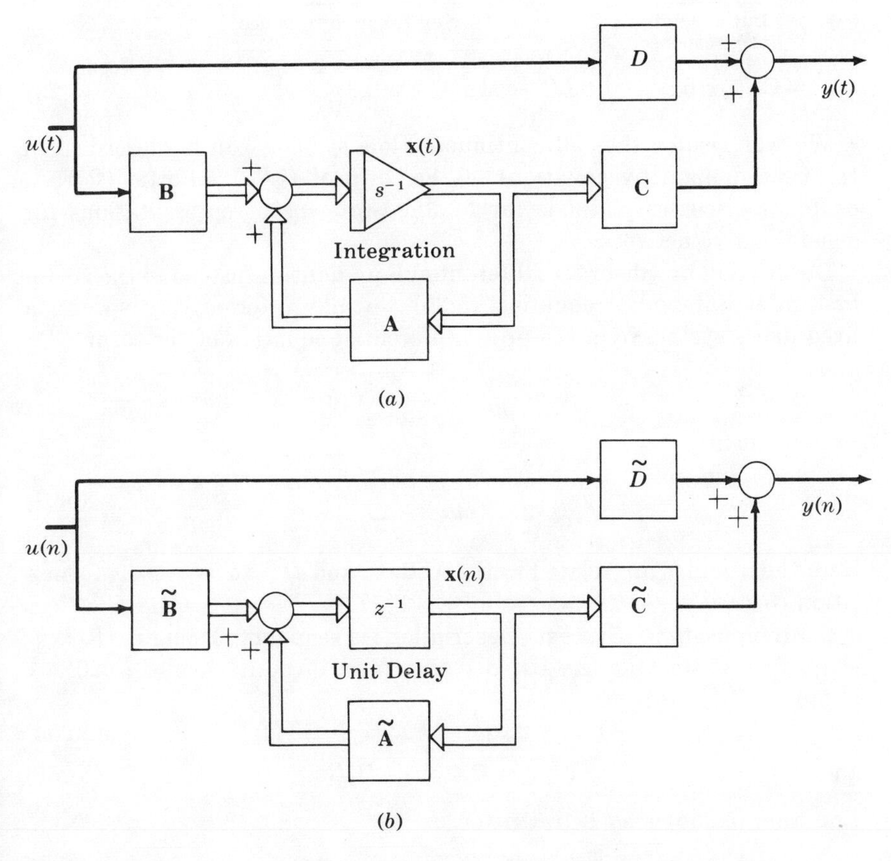

FIG. 5.3 General structure of linear systems. (*a*) Continuous-time; (*b*) discrete-time.

In those cases where the prototype continuous-time system does not have a positive pole-zero excess, then some form of low-pass filtering or smoothing should be provided as part of the design procedure. This may be done either before or after discretization has been made.

It is well known (Ref. 5, chap. 4) that if we define the system *fundamental* (transition) *matrix*

$$\begin{aligned}\boldsymbol{\phi}(t) &\triangleq e^{\mathbf{A}t}\\ &= \mathbf{I} + \mathbf{A}t + \tfrac{1}{2}(\mathbf{A}t)^2 + \cdots + (1/n!)(\mathbf{A}t)^n + \cdots \end{aligned} \tag{5.21}$$

then we find that the overall transfer operator, or impulse-reponse function, is given by

$$h(t) = \mathbf{C}\boldsymbol{\phi}(t)\mathbf{B} + D\delta(t) \tag{5.22}$$

and, of course,

$$\begin{aligned} y(t) &= \underbrace{\mathbf{C}\boldsymbol{\phi}(t)\mathbf{x}(0_+)}_{\text{Initial condition response}} + \underbrace{\int_0^\infty [\mathbf{C}\boldsymbol{\phi}(t-\lambda)\mathbf{B} + D\delta(t-\lambda)]u(\lambda)\,d\lambda}_{\text{Forcing function response}} \\ &= \mathbf{C}\boldsymbol{\phi}(t)\mathbf{x}(0_+) + \int_0^\infty h(t-\lambda)u(\lambda)\,d\lambda \end{aligned} \tag{5.23}$$

We will assume that all continuous-time systems can be described in the time domain by equations of the form of (5.20). Pottle (Ref. 1, chap. 3) discusses methods for finding state-space representations for general active networks.

Description by kth-order Differential Equations. Instead of the vector first-order differential equation (5.20) we may alternatively specify a fixed linear system by a kth-order differential equation of the form

$$y + b_1\dot{y} + \cdots + b_k\frac{d^k y}{dt^k} = a_0 u + a_1\dot{u} + \cdots + a_m\frac{d^m u}{dt^m}$$

or equivalently,

$$y + \sum_{i=1}^{k} b_i \frac{d^i y}{dt^i} = \sum_{j=0}^{m} a_j \frac{d^m u}{dt^m} \tag{5.24}$$

Here the a_j and b_i are related to the **A**, **B**, **C**, and D matrices in the formulation (5.20).

Continuous-time System Description: Frequency Domain (Ref. 5, chap. 7). If we take the Laplace transform of both sides of (5.20) we obtain

$$s\mathbf{X}(s) - \mathbf{x}(0_+) = \mathbf{A}\mathbf{X}(s) + \mathbf{B}U(s) \tag{5.25a}$$

and

$$Y(s) = \mathbf{C}\mathbf{X}(s) + DU(s) \tag{5.25b}$$

Equation (5.25a) may be rewritten

$$(s\mathbf{I} - \mathbf{A})\mathbf{X}(s) = \mathbf{x}(0_+) + \mathbf{B}U(s) \tag{5.26}$$

where $\mathbf{I}$ is the $k \times k$ identity matrix. From this we find

$$\mathbf{X}(s) = (s\mathbf{I} - \mathbf{A})^{-1}\mathbf{x}(0_+) + (s\mathbf{I} - \mathbf{A})^{-1}\mathbf{B}U(s) \tag{5.27}$$

and finally

$$Y(s) = [\mathbf{C}(s\mathbf{I} - \mathbf{A})^{-1}\mathbf{B} + D]U(s) + \mathbf{C}(s\mathbf{I} - \mathbf{A})^{-1}\mathbf{x}(0_+) \tag{5.28}$$

The transform of the impulse-response function is thus found by setting the initial conditions equal to zero, obtaining

$$H(s) = \mathcal{L}\{h(t)\} = \mathbf{C}(s\mathbf{I} - \mathbf{A})^{-1}\mathbf{B} + D \tag{5.29}$$

Also, we can take the Laplace transform of the general time-domain solution (5.23), to obtain

$$\mathbf{X}(s) = \mathbf{\Phi}(s)\mathbf{x}(0_+) + \mathbf{\Phi}(s)\mathbf{B}U(s) \tag{5.30}$$

thus we see that

$$\mathbf{\Phi}(s) = \mathcal{L}\{e^{\mathbf{A}t}\} = (s\mathbf{I} - \mathbf{A})^{-1} \tag{5.31}$$

is the Laplace transform of the fundamental matrix. Thus we have

$$H(s) = \mathbf{C}\mathbf{\Phi}(s)\mathbf{B} + D \tag{5.32}$$

Continuous-time Rational Transfer Functions. The transfer function $H(s)$ of a finite linear lumped-parameter network may be shown to be a rational function of the complex frequency variable; that is, if we evaluate the expression

$$\mathbf{C}(s\mathbf{I} - \mathbf{A})^{-1}\mathbf{B} + D$$

we find that it is a rational function $N(s)/D(s)$, with a denominator polynomial determined by $(s\mathbf{I} - \mathbf{A})^{-1}$; the roots of the denominator polynomial

$$D(s) = 1 + b_1 s + \cdots + b_k s^k \tag{5.33}$$

are commonly termed the *poles* of the system transfer function in the frequency domain. Similarly the roots of the numerator polynomial are called the *zeros*. Thus we have a rational transfer function

$$H(s) = \frac{Y(s)}{U(s)} = \frac{a_0 + a_1 s + \cdots + a_m s^m}{1 + b_1 s + \cdots + b_k s^k} \tag{5.34}$$

relating the Laplace transforms of the input and output. Implicit with this frequency-domain representation is the time-domain kth-order differential equation (5.24) relating the input $u(t)$ and the output $y(t)$. Of course, (5.34) may be obtained from (5.24) directly. In this chapter we will assume that the three methods of specifying a kth-order fixed linear system, (5.24), and (5.34), are equivalent, and any one of them may be used as the starting point for further analysis.

Sinusoidal Frequency Response. The response of fixed linear stable continuous-time systems to a sinusoidal forcing function

$$u(t) = A \sin(\omega t + \theta) \tag{5.35}$$

includes a unique *steady-state component*

$$y_{ss}(t) = B \sin(\omega t + \phi) \tag{5.36}$$

[corresponding to the particular solution of (5.24) to the forcing function (5.35)]. This sinusoidal function is the unique steady-state solution after all transients have subsided. In this case, it is possible to relate the input and output sinusoids by a *frequency response (operator)* $H(\omega)$. This is normally treated by the use of phasor notation, wherein we introduce a reciprocal one-to-one representation of the sinusoid

$$u(t) = A \sin(\omega t + \theta) \Leftrightarrow \frac{A}{\sqrt{2}} e^{j\theta} = \vec{U} = \frac{A}{\sqrt{2}} \underline{/\theta} \tag{5.37}$$

and

$$y_{ss}(t) = B \sin(\omega t + \phi) \Leftrightarrow \frac{B}{\sqrt{2}} e^{j\phi} = \vec{Y} = \frac{B}{\sqrt{2}} \underline{/\phi} \tag{5.38}$$

Conventionally, the absolute value of each phasor equals the rms value of the corresponding sinusoid; the phasor argument (angle) defines the phase of the sinusoid. We know that in phasor notation if

$$A \sin(\omega t + \theta) \Leftrightarrow \vec{U} \tag{5.39}$$

then

$$\frac{d^n}{dt^n} A \sin(\omega t + \theta) \Leftrightarrow (j\omega)^n \vec{U} \tag{5.40}$$

Thus we may substitute $j\omega$ for s in (5.34) to obtain

$$H(\omega) = \frac{a_0 + a_1 j\omega + \cdots + a_m (j\omega)^m}{1 + b_1 j\omega + \cdots + b_k (j\omega)^k} \tag{5.41}$$

relating the input and output phasors, i.e.,

$$\frac{\vec{Y}(\omega)}{\vec{U}(\omega)} = H(\omega) \tag{5.42}$$

Often of more importance, if we can determine $H(\omega)$ either empirically or using phasor analysis techniques, then by the principle of analytic continuation we may assume $H(s)$ is uniquely defined from the knowledge of $H(s = j\omega)$ along the $j\omega$ axis. Of course, in this context, we must be aware of the fact that non-minimum-phase systems will not be uniquely defined, unless we know exactly in which quadrant $H(\omega)$ lies, including any additional $2n\pi$ radians of phase shift. Further, $H(\omega)$ only defines a unique $H(s)$ if it is known for all frequencies in the range $0 \leq \omega < \infty$. Nevertheless, specifications of many continuous-time systems are commonly made in terms of the magnitude and phase of the function $H(\omega)$, and this will form one acceptable starting point in our discussions of design methods for suitable approximating discrete-time digital filters.

Discrete-time System Description: Time Domain (Ref. 5, p. 56). We assume that in a fashion similar to the above we can describe a linear fixed discrete-time causal system by the vector difference equations

$$\mathbf{x}(n+1) = \tilde{\mathbf{A}}\mathbf{x}(n) + \tilde{\mathbf{B}}u(n) \tag{5.43a}$$

$$y(n) = \tilde{\mathbf{C}}\mathbf{x}(n) + \tilde{D}u(n) \tag{5.43b}$$

where $$\mathbf{x}(n) = \mathbf{x}(t_n) = \mathbf{x}(t = nT)$$

(The tilde is used with the coefficient matrices to remind us that we are considering a discrete-time system.) The general structure is illustrated in Fig. 5.3*b*. Again, it is well known (Ref. 5, chap. 4) that the fundamental matrix should now be

$$\tilde{\boldsymbol{\phi}}(n) = \tilde{\mathbf{A}}^n \tag{5.44}$$

and thus we find that the overall discrete-time transfer operator (Ref. 5, p. 250)

$$h(n) = \begin{cases} \tilde{\mathbf{C}}\tilde{\boldsymbol{\phi}}(n-1)\tilde{\mathbf{B}} & n \geq 1 \\ \tilde{D} & n = 0 \end{cases} \tag{5.45a}$$

or

$$h(n) = \begin{cases} \tilde{\mathbf{C}}\tilde{\mathbf{A}}^{n-1}\tilde{\mathbf{B}} & n \geq 1 \\ \tilde{D} & n = 0 \end{cases} \tag{5.45b}$$

and also

$$y(n) = \underbrace{\tilde{\mathbf{C}}\tilde{\boldsymbol{\phi}}(n)\mathbf{x}(0)}_{\substack{\text{Initial} \\ \text{condition} \\ \text{response}}} + \underbrace{\sum_{k=0}^{n-1} \tilde{\mathbf{C}}\tilde{\boldsymbol{\phi}}(n-k-1)\tilde{\mathbf{B}}u(k) + \tilde{D}u(n)}_{\text{Forcing function response}}$$

$$= \tilde{\mathbf{C}}\tilde{\mathbf{A}}^n\mathbf{x}(0) + \sum_{k=0}^{n-1} \tilde{\mathbf{C}}\tilde{\mathbf{A}}^{n-k-1}\tilde{\mathbf{B}}u(k) + \tilde{D}u(n) \tag{5.46}$$

where $\mathbf{x}(0)$ is the initial state of the system at $n = 0$. We will assume that all discrete-time systems of interest here can be described in the time domain by equations of the form of (5.43) or (5.45).

Description by *k*th-order Difference Equations. Instead of the vector first-order difference equations (5.43), we again may alternatively specify a discrete-time fixed linear system by a kth-order difference equation of the form

$$y(n) + b_1 y(n-1) + \cdots + b_k y(n-k) = a_0 u(n) + a_1 u(n-1) + \cdots + a_m u(n-m)$$

or equivalently,

$$y(n) = -\sum_{i=1}^{k} b_i y(n-i) + \sum_{j=0}^{m} a_j u(n-j) \tag{5.47}$$

Here the a_j and b_i are related to the $\tilde{\mathbf{A}}$, $\tilde{\mathbf{B}}$, $\tilde{\mathbf{C}}$, and $\tilde{D}$ matrices of the formulation given in (5.43). Appendix C describes some of the computational operations involved in going from the state-space form of (5.43) to the difference equation form (5.47).

Discrete-time System Description: z Domain (Ref. 5, sec. 8.6). If we take the Z transform of both sides of (5.43) we obtain

$$z\mathbf{X}(z) - z\mathbf{x}(0_+) = \tilde{\mathbf{A}}\mathbf{X}(z) + \tilde{\mathbf{B}}\tilde{U}(z) \tag{5.48a}$$

$$Y(z) = \tilde{\mathbf{C}}\mathbf{X}(z) + \tilde{D}U(z) \tag{5.48b}$$

We may then solve these equations to obtain

$$\mathbf{X}(z) = (z\mathbf{I} - \tilde{\mathbf{A}})^{-1}z\mathbf{x}(0_+) + (z\mathbf{I} - \tilde{\mathbf{A}})^{-1}\tilde{\mathbf{B}}\mathbf{X}(z) \tag{5.49}$$

where $\mathbf{I}$ is the $k \times k$ unit matrix, and finally

$$Y(z) = \tilde{\mathbf{C}}(z\mathbf{I} - \tilde{\mathbf{A}})^{-1}z\mathbf{x}(0_+) + [\tilde{\mathbf{C}}(z\mathbf{I} - \tilde{\mathbf{A}})^{-1}\tilde{\mathbf{B}} + \tilde{D}]U(z) \tag{5.50}$$

The Z transform of the impulse response is thus found by setting the initial conditions equal to zero, and

$$\begin{aligned} H(z) = \mathcal{Z}\{h(n)\} &= \tilde{\mathbf{C}}(z\mathbf{I} - \tilde{\mathbf{A}})^{-1}\tilde{\mathbf{B}} + \tilde{D} \\ &= \tilde{\mathbf{C}}\mathbf{\Phi}(z)z^{-1}\tilde{\mathbf{B}} + \tilde{D} \end{aligned} \tag{5.51}$$

It can also be shown that

$$\hat{\mathbf{\Phi}}(z) = \mathcal{Z}\{\phi(n)\} = \mathcal{Z}\{\tilde{\mathbf{A}}^n\} = (z\mathbf{I} - \tilde{\mathbf{A}})^{-1} \tag{5.52}$$

is the Z transform of the fundamental matrix.

Discrete-time Rational Transfer Functions. The transfer function $H(z)$ may also be shown to be a rational function of the complex frequency variable z; that is, if we evaluate the terms in the matrix equation of (5.51) we find that it is of the form of a rational function $N(z)/D(z)$, with a denominator polynomial, as determined by $(s\mathbf{I} - \tilde{\mathbf{A}})^{-1}$; the roots of

$$D(z) = 1 + b_1 z^{-1} + b_2 z^{-2} + \cdots + b_k z^{-k} \tag{5.53}$$

are commonly termed the poles of the system transfer function in the z domain (see Appendix C). Similarly the roots of the numerator polynomial are termed the zeros. Thus we have a single scalar transfer function $H(z)$ of the form

$$H(z) = \frac{a_0 + a_1 z^{-1} + \cdots + a_m z^{-m}}{1 + b_1 z^{-1} + \cdots + b_k z^{-k}} \tag{5.54a}$$

or equivalently

$$H(z) = \frac{a_0 z^k + a_1 z^{k-1} + \cdots + a_m z^{k-m}}{z^k + b_1 z^{k-1} + \cdots + b_k} \tag{5.54b}$$

relating the Z transforms of the input and output. Implicit with this frequency-domain (z-domain here) representation is the time-domain *difference* equation (5.47). Also, (5.54) may be obtained from (5.47) directly. It is again assumed that all the forms of specifying a kth-order fixed linear discrete-time system, (5.43), (5.47), and (5.54), are equivalent, and any may be used as the starting point for further analysis.

Steady-state Sinusoidal Frequency Response of Discrete-time Fixed Linear Systems (Appendix A and Ref. 4, p. 238). If a (scalar) fixed linear discrete-time system (digital filter) with transfer function $H(z)$ is provided with samples from a sinusoidal signal, i.e.,

$$u(n) = E \cos (n\omega T) \tag{5.55}$$

then the output sequence will contain a *steady-state component*

$$\begin{aligned} y(n) &= G \cos (n\omega T + \theta) \\ &= |H(e^{j\omega T})| E \cos (n\omega T + \theta) \end{aligned} \tag{5.56}$$

where

$$\theta = \operatorname{Arg} [H(e^{j\omega T})] \tag{5.57}$$

Thus by analogy to the continuous-time case, we see that the complex number $H(e^{j\omega T})$ describes the steady-state frequency response of the system. We can establish the same correspondence with phasors and phasor algebra in the continuous-time case. Thus an analysis of $H(e^{j\omega T})$ provides us with analogous information about the spectral shaping produced by a given digital filter $H(z)$. Much of our later descriptions of filter behavior will be based on this viewpoint. Appendix A provides a z-domain proof of the above (see also Ref. 4).

It is also useful to note that $H(e^{j\omega T})$ is periodic in the frequency domain (see Sec. 5.4); that is

$$H(e^{j\omega T}) = H(e^{j(\omega + n\omega_s)T}) \qquad n \text{ an integer} \tag{5.58}$$

where ω_s is the angular sampling frequency

$$\omega_s = \frac{2\pi}{T}$$

This property is important to the understanding of the sampling theorem discussed in the next section.

5.4 The Sampling Theorem (Ref. 5, p. 158): A Link between Continuous- and Discrete-time Systems

We shall begin our discussion of the analytical links between continuous- and discrete-time systems by considering the properties of ideal, band-limited signals whose Fourier transform is identically zero outside a finite frequency range; in this case we may write†

$$X(\omega) \approx 0 \qquad |\omega| \geq W = 2\pi B \text{ rad/sec} \qquad |f| \geq B \text{ Hz} \tag{5.59}$$

The corresponding signal $x(t)$ is given by

$$x(t) = \frac{1}{2\pi} \int_{-2\pi B}^{2\pi B} X(\omega) e^{j\omega t} \, d\omega \tag{5.60}$$

Since $X(\omega)$ is zero outside of the interval $-2\pi B < \omega < 2\pi B$, it can be represented by a Fourier series valid for that range. Using the complex exponential form we obtain

$$X(\omega) = \sum_{n=-\infty}^{\infty} C_n e^{j\omega n T} \qquad T = \frac{1}{2B} = \frac{\pi}{W} \tag{5.61}$$

with the C_n given by

$$C_n = \frac{T}{2\pi} \int_{-2\pi B}^{2\pi B} X(\omega) e^{-jn\omega T} \, d\omega \tag{5.62}$$

By comparing (5.60) and (5.62), we see that

$$C_n = Tx(-nT) \tag{5.63}$$

so that we can write

$$x(t) = \frac{T}{2\pi} \sum_{n=-\infty}^{\infty} x(nT) \int_{-2\pi B}^{2\pi B} e^{j\omega(t-nT)} \, d\omega \tag{5.64}$$

Noting that

$$\int_{-2\pi B}^{2\pi B} e^{j\omega(t-nT)} \, d\omega = \frac{\sin 2\pi B(t-nT)}{2\pi B(t-nT)} = p(t-nT) \tag{5.65}$$

we have finally

$$x(t) = \sum_{n=-\infty}^{\infty} x(nT) p(t-nT) = \sum_{n=-\infty}^{\infty} X(nT) \frac{\sin 2\pi B(t-nT)}{2\pi B(t-nT)} \tag{5.66}$$

This last result is often called the *sampling theorem;* it may be stated as follows:

> If a signal $x(t)$ has a spectrum (Fourier transform) that is identically zero for $|f| \geq B$ Hz it can be completely reconstructed from its values at a countable set of equally spaced

† Note that this can only be an approximation to any real-world signal, since it can be shown (Ref. 9, p. 222) that a truly band-limited function cannot also be truly time-limited as well; on the other hand, all real-world signals begin at some point in time. Nevertheless, the concept of a band-limited function is very useful for theoretical treatments, and it can be satisfactorily approximated by a suitably filtered real-world signal.

sampling times spaced $T = \frac{1}{2}B$ seconds apart, i.e., from samples taken at a rate of $2B$ samples per second.

The above indicates that if we sample at a sufficiently fast rate ($1/T$ samples/sec) then we can retrieve the original time function $x(t)$ by the use of the ideal interpolation or data-reconstruction operator with impulse response†

$$p(t) = \frac{\sin[(\pi/T)(t - nT)]}{(\pi/T)(t - nT)} \tag{5.67}$$

Note that the sampling rate can be faster than $2B$ samples per second. The function $p(t)$ is unity for $t = 0$ and is zero for $t = kT$. Also, as indicated in Fig. 5.4*a*, $p(t)$ is otherwise nonzero for all time, and does not vanish for any other finite time. Thus we are reminded that all samples going back to $t = -\infty$ contribute to the reconstruction of $x(t)$. Immediately we see one problem in a physical implementation of the sampling theorem; viz., $p(t)$ is noncausal and thus not physically realizable. For this reason, plus the fact that a truly band-limited function requires an infinitely long sequence of samples for exact representation, any real-

† Note that impulse response in the context of the present discussion implies the response to a unit-valued data sample at time $nT = 0$ (Kronecker delta function response).

(*a*)

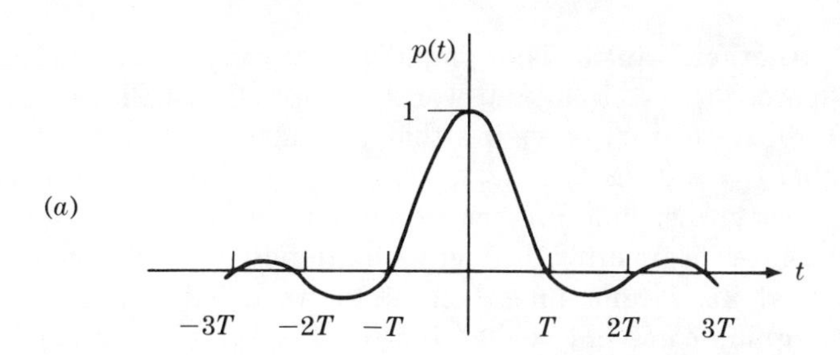

(*b*)

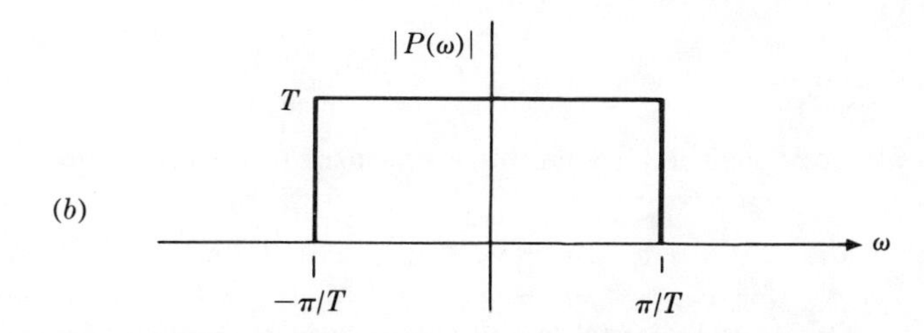

FIG. 5.4 Ideal data-reconstruction operator. (*a*) Impulse response $p(t)$; (*b*) Fourier transform of $p(t)$.

world data-reconstruction operation will yield only an approximate replica of the original signal. As we will see later, it is possible to obtain high accuracy by the use of hold circuits that approximate $p(t)$, and sampling faster than the sampling theorem requires.

Further insight may be obtained by looking at the Fourier transform of $p(t)$; we obtain (Fig. 5.4*b*)

$$P(\omega) = \begin{cases} T & \text{for } |\omega| \leq \frac{\pi}{T} \\ 0 & \text{for } |\omega| > \frac{\pi}{T} \end{cases} \tag{5.68}$$

We can deduce that physical hold circuits designed to approximate $p(t)$ should have the following properties:

1. The area under the impulse response should equal T, and the impulse response should equal 1 at $t = 0$ and 0 at $t = nT$, $n \neq 0$.
2. The Fourier transform of $p(t)$ should equal T at $\omega = 0$, and be as close to zero in magnitude as is possible for $|\omega| > \pi/T$ rad/sec.
3. $P(\omega)$ has zero phase shift (where acceptable, a physical hold circuit may introduce a constant additional delay, but should otherwise exhibit flat phase response). We shall later see that the above guidelines are useful for the design and evaluation of data-reconstruction operators.

Another Link: The Impulse-sampling Model (Laplace-transform Approach). Actual real-world sampling usually involves deriving a sequence of sample values that is realistically modeled by the sequence $x(nT) = x(0), x(1), \ldots, x(nT), \ldots$; Z-transform techniques apply directly to such a number sequence.

Since we are often dealing with samples derived from a continuous-time signal, it is sometimes helpful to work with an alternative model of sampling commonly called the impulse-sampling model. Here we treat the sampling operation as modulating the magnitude (or area) of a series of Dirac impulses to yield the *sampled-data impulse train* $x^*(t)$ defined by

$$x^*(t) = x(t) \sum_{n=0}^{\infty} \delta(t - nT) \tag{5.69}$$

Here $\delta(t)$ is the Dirac impulse operator (see Ref. 5, sec. 3.2); thus we have†

$$x^*(t) = \sum_{n=0}^{\infty} x(nT)\, \delta(t - nT) \tag{5.70}$$

† We must be careful to note that we seldom encounter physical pulse trains that in themselves approximate $x^*(t)$ by a train of pulses of short duration compared to the sampling interval. Rather we normally encounter discrete-time variables that

The Laplace transform of (5.70) yields (Ref. 6, p. 34)

$$X^*(s) = \sum_{n=-\infty}^{\infty} X(s + jn\omega_s) + \frac{T}{2} x(0_+) \tag{5.71}$$

where
$$\omega_s = \frac{2\pi}{T}$$
$$x(0_+) = x(t = 0_+) = \lim_{\epsilon \to 0} x(\epsilon) \qquad \epsilon > 0$$
$$X(s) = \mathcal{L}\{x(t)\}$$

Equation (5.71) shows the effect of sampling in the frequency domain. It will be recalled that the spectrum of $x(t)$ is given by $X(\omega)$. Therefore, the effect of sampling is to generate an infinite number of sidebands separated by integral multiples of the sampling frequency and proportional in amplitude to the input spectrum. This spectrum is illustrated in Fig. 5.5*a*. The sidebands generated by sampling are termed the "complementary signal" while the central part, which is proportional to the Fourier transform of $x(t)$, might be called the pure or baseband signal.

The reconstruction of the continuous signal from the sampled signal then involves essentially the separation of the pure signal from the complementary signals. It is therefore necessary to provide a low-pass filter to eliminate all complementary signals, while leaving the pure signals entirely unaffected. The desired transfer characteristics $P(\omega)$ of this filter in the Fourier domain are shown in Fig. 5.5*b*. If the actual response characteristics differ from the ideal, the filter either alters the pure signal or permits some of the complementary frequency components to pass through.

Again we see a basis for the sampling theorem. To keep the first lobes of the complementary signal from overlapping the center baseband, it is necessary for the upper limit of $X(\omega)$ to be limited to one-half of ω_s. Thus again if B is the upper limit of the spectrum of $x(t)$ in Hz, then the sampling interval T must be chosen so that

$$2\pi B < \frac{\omega_s}{2} = \frac{2\pi}{2T} = \frac{\pi}{T}$$

---

change in value periodically, or are transmitted periodically, etc. Thus although $x^*(t)$ is a useful ideal model function, we seldom see anything even close to such a pulse train in most signal-processing systems. If we use a model that includes a desampler (data hold or reconstruction operator) following the impulse train, then we have a model that may approximate real signals.

Nevertheless we proceed with an analysis of the characteristics of $x^*(t)$ in the frequency domain, since such a treatment provides considerable insight into the nature of the sampling operation. We must keep in mind the fact that real-world sampling and data-reconstruction operations differ in important ways from the ideal impulse-sampling and "brick-wall" reconstruction filter operations described next.

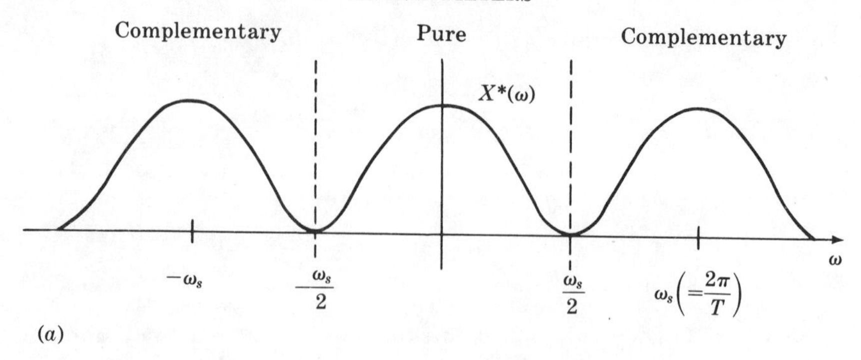

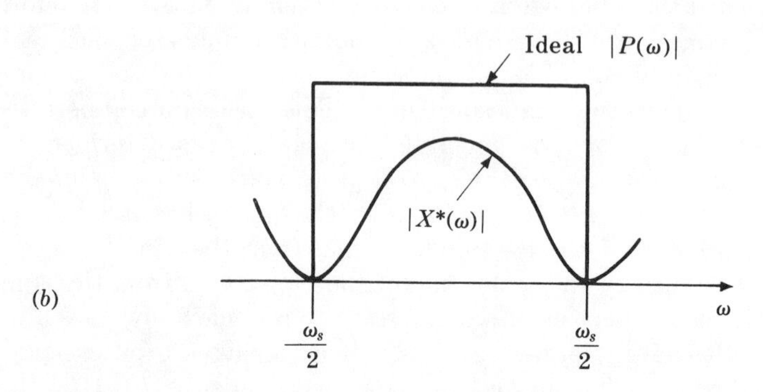

FIG. 5.5 Spectrum of $x^*(t)$. (a) $X^*(\omega)$; (b) filtering effect of $P(\omega)$.

or

$$T < \frac{1}{2B}$$

or

$$B < \frac{1}{2T}$$

that is, the sampling rate (samples/sec) must be twice the upper limit of the spectrum in Hz.

Aliasing Errors. Suppose we sample slower than the sampling theorem says we should; what effect does this have on the resultant information in the samples? We can visualize the effects by looking at the spectrum that results from sampling too slowly, or we can reconstruct the signal and compare it to the original. Figure 5.6*b* shows the spectrum that results from sampling the signal in *a*; we see the "folding" effect which leads to a distorted, and oftentimes meaningless result when we reconstruct, since reconstruction treats all spectral components below π/T as though they belonged to the original signal. Again, Fig. 5.7 shows the result of sampling a sine wave too slowly. The successive samples follow a sinusoidal variation, but at too low a frequency! Hence, the reconstructed data appear correct, unless one checks to see that the

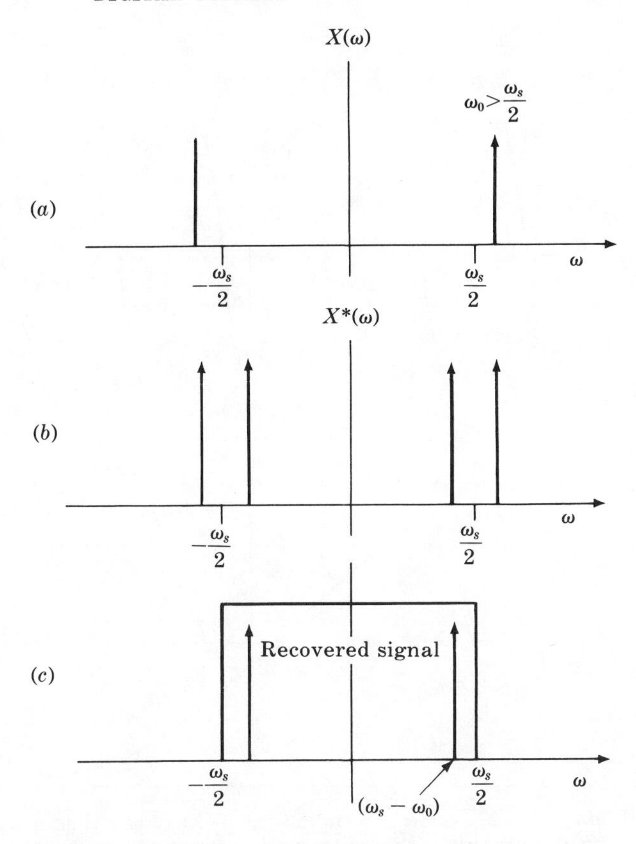

FIG. 5.6 Effect of sampling too slowly. (*a*) Spectrum of original sine wave, $\cos \omega_0 t$; (*b*) spectrum of sampled sine wave; (*c*) spectrum of recovered signal, a sine wave of incorrect frequency.

frequency is wrong. Note in particular that if we sample a sine wave of frequency f at a rate f we get constant sample values; in fact, the samples all have the same value for any sampling rate of f/n, n an integer.

Relationships between Laplace and Z Transforms. Consider the Z-transform relationship of (5.13), repeated here†

$$X_z(z) \triangleq \sum_{n=0}^{\infty} x(n)z^{-n} = \sum_{n=0}^{\infty} x(nT)z^{-n} \tag{5.72}$$

Now make the substitution

$$z = e^{sT}$$

† In this section, we will write the Z transform as $X_z(z)$ with the subscript z to emphasize the argument, and to distinguish $X(z)$ from $X^*(s)$ or $X(s)$.

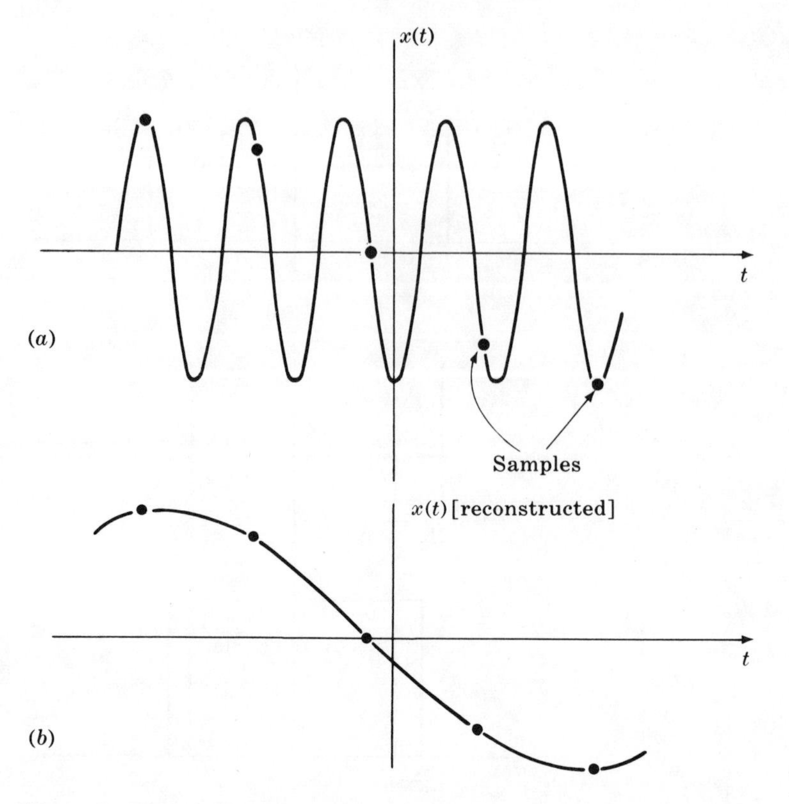

FIG. 5.7 Time-domain sketch of the effect of sampling a sine wave too slowly. (a) Original wave form; (b) resulting reconstructed wave form.

We will call the resulting function of s:

$$X^*(s) = X_z(z = e^{sT}) = \sum_{n=0}^{\infty} x(n)e^{-nsT} \tag{5.73}$$

We recall that this is equivalently (time-domain shifting theorem of Laplace transform) the Laplace transform of the continuous-time function $x^*(t)$, viz.,

$$X^*(s) = \mathcal{L}\left\{\sum_{n=0}^{\infty} x(n)\delta(t - nT)\right\} = \mathcal{L}\{x^*(t)\} \tag{5.74}$$

with $$x^*(t) = \sum_{n=0}^{\infty} x(n)\delta(t - nT)$$

and again $\delta(t - nT)$ is the usual Dirac impulse.

Note the implication that $x^*(t)$ is a continuous-time function. This leads quite naturally to the viewpoint that $X_z(z)$ is related to $x^*(t)$; viz.,

$$\mathcal{Z}\{x(nT)\} = X_z(z) \qquad z \Leftrightarrow e^{sT} \qquad X^*(s) = \mathcal{L}\{x^*(t)\} \tag{5.75}$$

Note also that $X^*(s)$ is the Laplace transform of a continuous-time function whereas $X_z(z)$ is the Z transform of a discrete-time function. Thus the conformal transformation $z = e^{sT}$ provides a link between continuous-time and discrete-time functions in the sense that identical values of $x(t = nT)$ in the continuous-time case and $x(n)$ in the discrete-time case are implied when $X^*(s)$ and $X_z(z)$ are thus related. We emphasize that

$$x^*(t) = \sum_{n=0}^{\infty} x(nT)\delta(t - nT)$$

is not *itself* a valid approximation to the continuous-time function $x(t)$ and thus $X_z(z)$ is not a legitimate z-domain representation of a continuous-time function $x(t)$.

Two Analytical Viewpoints of Mixed (Continuous/Discrete-time) Systems. To further clarify this point, let us refer again to Fig. 5.2. We may adopt two equivalent viewpoints:

1. The signal at point B can be treated as a number sequence (discrete-time function) $x(nT)$. It is operated upon by the digital processor to produce a second number sequence $y(nT)$. Subsequently the sequence $y(nT)$ drives a hold circuit or data-reconstruction system that generates a continuous-time output $y(t)$. Accompanying this viewpoint, we specify the response of the digital processor by a discrete-time transfer operator $h(n)$, described more fully in Sec. 5.7, or its z-domain equivalent $H(z)$.†
2. Alternatively, the signal at point B can be treated as a Dirac impulse train $x^*(t)$. Note that this signal model is a function of a continuous-time argument. The digital processor produces a second Dirac impulse train $y^*(t)$. We then assume that the data-reconstruction circuit responds to a *Dirac* impulse to give the same output as was obtained from a Kronecker delta function in the previous model. In this framework, we can talk about the continuous-time description of the data-reconstruction circuit in terms of an impulse response $p(t)$ and a Fourier transform $P(\omega)$.

† In this context, we think of the discrete-time system impulse response $h(n)$ as the response to a Kronecker delta function (Ref. 5, sec. 3.4)

$$\delta_K(n - k) = \begin{cases} 0 & n \neq k \\ 1 & n = k \end{cases}$$

Similarly, the response of a given data-reconstruction circuit $p(t)$ is described as the continuous-time output resulting from a data value of 1 delivered at $t = 0$, that is, $n = 0$.

We have stressed these two viewpoints because they both are common in the literature, and often some confusion arises when they are mixed. We prefer to look upon the operation of sampling as the interfacing operation between continuous-time and discrete-time signals, digital filtering as an operation on a discrete-time data sequence, and the final data reconstruction as the interfacing operation between a discrete-time signal and a continuous-time output. The Dirac-impulse-train viewpoint is nevertheless quite useful, particularly in providing insight into the spectral properties of sampled functions. We will not be led astray if we remember that $x^*(t)$ has a Laplace transform that can be mapped into the z domain by the conformal transformation $z = e^{sT}$ to *yield the same* $X(z)$ as the Z transform of $x(nT)$. This equivalence between the $X^*(s)$ and $X(z)$ representations must not, however, be misconstrued to indicate that $x^*(t)$ is a valid approximation to $x(t)$. In order to derive a valid approximation to $x(t)$, $x^*(t)$ must be used with some type of approximation operator $p(t)$. Thus although impulse trains of the form of $x^*(t)$ never appear at a physical port of a real-world system, we can nevertheless treat hybrid systems of the type shown in Fig. 5.2 as though such signals were being manipulated internally. In actuality, one cannot distinguish between the two viewpoints following the data-reconstruction operation.

5.5 Data Reconstruction: Generating a Continuous-time Function from a Discrete-time Data Train

It is easy to develop the habit of viewing a discrete-time data train as being an approximation to a continuous-time function, if the successive data values are close enough together. Similarly, we often think of a discrete-time operator as approximating a continuous-time operator for the same reasons. It is important to have the implications of such approximations well in mind, and to this end, let us consider some ways of formally making such approximations in a valid and consistent manner.

First of all, consider the data sequence of $x(nT)$ of Fig. 5.1*b*, which we assume was derived by sampling the continuous-time parent signal $x(t)$ in Fig. 5.1*a*. As mentioned earlier, we often represent the sampled wave form by the impulse train

$$x^*(t) = \sum_{n=0}^{\infty} x(t)\delta(t - nT)$$

for the purpose of discussing Laplace transforms, frequency spectra, etc. Note however that the continuous-time (quasi-discrete-time) function defined as $x^*(t)$ is *not* an approximation to $x(t)$, in that $x^*(t)$ would not affect a given continuous-time system in the same manner, regardless of how closely spaced the data points were. On the other hand, if we construct a train of pulses whose area is properly related to the area under the original function $x(t)$ during each sampling interval, then we may legitimately expect that this pulse train will serve as a useful approximation in the continuous-time sense. This point is carefully made in Refs. 11 and 12.

The properties of the ideal reconstruction filter described above by (5.68) may be used to suggest means for generating suitable approximations that can be implemented by realizable data-reconstruction operators.

Of course the reconstruction operator should be linear and time-invariant in the sense that a particular sample value $x(nT)$ will produce the same pulse shape regardless of the value of the index n. In this section we will describe some generally useful approximating methods. We are presently concerned with these operators as mathematical models of the approximation process. In the next section we will consider the performance of the real-world counterparts of some of these approximation operators; in that context, they are more commonly called hold circuits.

The Impulse Approximation (H_{-1}). Although not useful as a physical data-reconstruction operator, one can postulate the impulse approximation operator H_{-1} (Fig. 5.8*a*). Its operation is described by

$$H_{-1}: \qquad x(t) = Tx(n)\delta(t) \qquad nT < t \leq (n+1)T \tag{5.76}$$

A train of pulses from this operator approximates $x(t)$

$$x(t) \approx T \sum_{n=0}^{\infty} x(t)\delta(t - nT) = T \sum_{n=0}^{\infty} x(nT)\delta(t - nT) \tag{5.77}$$

Thus the resulting wave form has the proper area under each interval $nT < t \leq (n+1)T$. Otherwise the resulting approximation is seemingly rather poor. Nevertheless, the impulse approximation is a useful one for mathematical modeling of continuous-time systems by discrete-time blocks; it will be used later for the development of the impulse-invariant method of designing digital filters (Sec. 5.9).

Zero-order Hold (H_0). The most common data-reconstruction operator is the zero-order hold or box-car circuit (Fig. 5.8*b*). Its operation is described by

$$H_0: \qquad x(t) = x(n) \qquad nT < t \leq (n+1)T \tag{5.78}$$

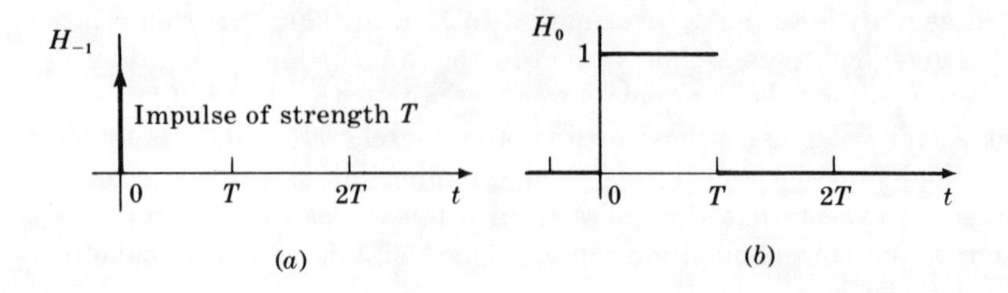

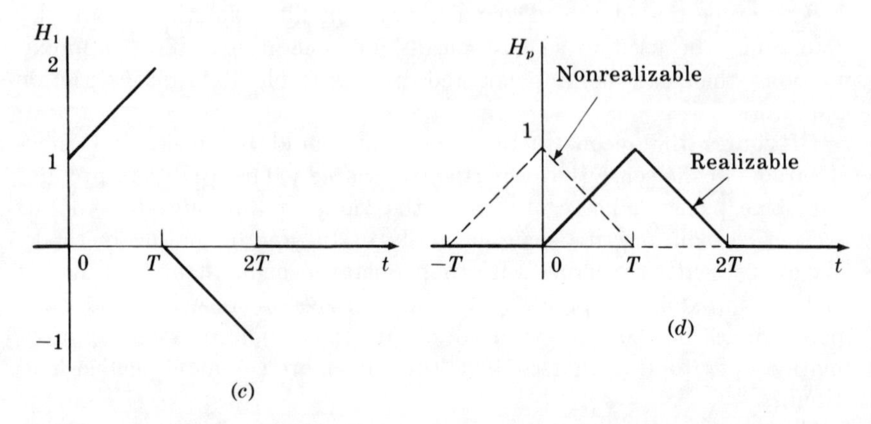

FIG. 5.8 Data-reconstruction operators. (a) Impulse approximation H_{-1}; (b) zero-order hold H_0; (c) first-order (extrapolative) hold H_1; (d) polygonal hold (realizable and nonrealizable) H_p.

Extrapolative First-order Hold (H_1). The operation of this circuit is described by

$$H_1:\quad x(t) = x(n) + [x(n) - x(n-1)]\frac{t - nT}{T} \qquad nT < t \leq (n+1)T \tag{5.79}$$

Note that we now must store one additional past value of the input sequence, and use the difference $x(n) - x(n-1)$ to estimate the slope during each extrapolation interval (Fig. 5.8c).

Polygonal Hold (H_p). This is a linear interpolator that performs the operation (Fig. 5.8d)

$$H_p:\quad x(t) = x(n-1) + [x(n) - x(n-1)]\frac{t - nT}{T} \qquad nT \leq t \leq (n+1)T \tag{5.80}$$

Note that we have presented a physically realizable version of the polygonal hold. The *nonrealizable* version of the circuit

$$x(t) = x(n) + [x(n+1) - x(n)]\frac{t - nT}{T} \qquad nT \leq t \leq (n+1)T \tag{5.81}$$

may be used for input approximation in the design of a digital filter (e.g., in Sec. 5.9), but it is not capable of being implemented physically.

We note that all of the above are linear, homogeneous time-invariant operators. As such, we may analyze their response in either the frequency or time domain, if we assume the input sequence is derived from a wide-sense stationary source, and that the sampling times are independent of the input-signal phase.

Frequency Response of the Hold Operators. The spectral behavior of the hold operators may be examined by finding the Laplace transform of the operator impulse response and then substituting $j\omega$ for s. The operators' transforms are easily found to be

$$H_{-1}(s) = T \tag{5.82a}$$

$$H_0(s) = \frac{1 - e^{-sT}}{s} \tag{5.82b}$$

$$H_1(s) = \frac{1 + Ts}{T}\left(\frac{1 - e^{-sT}}{s}\right)^2 \tag{5.82c}$$

$$H_p(s) = \frac{1}{T}\left(\frac{1 - e^{-sT}}{s}\right)^2 = \frac{1}{T}[H_0(s)]^2 \tag{5.82d}$$

From the above, the frequency-response functions are found to be (i.e., by setting $s = j\omega$)

$$H_{-1}(\omega) = T \tag{5.83a}$$

$$H_0(\omega) = \frac{2}{\omega}\sin\left(\frac{\omega T}{2}\right)e^{-\frac{j\omega T}{2}} = T\,\frac{\sin \omega T/2}{\omega T/2}\;\underline{/-\omega T/2} \tag{5.83b}$$

$$H_1(\omega) = T\sqrt{1 + (\omega T)^2}\left(\frac{\sin \omega T/2}{\omega T/2}\right)^2 \underline{/\tan^{-1}(\omega T) - \omega T} \tag{5.83c}$$

$$H_p(\omega) = T\left(\frac{\sin \omega T/2}{\omega T/2}\right)^2 \underline{/-\omega T} \tag{5.83d}$$

Plots of the magnitude and phase characteristics of the above operators are shown in Fig. 5.9. One finds that the polygonal hold H_p has a flat phase characteristic, with a delay of T seconds, and thus it is normally suited to open-loop reconstruction. Similarly, the zero-order hold H_0 introduces a delay of $T/2$. The extrapolative first-order hold H_1 is potentially useful for data reconstruction where signal delay must be minimized, e.g., closed-loop control and hybrid computation (however, some delay is introduced). Although the impulse-approximation operator has ideal response in the range $0 \leq \omega \leq \pi/T$, it does not have the desired zero magnitude beyond $\omega = \pi T$.

The frequency-response curves will later be seen to be somewhat misleading, and one cannot in general assume that the hold operators

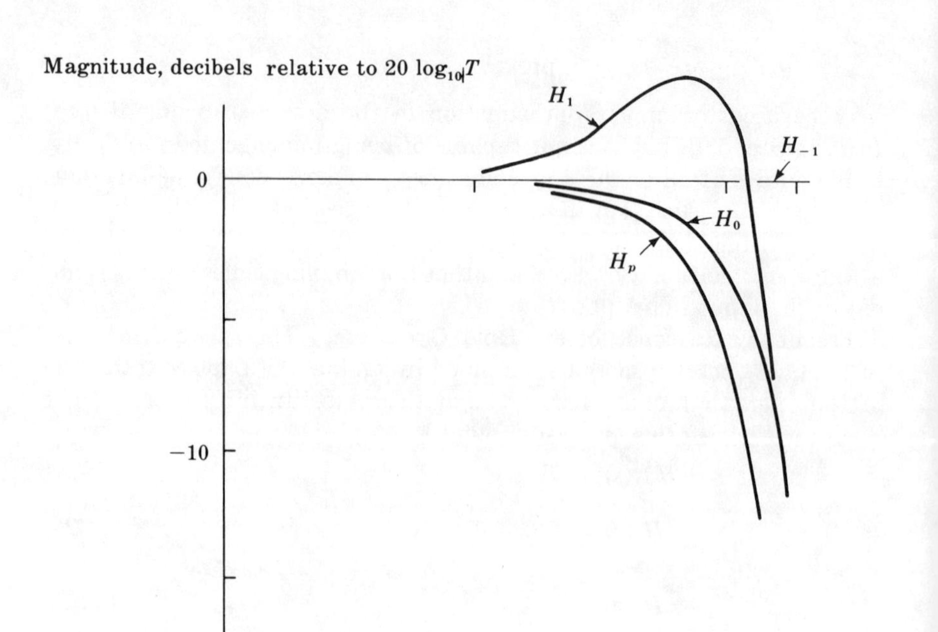

(a)

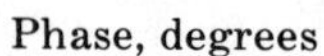

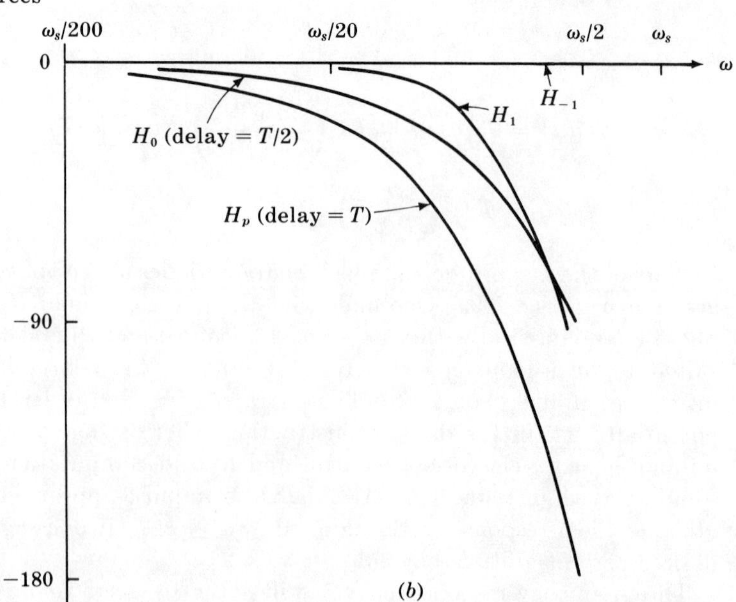

(b)

Fig. 5.9 Data-reconstruction-operator frequency response. (a) Magnitude in db; (b) phase in degrees.

will behave in a given sampled data system exactly as predicted by these curves.

5.6 Data-reconstruction Errors

Hold Circuits. This section summarizes some results of empirical and analytical studies of practical sampled-data reconstruction filters. It describes the errors associated with reconstructing *sine waves* and *band-limited noise.* In the first case, the error criterion is peak sine-wave reconstruction error; in the second case, two criteria are used, average rms error, averaged over an entire reconstruction interval, and maximum rms error at any given point during a reconstruction interval.[14] Three common types of data-reconstruction hold circuits are discussed, the zero-order hold, the extrapolative first-order hold, the polygonal (interpolative first-order) hold, designated H_0, H_1, and H_p respectively. It will be seen that the choice of sampling rate is related to the accuracy and ease of data reconstruction, and not only to the spectral characteristics of the original signal source.

Our analysis assumes that the necessary D/A conversion involved in providing the driving signal for the hold circuit is performed as part of the hold operation in a time that is negligibly small compared to the sampling time. Amplitude quantization errors are also not treated here; i.e., we assume the input to the hold circuit is the discrete-time signal of Fig. 5.1*b*.

In this section we assume the hold circuit receives input values $x(n) = x(t = nT)$, and produces a continuous-time output signal $x(t)$.

Worst-case Reconstruction Errors—Sine-wave Source Signals. The analysis for the zero-order hold and the extrapolative first-order hold follow directly from Karplus.[13] We assume that the data samples are from a sinusoid of peak amplitude A and frequency f Hz; that is, the desired reconstructed signal is

$$x = A \sin 2\pi ft = A \sin \omega t \qquad \omega = 2\pi f \tag{5.84}$$

Also we designate y = actual reconstructed signal.

Zero-order Hold. We can see by inspection that the actual signal obtained from a zero-order hold circuit will have instantaneous error near the zero crossings of the sinusoid; thus, we can find the maximum error by considering the case where one sample of $A \sin \omega t$ is taken at $t = 0$ and the next at $t = T$.

The error $\epsilon = x - y$ is therefore

$$\epsilon = A \sin \omega t \qquad 0 < t < T \tag{5.85}$$

and clearly ϵ max $= A\omega T$. Let δ = maximum error in term of percent of half-scale

$$\delta = 628\, fT \text{ percent} \tag{5.86}$$

Then if we designate the number of samples per cycle

$$p = \frac{1}{fT} \text{ samples/cycle} \tag{5.87}$$

we have the relationship

$$\delta = \frac{628}{p} \text{ percent} \tag{5.88}$$

or

$$p = \frac{628}{\delta} \text{ samples/cycle} \tag{5.89}$$

Table 5.2 shows the number of samples per cycle required for various percent accuracies, as calculated from (5.89).

TABLE 5.2 NUMBER OF SAMPLES PER CYCLE REQUIRED FOR A GIVEN SINE-WAVE RECONSTRUCTION ACCURACY

	Samples per cycle		
Accuracy %	*Zero-order*	*First-order*	*Polygonal*
0.01	62,800	628	223
0.1	6,280	199	71
0.5	1,256	89	32
1.0	628	63	22

Extrapolative First-order Hold. Here we can see that the maximum error occurs near the peak of a cycle, and so for convenience, let us assume

$$x = A \cos \omega t \tag{5.90}$$

then for the extrapolative first-order hold

$$\begin{aligned}\epsilon &= 2A(1 - \cos \omega t)\\ &\approx A\omega^2 t^2\end{aligned} \tag{5.91}$$

and thus

$$\epsilon_{\max} = A\omega^2 T^2$$

or

$$\delta = 100(\omega T)^2 \text{ percent} \tag{5.92}$$

Again if we let $p = 1/fT$ be the number of samples per cycle we have

$$\begin{aligned}\delta &= 100(2\pi fT)^2\\ &= \frac{3{,}940}{p^2} \text{ percent}\end{aligned} \tag{5.93}$$

or

$$p = \frac{62.8}{\sqrt{\delta}} \text{ samples/cycle} \tag{5.94}$$

Again, Table 5.2 presents typical values of p for the extrapolative first-order hold. We see that the required sampling rate is markedly reduced, when compared to the zero-order hold.

Polygonal Hold. Again, the maximum absolute error occurs near the peak of the sine wave. This can most easily be estimated if we assume a cosine wave, and assume one sample occurs at $t = -T/2$ and the other at $+T/2$. Then the error is

$$\begin{aligned} \epsilon &= A \cos \omega t - A \cos \frac{\omega T}{2} \\ &= A\left(\cos \omega t - \cos \frac{\omega T}{2}\right) \end{aligned} \tag{5.95}$$

or, at $t = 0$,

$$\begin{aligned} \epsilon_{\max} &= A\left(1 - \cos \frac{\omega T}{2}\right) \\ &\approx A \frac{(\omega T)^2}{8} \end{aligned} \tag{5.96}$$

thus

$$\begin{aligned} \delta &= \frac{100\, \epsilon_{\max}}{A} = 495\, (fT)^2 \\ &= \frac{495}{p^2} \text{ percent} \end{aligned} \tag{5.97}$$

$$p = \frac{22.3}{\sqrt{\delta}} \text{ samples/cycle} \tag{5.98}$$

Equation (5.97) indicates that for a given maximum error the number of samples per cycle can be reduced by a factor of $\sqrt{8}$ if one uses the polygonal interpolative rather than the extrapolative first-order hold; also, for a given value of p, the error is reduced by a factor of 8, which is a decided improvement over both conventional hold operators. It is emphasized that one must be able to tolerate the extra delay of one sample period incurred.

Summary of Sine-wave Reconstruction Errors. An examination of Table 5.2 indicates that the polygonal hold is clearly better for situations where the one-sample delay in reconstruction can be tolerated. Otherwise, the first-order hold is the best of the three holds with regard to peak error in sine-wave reconstruction. It also should be noted that the error of the zero-order hold is only one-half of that indicated by (5.88) when the output is compared to the original sine wave delayed by $T/2$. Nevertheless, the other two hold circuits are generally superior.

Data-reconstruction Errors: Random Signals. We will use the following relationship.

Given $Z = \sum_{i=1}^{N} a_i x_i$, the x_i have zero mean and statistical variance σ^2. Then

$$\text{Var} = \{Z\} = \sigma^2 \sum_{i=1}^{N} a_i^2 + \sum_{i \neq j} a_i a_j \tag{5.99}$$

We assume that the sampled signal is wide-sense stationary with rms value σ and zero mean. Let us designate the normalized auto covariance

$$\rho_{xx}(\tau) = \frac{1}{\sigma^2} R_{xx}(\tau) = \frac{1}{\sigma^2} \lim_{T \to \infty} \frac{1}{2T} \int_{-T}^{T} x(t)x(t + \tau)\, dt \tag{5.100}$$

Zero-order Hold. Consider the error during a typical reconstruction interval: $nT < t \leq (n + 1)T$. Defining

$$u = \frac{t - nT}{T} \qquad 0 \leq u \leq 1 \tag{5.101}$$

Then the zero-order hold error is

$$\epsilon = x(nT + uT) - x(nT) \tag{5.102}$$

and

$$\text{Var}\{\epsilon\} = 2\sigma^2[1 - \rho(uT)] \tag{5.103}$$

Also the average mean-square error is

$$\overline{\epsilon^2} = \int_0^1 \text{Var}\,\{\epsilon\}\, du = 2\sigma^2 \int_0^1 [1 - \rho(uT)]\, du \tag{5.104}$$

First-order Hold. A similar analysis yields

$$\text{Var}\{\epsilon\} = 2\sigma^2\{(1 + u + u^2) - (1 + u)\rho(uT) + u\rho[(1 + u)T] - u(1 + u)\rho(T)\} \tag{5.105}$$

and

$$\overline{\epsilon^2} = \sigma^2\left(\frac{11}{3} - \frac{5}{3}\rho(T) + 2\int_0^1 \{u\rho[(1 + u)T] - (1 + u)\rho(uT)\}\, du\right) \tag{5.106}$$

Polygonal Hold. Again we find

$$\text{Var}\{\epsilon\} = 2\sigma^2\{(u^2 - u + 1) + (u - 1)\rho(uT) - u(u - 1)\rho(T) - u\rho[(1 - u)T]\} \tag{5.107}$$

and

$$\overline{\epsilon^2} = \sigma^2\left[\frac{5}{3} + \frac{\rho(T)}{3} - 4\int_0^1 (1 - u)\rho(uT)\, du\right] \tag{5.108}$$

The latter result agrees with Leneman and Lewis.[16]

Ideally Band-limited Random Noise. Let us assume

$$\rho_{xx}(\tau) = \frac{\sin (2\pi B\tau)}{2\pi B\tau} \tag{5.109}$$

TABLE 5.3 DATA-RECONSTRUCTION ERRORS, IDEALLY BAND-LIMITED NOISE†

Hold circuit	*Maximum rms error*	*Average rms error*
Zero-order	$362/p$	$209/p$
First-order	$1{,}850/p^2$	$894/p^2$
Polygonal	$220/p^2$	$161/p^2$

† p = number of samples/Hz of noise bandwidth.

where B = noise bandwidth in Hz. Let $2\pi BT = V$, then

$$\rho_{xx}(uT) = \frac{\sin Vu}{Vu} \approx 1 - \frac{(Vu)^2}{6} + \frac{(Vu)^4}{120} + \cdots \tag{5.110}$$

We can then evaluate the above error estimates for small V. For the zero- and first-order hold we assume that Var $\{\epsilon\}_{\max}$ occurs at $u = 1$, and for the polygonal hold at $u = \frac{1}{2}$. Table 5.3 summarizes the results with $p = 1/BT$, that is, the number of samples/Hz of noise bandwidth; Table 5.4 indicates the required sampling rates for a given rms accuracy.

5.7 Classes of Digital Filters: Realizations

The general state-space description of a discrete-time system (5.43) specifies a set of algorithms that may be used directly for designing a digital filter or programming a digital computer. In many digital-simulation applications it is important to preserve the identity of various internal state variables, and the formulation of (5.43) does so by specifying the calculation of a specific choice of variables, $\mathbf{x}$. We shall see in Sec. 5.9 that a variety of digital-filter-design methods lead to algorithms of this general form; associated with these algorithms is the general block diagram of Fig. 5.2*b*. Note that if we do not wish to preserve the identity of a given set of state variables, but only the input-output

TABLE 5.4 REQUIRED NUMBER OF SAMPLES PER HZ OF NOISE BANDWIDTH

	Average rms error %			*Maximum rms error %*		
Hold	1	0.5	0.1	1	0.5	0.1
Zero-order	209	418	2,090	362	724	3,620
First-order	30	43	95	43	60	136
Polygonal	13	19	41	15	21	48

relationship, then we may select a different set of state variables that may lead to a more efficient implementation.

Suppose that we are interested in implementing a given input-output relationship, without requiring that specific state variables be calculated; then a more convenient representation is the difference equation form (5.47) or the corresponding z-domain transfer function (5.54). In this section, let us assume that only a given digital-filter input-output relationship is to be implemented, specified by

$$y(n) = \sum_{j=0}^{m} a_j\, u(n-j) - \sum_{i=1}^{k} b_i\, y(n-i) \tag{5.111}$$

The input-output relationship of (5.111) is equivalent to a state-variable form, (5.43). The equivalence may be established by means of appropriate transformations (see Appendix C). Moreover, the choice of state variables to be used in the actual implementation is not in general unique; several of the more commonly used structures are illustrated in Fig. 5.10. They are easily obtained from the difference equation form of (5.112). Taking the Z transform of (5.111), we obtain the z-domain transfer function

$$\frac{Y(z)}{U(z)} = H(z) = \frac{a_0 + a_1 z^{-1} + \cdots + a_m z^{-m}}{1 + b_1 z^{-1} + \cdots + b_k z^{-k}} \tag{5.112}$$

Equation (5.112) suggests the *direct* form of realization of Fig. 5.10*a*. Three other common forms of realization are shown, the *canonical* (*b*), the *cascade* (*c*), and the *parallel* (*d*). The direct and canonical forms are easily obtained from inspection of the coefficients of $H(z)$. In obtaining the canonical form it is useful to note (see Fig. 5.10*b*) that (5.112) can be rewritten in the form

$$F(z) = \frac{1}{1 + b_1 z^{-1} + b_2 z^{-2} + \cdots + b_k z^{-k}} \tag{5.113}$$

$$Y(z) = (a_0 + a_1 z^{-1} + a_2 z^{-2} + \cdots + a_m z^{-m})\, F(z) \tag{5.114}$$

The cascade configuration of Fig. 5.10*c* is based upon the fact that we can factor the transfer function into pole-zero pairs, i.e.,

$$H(z) = a_0 \frac{1 + \alpha_1 z^{-1}}{1 + \beta_1 z^{-1}} \cdots \frac{1 + \alpha_m z^{-1}}{1 + \beta_m z^{-1}} \cdots \frac{1}{1 + \beta_k z^{-1}}$$

$$= \frac{a_0 \prod_{i=1}^{m} (1 + \alpha_i z^{-1})}{\prod_{i=1}^{k} (1 + \beta_i z^{-1})} \tag{5.115}$$

The parallel form follows directly from a partial fraction expansion of (5.112) in the form

$$H(z) = \sum_{j=1}^{k} \frac{\gamma_j}{1 + \beta_j z^{-1}} \tag{5.116}$$

Note that the cascade and parallel forms (5.115 and 5.116) imply that all z-domain poles and zeros are real. In many practical cases, some poles and/or zeros may occur in complex-conjugate pairs. In these cases, one must use cascade or parallel structures involving second-order difference equation blocks (see Ref. 2, p. 150) in order to obtain an implementation using only real-valued coefficients.

Recursive and Nonrecursive Filters. In general, when the b_k are nonzero, we have a *recursive* filter, viz., one in which past values of the output are used in calculating the new output. *Recursive* filters therefore have *infinite memory* of the past input sequence and internal state of the processing system.

When the b_k are all zero, we then have a *nonrecursive, transversal,* or *finite-memory* filter of the general type

$$y(n) = \sum_{j=0}^{m} a_j\, u(n - j) \tag{5.117}$$

which specifies a direct implementation in the form of Fig. 5.10a (with the $b_i = 0$).

Signal-processing Delays: Realizability. Strictly speaking, the general transfer function of (5.112) is not physically realizable if the numerator constant term a_0 is nonzero, since this implies the next value of the output is calculated at the instant when the corresponding input value is received. There will always be a digital-processing time after $u(n)$ is received before $y(n)$ may be delivered to an output device. If the digital-processing time T_c is a significant fraction of the total sampling time T, then this additional delay can create serious problems in closed-loop systems. If it is necessary to deliver $y(n)$ as closely as possible to the corresponding time $t = nT$ then one may have to use a filter with $a_0 = 0$. Then the new output value can be transferred to the output device (e.g., a D/A converter buffer register) at the beginning of the next sampling period with a minimum of delay.

If a_0 is not zero, then the output value $y(n)$ must be delayed with respect to the receipt of $u(n)$ by at least the time to form the product $a_0\, u(n)$ and add its contribution to the output. Of course in some cases where T is relatively large compared to the digital-processing speed, this delay may not be important.

In the following discussions of general design methods (particularly in Sec. 5.9) we will designate filter algorithms where $a_0 = 0$ as being *real-*

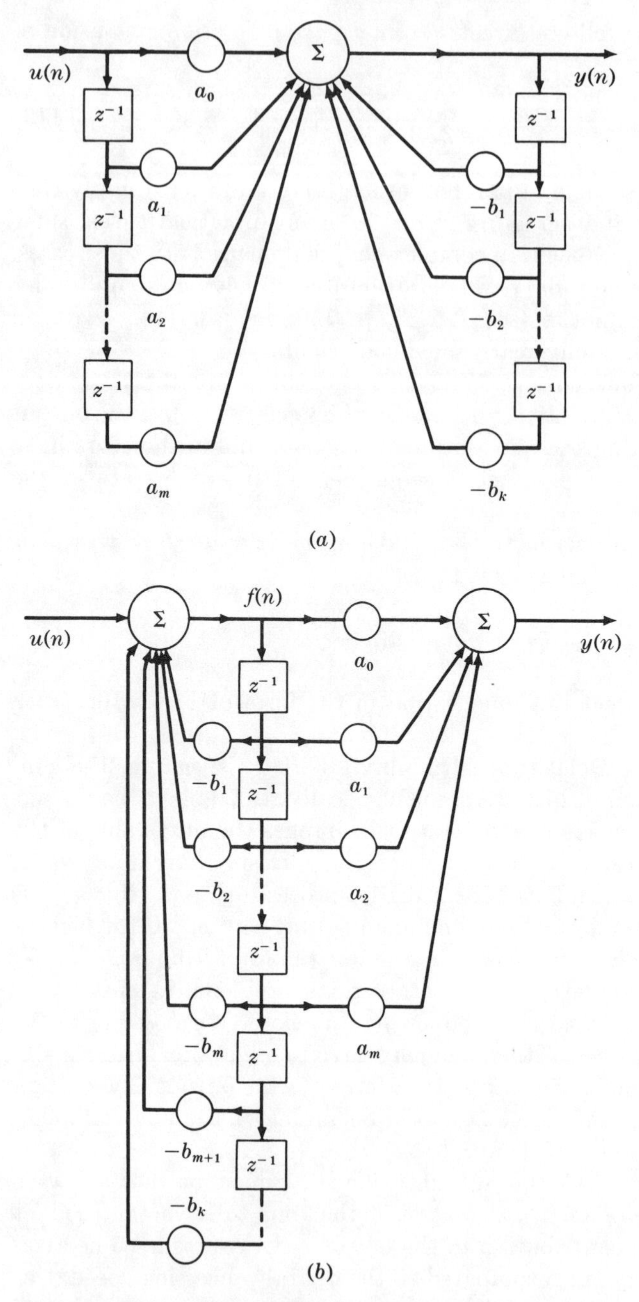

FIG. 5.10 Digital-filter implementations. (*a*) Direct realization; (*b*) canonical; (*c*) cascade; (*d*) parallel.

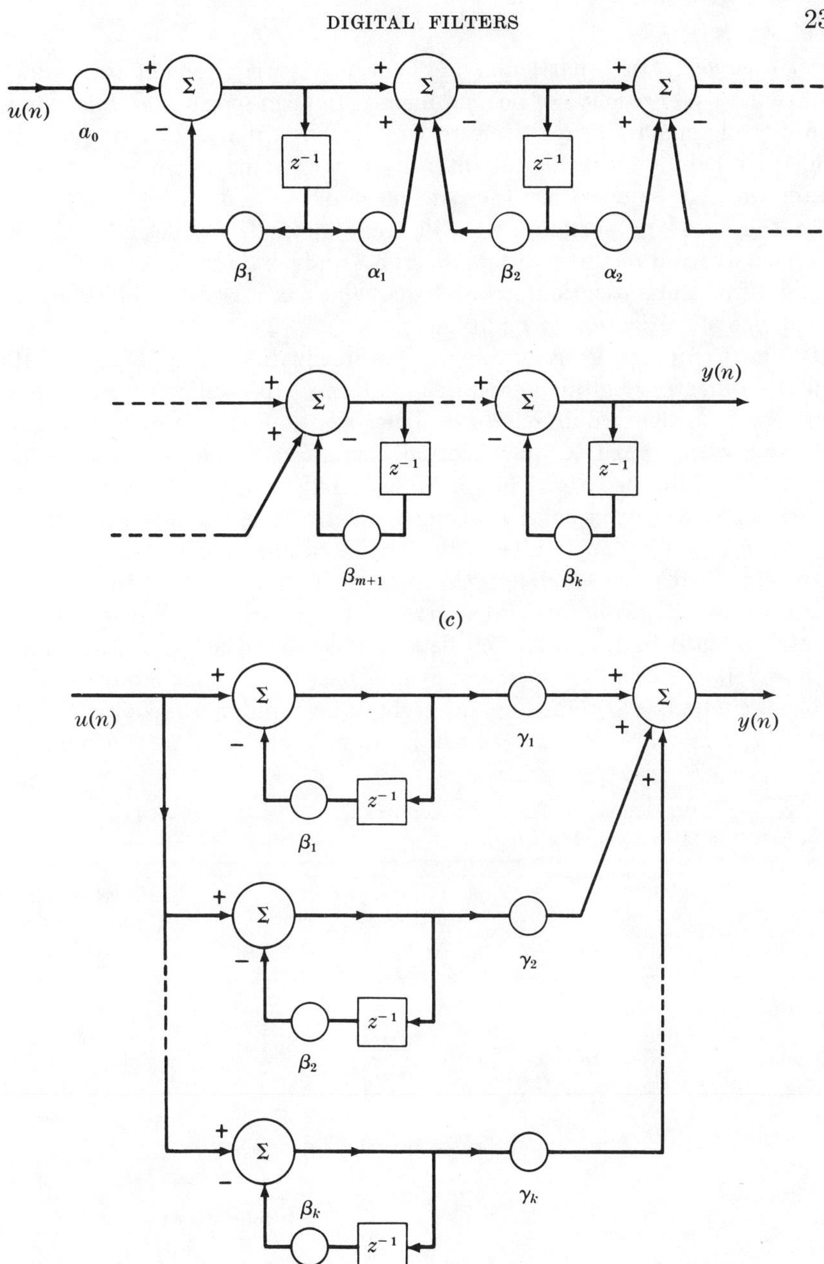

FIG. 5.10 Continued.

time realizable. In a particular application the requirement for realizability in this sense may not be a problem. In open-loop signal processing and digital simulation, real-time realizability is usually not required; on the other hand, if the digital filter is part of some larger closed-loop system, it may be necessary for a_0 to be zero.

In the following sections, we will treat methods for designing both nonrecursive and recursive filters (Secs. 5.8 and 5.9); Sec. 5.11 will discuss round-off and quantization error effects, which in many cases bear upon the choice of realization to be used.

Digital Computer Programming. The realization forms of Fig. 5.10 lead to different digital computer programs, with different execution times and storage requirements. (They also have inherently different behavior with regard to quantization errors; this topic is discussed in Sec. 5.11.) Monroe (Ref. 3, chap. 25) has analyzed the computer requirements for scaled fixed-point implementation of each of the realizations of Fig. 5.10. Similar results, using the notation of (5.111), are summarized in Table 5.5, which lists the number of additions, multiplications, and data transfers required for a given filter order, as well as the total storage requirement. Table 5.6 lists relative execution times in terms of numbers of computer memory cycles, based on the capabilities of a typical, contemporary, medium-sized digital computer, the DEC PDP-9 (Ref. 17, chap. 7). We can conclude from Table 5.5 that the relative

TABLE 5.5 COMPARISON OF DIGITAL COMPUTER PROGRAMMING METHODS (See Fig. 5.10)†

Method	*Number of operations*			*Total storage*
	Additions	Multiplications	Data transfers‡	(Constants and variables)
Direct (Fig. 5.10*a*)	$k + m$	$k + m + 1$	$k + m + 2$	$2k + 2m + 3$
Canonical (Fig. 5.10*b*)	$k + m$	$k + m + 1$	$k + 5$	$2k + m + 4$
Cascade (Fig. 5.10*c*)	$k + m + 2$	$k + m + 3$	$k + 5$	$2k + m + 6$
Parallel (Fig. 5.10*d*)	$2k + 1$	$2k + 2$	$k + 5$	$3k + 5$
Nonrecursive (Fig. 5.10*a*, $k = 0$)	m	$m + 1$	$m + 2$	$2m + 3$

† k is the number of nonzero, nonunity denominator coefficients; in (5.111), $m + 1$ is the number of nonzero, nonunity numerator coefficients; totals assume no missing coefficients and $m \leq k$, hence they are upper bounds for recursive filters.

‡ Data transfers include one input and one output transfer.

TABLE 5.6 TYPICAL EXECUTION TIMES

Method	*Number of memory cycles*†
Direct	$29k + 29m + 29$
Canonical	$29k + 25m + 41$
Cascade	$29k + 25m + 91$
Parallel	$54k + 66$
Nonrecursive	$25m + 29$

† Based upon the DEC PDP-9 digital computer with hardware multiply; add and transfer times are each assumed to take 4 cycles, multiply time is 21 cycles. PDP-9 cycle time is 1 μsec.

execution speed of a particular realization depends upon the number of numerator coefficients. In general, the direct method is faster than the canonical method if m is less than 3. The cascade is slower than the canonical method; it is also slower than the direct method if $m < 15$. The parallel method is slower than the cascade if $k > m + 1$. This gross comparison applies only to the particular computer used; using the expressions in Table 5.5, a similar comparison can be made for a different machine.

The direct method requires generally more storage than the canonical or cascade method; the parallel method requires more than the canonical or cascade method if m is small.

5.8 Nonrecursive-filter-design Methods

The nonrecursive filter (5.117) implies the z-domain transfer function

$$H(z) = \sum_{k=0}^{m} a_k z^{-k} \tag{5.118}$$

(this type of filter is also referred to as a *finite-memory* or *transversal* filter). Since it does have only a finite memory of past input values, it is not a suitable filter for integration, but it is useful for such operations as smoothing (low-pass filtering), interpolation and extrapolation (delay and prediction), differentiation, and combinations of the above. Generally the frequency response of the desired operation should change gradually. It is not normally as economical of machine time as the recursive filter (Sec. 5.9). The design methods presented here rely upon polynomial curve fitting (for time-domain synthesis) and Fourier-series expansion techniques (frequency-domain synthesis).

Time-domain Design (Polynomial Fit). In some cases, the desired filter response may be specified in terms of a time-domain impulse response, $h(n)$, (5.45). Recognizing that

$$\mathcal{Z}\{h(n)\} = h(0) + h(1)z^{-1} + \cdots \tag{5.119}$$

it is possible to approximate $h(n)$ by a finite set of coefficients; i.e., we set

$$a_k = h(k) \qquad k = 0, 1, 2, \ldots, m \tag{5.120}$$

Obviously the accuracy of the approximation depends upon the number of terms used, $m + 1$. This in turn depends upon the bandwidth over which the approximation is to be effective.

In many traditional applications of nonrecursive filters, the filter's input is assumed to be a polynomial of degree m or less. In this case, one can apply the classical, curve-fitting methods of numerical analysis, recognizing the correspondence between the Z-transform operator z and the unit displacement operator, defined by

$$E^k\{x(n)\} = x(n + k) \tag{5.121}$$

and also noting that

$$E^{-1} = 1 - \nabla \tag{5.122}$$

where ∇ is the classical backward difference operator defined by

$$\nabla\{x(n)\} = x(n) - x(n - 1) \tag{5.123}$$

Monroe has treated this approach in considerable detail (Ref. 3, chaps. 8 and 9), and has derived formulas for smoothing, predicting, and differentiating polynomials of a given order by means of an m-term nonrecursive filter (in Monroe's notation, our m is designated $N - 1$). Monroe also develops the use of the variance reduction factor

$$R = \sum_{k=0}^{m} a_k^2 \tag{5.124}$$

as a measure of the noise reduction or "smoothing" capabilities of a nonrecursive filter. He then applies polynomial-fitting constraints by means of lagrangian multipliers to achieve a design of an mth-order filter that fits a qth-order polynomial with minimum value for the variance reduction factor (Ref. 3, chap. 9).

Brubaker and Stevens[18] and Brubaker and Peterson[19] present the results of some empirical studies of nonrecursive designs based upon the use of orthogonal (Legendre) polynomials, and they compare their performance to filters designed by the above method.

For a given number of coefficients, Kuo and Kaiser (Ref. 1, secs. 7.4 and 7.5) suggest that designs based on polynomial-fitting methods generally are not as efficient as those based on a frequency-domain approach in the sense that the useful filter bandwidth is a smaller fraction of the theoretical upper limit $\omega = \pi/T = \omega_s/2$. We therefore turn our attention to frequency-domain methods, which have been proved to be useful for accurate broadband design.

Frequency-domain Design (Fourier-series Expansion). The frequency response of the z-domain transfer function $H(z)$ of (5.118) may be found by the substitution

$$z = e^{sT}\Big|_{s=j\omega} = e^{j\omega T}$$

from which we obtain

$$H(\omega) = \sum_{k=0}^{m} a_k e^{-jk\omega T} \tag{5.125}$$

where $H(\omega)$ is the s-domain transfer function obtained from $H(z)$. This is recognized as a Fourier exponential series expansion. As is evident in (5.71), $H(\omega)$ is periodic, i.e.,

$$H(\omega \pm jn\omega_s) = H(\omega) \tag{5.126}$$

That is, it is periodic along the $j\omega$ axis with *period* ω_s. Thus we see that (5.125) can be used to represent a desired $H(\omega)$ by a Fourier-series approximation that applies to the interval $|\omega| \leq \omega_s/2$. Care must be taken to avoid discontinuities in $H(s)$ and its first derivative which would result in a Gibb's-phenomenon oscillation characteristic of truncated Fourier series at points of discontinuity.

Nonrecursive filters inherently introduce delay, which may not be part of the desired transfer-function specification. If this delay is acceptable, one can obtain an approximation for a delayed version of $H(\omega)$ by the following method:

1. Split $H(\omega)$ into its real and imaginary components, i.e.,

$$H(\omega) = U(\omega) + jV(\omega) \tag{5.127}$$

We know for realizable $H(\omega)$ that $U(\omega)$ is even and $V(\omega)$ is odd, i.e.,

$$U(\omega) = U(-\omega) \tag{5.128a}$$

and

$$V(\omega) = -V(-\omega) \tag{5.128b}$$

2. Find Fourier-series expansions for U and V along the $j\omega$ axis. Using the above properties, we know the series will be of the form

$$U(\omega) = \alpha_0 + \sum_{n=1}^{\infty} \alpha_n \cos n\omega T \tag{5.129a}$$

and

$$V(\omega) = \sum_{n=1}^{\infty} \beta_n \sin n\omega T \tag{5.129b}$$

3. From (5.129), noting that

$$\cos x = \frac{e^{jx} + e^{-jx}}{2} \tag{5.130a}$$

and

$$\sin x = \frac{e^{jx} - e^{-jx}}{2j} \tag{5.130b}$$

we can write

$$jV(\omega) = \frac{1}{2} \sum_{n=1}^{\infty} \beta_n(e^{jn\omega T} - e^{-jn\omega T}) \tag{5.131a}$$

and

$$U(\omega) = \alpha_0 + \frac{1}{2} \sum_{n=1}^{\infty} \alpha_n(e^{jn\omega T} + e^{-jn\omega T}) \tag{5.131b}$$

4. Then by analytic continuation, with $z = e^{sT}$, we can write

$$H(z) = \alpha_0 + \frac{1}{2}\left[\sum_{n=1}^{\infty} (\alpha_n - \beta_n)z^{-n} + \sum_{n=1}^{\infty} (\alpha_n + \beta_n)z^n\right] \tag{5.132}$$

5. The series (5.132) is normally truncated symmetrically, for a total of $2J + 1$ terms. We then have

$$H(z) = \alpha_0 + \frac{1}{2}\left[\sum_{n=1}^{J} (\alpha_n - \beta_n)z^{-n} + \sum_{n=1}^{J} (\alpha_n + \beta_n)z^n\right] \tag{5.133}$$

6. To put (5.133) in the form of (5.118) we multiply each term by z^{-J} to obtain

$$\begin{aligned} H(z) &= z^{-J}H_z(z) \\ &= \alpha_0 z^{-J} + \frac{1}{2}\sum_{n=1}^{J} (\alpha_n - \beta_n)z^{-n-J} + \frac{1}{2}\sum_{n=1}^{J} (\alpha_n + \beta_n)z^{n-J} \\ &= \sum_{k=0}^{2J} a_k z^{-k} \end{aligned} \tag{5.134}$$

where $2J = m$, and

$$\begin{aligned} a_0 &= \tfrac{1}{2}(\alpha_J + \beta_J) \\ a_1 &= \tfrac{1}{2}(\alpha_{J-1} + \beta_{J-1}) \\ &\cdots\cdots\cdots\cdots \\ a_{J-1} &= \tfrac{1}{2}(\alpha_1 + \beta_1) \\ a_J &= \alpha_0 \\ a_{J+1} &= \tfrac{1}{2}(\alpha_1 - \beta_1) \\ &\cdots\cdots\cdots\cdots \\ a_{m-1} &= a_{2J-1} = \tfrac{1}{2}(\alpha_{J-1} - \beta_{J-1}) \\ a_m &= a_{2J} = \tfrac{1}{2}(\alpha_J - \beta_J) \end{aligned} \tag{5.135}$$

Associated with the above, of course, is the time-domain impulse response

$$h(n) = \sum_{k=0}^{m} a_k \delta(n - k) \tag{5.136}$$

An example of a differentiator $H(s) = s$ design based on this Fourier-series method follows in Sec. 5.10 (Fig. 5.13 and the accompanying text). The frequency response exhibits the Gibb's oscillation that results from the discontinuity in $|H(\omega)|$ at $\omega = \omega_s/2$.

Modified Fourier-series Method. In order to reduce the Gibb's oscillation, one can modify the transfer function in the neighborhood of discontinuities (Ref. 1, sec. 7.4.3, and Ref. 20). This is conveniently done by multiplying the impulse response $h(n)$ derived by the Fourier-series method (above) by a properly chosen weighting function $w(n) = w(t = nT)$. This multiplication in the time domain corresponds to convolving the original $H(\omega)$ with $W(\omega) = \mathfrak{F}\{w(t)\}$. We choose $w(t)$ to be an even function satisfying (see Fig. 5.11)

$$\begin{aligned} w(t) &= w(-t) \\ w(t) &= 0 \qquad |t| > JT \end{aligned} \tag{5.137}$$

and so that $W(j\omega)$ passes only frequencies in a narrow central lobe, i.e.,

$$W(\omega) \approx 0 \qquad |\omega| > \Delta\omega \ll \frac{\omega_s}{2} \tag{5.138}$$

We thus obtain a modified transfer function without discontinuities (Fig. 5.11*d*). The procedure is outlined below:

1. Perform the analysis for the Fourier-series method indicated in steps 1 to 5 above to obtain the transfer function of (5.133).
2. Select a weighting function satisfying the properties of (5.137) and (5.138); see below for typical choices.
3. Modify (5.133) to obtain

$$\begin{aligned} H_z(z) = \alpha_0 w(0) &+ \frac{1}{2} \sum_{n=1}^{J} (\alpha_n - \beta_n) w(n) z^{-n} \\ &+ \frac{1}{2} \sum_{n=1}^{J} (\alpha_n + \beta_n) w(n) z^{n} \end{aligned} \tag{5.139}$$

4. Change (5.139) into the form of (5.118) by multiplying each term by z^{-J} to obtain

$$H(z) = \sum_{k=0}^{2J} a_k z^{-k}$$

where $2J = m$, and

$$\begin{aligned}
a_0 &= \tfrac{1}{2}w(J)(\alpha_J + \beta_J) \\
a_1 &= \tfrac{1}{2}w(J-1)(\alpha_{J-1} + \beta_{J-1}) \\
&\cdots\cdots\cdots\cdots\cdots \\
a_{J-1} &= \tfrac{1}{2}w(1)(\alpha_1 + \beta_1) \\
a_J &= \alpha_0 \\
a_{J+1} &= \tfrac{1}{2}w(1)(\alpha_1 - \beta_1) \\
&\cdots\cdots\cdots\cdots \\
a_{m-1} &= a_{2J-1} = \tfrac{1}{2}w(J-1)(\alpha_{J-1} - \beta_{J-1}) \\
a_m &= \tfrac{1}{2}w(J)(\alpha_J - \beta_J)
\end{aligned} \tag{5.140}$$

Choice of Weighting Function. A large variety of weighting functions have been studied for controlling the ripple in the frequency response of finite-length weighted sequences.[1] Of these, only two will be discussed here (the reader is referred to the work of Taylo[20] for a discussion of several others).

A relatively simple, but effective, weighting function is Hamming's window function defined by

$$W_H(n) = w(nT) = \begin{cases} 0.54 + 0.46 \cos\left(\pi \dfrac{n}{J}\right) & n \leq J \\ 0 & n > J \end{cases} \tag{5.141}$$

This weighting function has 99.96 percent of its energy in the frequency range below the first zero of $W(\omega)$ at

$$\omega = \frac{2\pi}{JT} = \Delta\omega \qquad \text{for the Hamming function} \tag{5.142}$$

A more general family of weighting functions is discussed by Kuo and Kaiser (Ref. 1, pp. 232ff.) given by

$$W_K(n) = \frac{I_0[R\sqrt{1 - (9n/J)^2}]}{I_0(R)} \tag{5.143}$$

where I_0 is the modified Bessel function of the first kind, zero order, and R is typically chosen in the range $4 < R < 9$ (here we use R to represent the parameter called $\omega_a\tau$ by Kaiser). The majority of the energy is concentrated below the first spectral zero at

$$\omega = \frac{\sqrt{R^2 + \pi^2}}{JT} = \Delta\omega \qquad \text{for the Kaiser function} \tag{5.144}$$

TABLE 5.7 PROPERTIES OF THE KAISER WEIGHTING FUNCTION [Eq. (5.143)]

R	*Overshoot, % amplitude*†	$\Delta\omega\, JT$‡
0	17.9	π
5.0	0.33	5.91
5.44	0.25	2π
6.0	0.14	6.78
6.5	0.83	7.23
7.0	0.050	7.68
7.5	0.029	8.14
8.0	0.017	8.60
8.5	0.010	9.08

† Based on Ref. 1, table 7.1.
‡ $\Delta\omega\, JT = \sqrt{R^2 + \pi^2}$.

Table 5.7 lists values of $\Delta\omega\, JT$ corresponding to a particular value of R. Large values of R yield a weighting function with a wider central lobe in the frequency domain, and greater reduction of the Gibb's oscillation. This table provides a means for estimating the values of R and J required to achieve a desired amount of smoothing of the Gibb's oscillation (based on Ref. 1, pp. 235–238).

Using the information in this table, one can estimate the number of terms required in the nonrecursive filter as follows:

1. Determine the allowable transition width (ω_1 to ω_2 in Fig. 5.11d) to be permitted in the neighborhood of the discontinuity. Note that when we convolve the spectral response of the weighting function with the desired $H(s)$ we obtain a transition region of width $2\Delta\omega$; hence

$$\omega_2 - \omega_1 = 2\Delta\omega \tag{5.145}$$

Thus we can estimate the required value of $\Delta\omega$.

2. Find J to achieve the necessary $\Delta\omega$. For example, using the Hamming weight, (5.141), we solve

$$\omega_2 - \omega_1 \approx 2\Delta\omega = \frac{4}{JT}$$

or

$$J = \frac{4\pi}{T(\omega_2 - \omega_1)} = \frac{2\pi}{T\Delta\omega} \tag{5.146}$$

Similarly, for the Kaiser weight, (5.143), select a value of R from Table

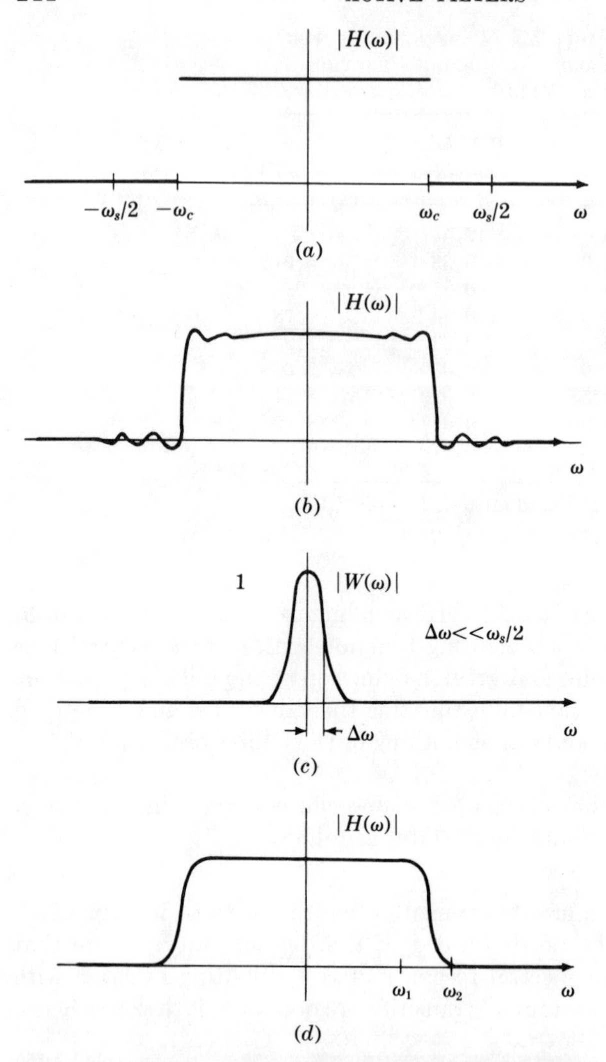

FIG. 5.11 Use of weighting function to reduce Gibb's-phenomenon oscillation. (*a*) Desired $|H(\omega)|$; (*b*) response resulting from direct Fourier-series design, Eq. (5.135); (*c*) typical weighting function $W(\omega)$; (*d*) response obtained by the use of the weighting function.

5.7 which yields the desired overshoot reduction; then solve

$$\omega_2 - \omega_1 = 2\Delta\omega = \frac{2\sqrt{R^2 + \pi^2}}{JT} \tag{5.147}$$

or

$$J = \frac{2\sqrt{R^2 + \pi^2}}{T(\omega_2 - \omega_1)} \tag{5.148}$$

Of course we must also choose T sufficiently high so that $\omega_s/2 = \pi/T$ exceeds the highest frequency component (in rad/sec) to be accommodated by the filter.

Some examples of this approach appear in Sec. 5.10.

Mean-square Design in the Frequency Domain. Fleischer[21] describes a procedure for optimal design of nonrecursive filters based upon minimizing a weighted mean-square error with respect to a desired frequency-response function. This is a generally powerful approach which, of course, requires considerable computer analysis to achieve the final design. Fleischer's method also permits including the effect of any output data-reconstruction filter that may accompany the digital operator, and is perhaps the most powerful general method for nonrecursive-filter design. The reader is referred to his original paper for further details.

5.9 Recursive-filter-design Methods

Figure 5.12 shows the viewpoint used in a general approach to designing a set of discrete-time algorithms for a given digital filter. In general we assume that we are attempting to match the behavior of a corresponding continuous-time system specified by a particular $H(s)$ or $h(t)$. There are no restrictions on the prototype continuous-time system except the usual one of physical realizability in the sense that $h(t)$ is a real-valued fixed linear causal impulse-response function. It is not necessary for $H(s)$ to have a positive pole-zero excess, but we will normally assume that it does, unless otherwise specified.

The digital filter is designed to provide the same behavior at the sample times as the system of Fig. 5.12. The principal variation in

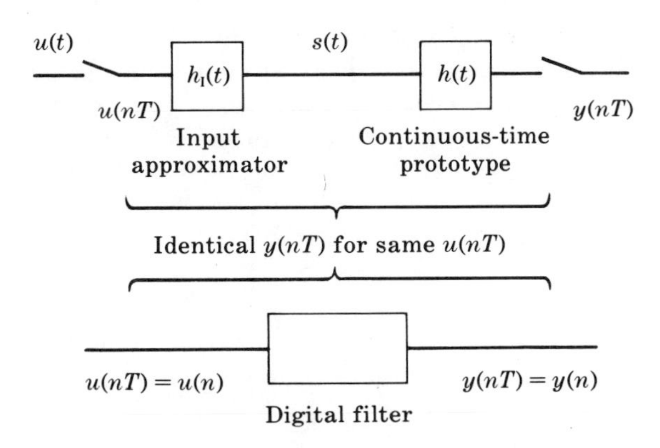

FIG. 5.12 The hold-invariant method.

design is the choice of an input approximation or hold operator h_I. Any one of the operators discussed in Secs. 5.4 and 5.5 may be used; others may also be appropriate in special situations. The present discussion will emphasize the considerations associated with making a selection from one of the hold operators; for discussion purposes, each has been given a letter designation A, B, etc.

The Impulse Approximation (Method O). A common input approximator is the simple Dirac impulse operator (Fig. 5.8a)

$$h_{IO}(t) = T\delta(t) \qquad \text{method } O \tag{5.149}$$

This operator is implicit in the "impulse-invariant" methods used by Kuo and Kaiser[1] and Rader and Gold.[2] Here the input samples are assumed to have a continuous-time equivalent function that is merely a train of Dirac impulse of value $Tu(nT)$; the continuous-time signal $s(t)$ provided to $h(t)$ is thus

$$s_O(t) = T\Sigma u(n)\delta(t - nT) \qquad \text{method } O\dagger \tag{5.150}$$

This seemingly crude approximation works surprisingly well in some cases. As we will later see, however, it has some limitations.

The Zero-order Hold (Method A). The next logical choice of an input approximator is the zero-order hold operator of Fig. 5.8b; which in this context leads to

$$h_{IA}(t) = \begin{cases} 1 & 0 < t \leq T \\ 0 & \text{elsewhere} \end{cases} \tag{5.151}$$

and correspondingly

$$s_A(t) = u(n) \qquad nT < t \leq (n+1)T \tag{5.152}$$

This approximator works quite well in most instances; its principal limitation is that it introduces a half-sample period delay, $T/2$.

First-order Extrapolative Hold (Method B). The first-order hold of Fig. 5.8c reduces the delay at frequencies below about $\omega_s/4$, when compared to method A, and also leads to a real-time realizable digital filter ($a_0 = 0$). It has impulse response

$$h_{IB}(t) = \begin{cases} 1 + \dfrac{t}{T} & 0 < t \leq T \\ 1 - \dfrac{t}{T} & T < t \leq 2T \\ 0 & \text{elsewhere} \end{cases} \tag{5.153}$$

† Note that the impulse-invariant method is implicit in the Z-transform definition used by Wilts[11] and Kuo and Kaiser (Ref. 1, sec. 7.2). We chose to use the definition of (5.13), as explained previously.

and correspondingly

$$s_B(t) = u(n) + \frac{t - nT}{T}[u(n) - u(n-1)] \qquad nT < t \leq (n+1)T \tag{5.154}$$

Polygonal Holds (Methods *C* and *D*). The polygonal hold of Fig. 5.8*d* may be used as an input approximator in either the physically realizable form (5.80) or the nonrealizable form (5.81). The realizable form produces a digital filter with a full sample period delay [a_0 in (5.111) will be zero]; however, the phase response will be otherwise quite good. If the nonrealizable form is used, the resulting filter will include the a_0 term in the numerator.

We will call the method that utilizes the physically realizable form method C, and the method that uses the undelayed "nonrealizable" form method D.

The realizable form has impulse response

$$h_{IC}(t) = \begin{cases} \dfrac{t}{T} & 0 < t \leq T \\ 2 - \dfrac{t}{T} & T < t \leq 2T \\ 0 & \text{elsewhere} \end{cases} \tag{5.155}$$

and correspondingly

$$s_C(t) = u(n-1) + \frac{t - nT}{T}[u(n) - u(n-1)] \qquad nT < t \leq (n+1)T \tag{5.156}$$

The nonrealizable form has impulse response

$$h_{ID}(t) = \begin{cases} 1 + \dfrac{t}{T} & -T < t \leq 0 \\ 1 - \dfrac{t}{T} & 0 < t \leq T \\ 0 & \text{elsewhere} \end{cases} \tag{5.157}$$

$$s_D(t) = u(n) + \frac{t - nT}{T}[u(n+1) - u(n)] \qquad nT < t \leq (n+1)T \tag{5.158}$$

Time-domain Design (State-space Methods). We will here treat the scalar-input, scalar-output case; this may easily be extended to the multiple-input, multiple-output case (see Ref. 22). The methods presented are convenient for designing digital filters with a digital computer, but the Z-transform methods presented later are generally more convenient for hand calculation.

We assume that the filter is to be designed to approximate a continuous-time prototype specified by (5.20), repeated here

$$\dot{\mathbf{x}}(t) = \mathbf{A}\mathbf{x}(t) + \mathbf{B}u(t)$$
$$y(t) = \mathbf{C}\mathbf{x}(t)$$

where $y(t)$ is a scalar-output function
$u(t)$ is a scalar-input function
$\mathbf{x}(t)$ is a $k \times 1$ state vector function
$\mathbf{A}$ is a $k \times k$ constant matrix
$\mathbf{B}$ is a $k \times 1$ constant matrix
$\mathbf{C}$ is a $1 \times k$ constant matrix

[For the time being, we assume the constant D in (5.20*b*) to be zero; we will later discuss means for including this term.] Thus we again assume the structure shown earlier in Fig. 5.3*a*; we know (Ref. 5, p. 125)

$$\mathbf{x}(t) = e^{\mathbf{A}t}\mathbf{x}(0_+) + \int_{0^+}^{t} e^{\mathbf{A}(t-\lambda)}\mathbf{B}u(\lambda)\, d\lambda \tag{5.159}$$

Therefore, at discrete times $t = nT$:

$$\mathbf{x}(n+1) = \mathbf{x}(nT+T) = e^{\mathbf{A}(n+1)T}\mathbf{x}(0_+) + \int_{0}^{(n+1)T} e^{\mathbf{A}[(n+1)T-\lambda]}\mathbf{B}u(\lambda)\, d\lambda \tag{5.160}$$

or

$$\mathbf{x}(n+1) = e^{\mathbf{A}T}\left[e^{\mathbf{A}nT}\mathbf{x}(0_+) + \int_{0}^{nT} e^{\mathbf{A}(nT-\lambda)}\mathbf{B}u(\lambda)\, d\lambda\right] + \int_{nT}^{(n+1)T} e^{\mathbf{A}[(n+1)T-\lambda]}\mathbf{B}u(\lambda)\, d\lambda \tag{5.161}$$

Since the term in the brackets equals $\mathbf{x}(nT) = \mathbf{x}(n)$, we have

$$\mathbf{x}(n+1) = e^{\mathbf{A}T}\mathbf{x}(n) + \int_{nT}^{(n+1)T} e^{\mathbf{A}[(n+1)T-\lambda]}\mathbf{B}u(\lambda)\, d\lambda \tag{5.162}$$

Equation (5.162) *is exact,* provided *we could evaluate the integral exactly;* however, here we must estimate its value, based on a knowledge of u at discrete sample times. It is often convenient to rewrite the last term of (5.162) using the variable substitution

$$\lambda = (n+1)T - v \qquad d\lambda = -dv \qquad v = (n+1)T - \lambda \tag{5.163}$$

thus obtaining

$$\mathbf{x}_{n+1} = e^{\mathbf{A}T}\mathbf{x}(n) + \int_{0}^{T} e^{\mathbf{A}v}\mathbf{B}u[(n+1)T - v]\, dv$$
$$= e^{\mathbf{A}T}\mathbf{x}(n) + \mathbf{J} \tag{5.164}$$

where
$$\mathbf{J} = \int_{0}^{T} e^{\mathbf{A}v}\mathbf{B}u[(n+1)T - v]\, dv \tag{5.165}$$

The various hold approximations yield different expressions for $\mathbf{J}$, which then determines the form of (5.165).

Method O (The Impulse-invariant Method). If we assume that between sample point s the input is approximated by

$$u(t) \approx s_O(t) = Tu(n)\delta(t - nT) \qquad nT < t \leq (n+1)T \quad (5.166)$$

here we intentionally choose to include the impulse on the right end of the interval; this choice leads us to an algorithm that is equivalent to the "impulse-invariant" method[2]; upon evaluating the right term of (5.164) using (5.166) we obtain

$$\begin{aligned} \mathbf{J} &= \int_0^T e^{\mathbf{A}v}\mathbf{B}Tu(n+1)\delta(v)\,dv \\ &= T\mathbf{B}u(n+1) \end{aligned} \quad (5.167)$$

and thus the discrete-time algorithms based on the impulse approximation become

$$\mathbf{x}(n+1) = e^{\mathbf{A}T}\mathbf{x}(n) + T\mathbf{B}u(n+1) \quad (5.168a)$$

$$y(n) = \mathbf{C}\mathbf{x}(n) \quad (5.168b)$$

This last equation is the time-domain equivalent form of the "impulse-invariant" method.[1,2] For sufficiently small T it yields a good approximation to the continuous-time prototype; however if T is too large it yields a digital filter with a noticeable overall gain error. The resulting digital filter will be real-time realizable ($a_0 = 0$) only if the pole-zero excess of the prototype is two or greater; it should be used with caution for approximating an $H(s)$ with a pole-zero excess of less than two (see, e.g., Ref. 6, sec. 4.7, and further discussion below). Note also from (5.168a) that $u(n+1)$ must be available to calculate the state-variable vector $\mathbf{x}(n+1)$.

Method A (Zero-order Hold). The preceding method is often called impulse-invariant; we now direct our attention to a more general class of methods which Mantey and Franklin[24] have called the hold circuit equivalent technique; referring again to Fig. 5.12, we postulate more effective hold circuits h_I. Consider next the zero-order hold described by (5.151); using this we approximate $u(t)$ by

$$u(t) \approx u(n) \qquad nT < t \leq (n+1)T \quad (5.169)$$

and thus using the right term of (5.164) we obtain

$$\mathbf{J} = \int_0^T e^{\mathbf{A}v}\mathbf{B}u(n)\,dv = \mathbf{A}^{-1}(e^{\mathbf{A}T} - \mathbf{I})\mathbf{B}u(n) = \mathbf{L}_1\mathbf{B}u(n) \quad (5.170)$$

where

$$\mathbf{L}_1 \triangleq \int_0^T e^{\mathbf{A}v}\,dv = \mathbf{A}^{-1}(e^{\mathbf{A}T} - \mathbf{I}) \quad (5.171)$$

This then yields the algorithms of method A

$$\mathbf{x}(n+1) = e^{\mathbf{A}T}\mathbf{x}(n) + \mathbf{L}_1\mathbf{B}u(n) \quad (5.172a)$$

$$y(n) = \mathbf{C}\mathbf{x}(n) \quad (5.172b)$$

Method B **(Extrapolative First-order Hold).** As we shall see later, the discrete-time algorithms provided by method A above contain an inherent delay of $T/2$. If we assume the extrapolative first-order hold described by (5.153), we thus represent $u(t)$ in the interval $nT < t \leq (n+1)T$ by linear extrapolation, viz.,

$$u \approx u(n) + \frac{u(n) - u(n-1)}{T}(t - nT) \tag{5.173}$$

Substituting (5.173) in (5.164) we obtain

$$\begin{aligned}\mathbf{J} &= \int_0^T e^{\mathbf{A}v}\mathbf{B}\left\{u(n) + \left[\frac{u(n) - u(n-1)}{T}\right](T - v)\right\} dv \\ &= \int_0^T e^{\mathbf{A}v}\mathbf{B}\left\{[2u(n) - u(n-1)] + [u(n-1) - u(n)]\frac{v}{T}\right\} dv \\ &= (2\mathbf{L}_1 - \mathbf{L}_2)\mathbf{B}u(n) + (\mathbf{L}_2 - \mathbf{L}_1)u(n-1)\end{aligned} \tag{5.174}$$

where

$$\mathbf{L}_1 \triangleq \int_0^T e^{\mathbf{A}v}\, dv = \mathbf{A}^{-1}(e^{\mathbf{A}T} - \mathbf{I}) \tag{5.175a}$$

$$\mathbf{L}_2 \triangleq \frac{1}{T}\int_0^T e^{\mathbf{A}v}v\, dv = \frac{(\mathbf{A}^{-1})^2}{T}[\mathbf{I} - (\mathbf{I} - \mathbf{A}T)e^{\mathbf{A}T}] \tag{5.175b}$$

Methods C **and** D **(Interpolative First-order Holds).** Method B is based upon linear extrapolation of the input; if instead we use the interpolative hold described by (5.155) we obtain a better approximation for $u(t)$, at the expense of additional delay. We will call this method C, wherein we represent $u(t)$ in the interval $nT < t \leq (n+1)T$ by

$$u \approx u(n-1) + \frac{u(n) - u(n-1)}{T}(t - nT) \tag{5.176}$$

Substituting (5.176) into (5.164) we obtain the expression for method C:

$$\begin{aligned}\mathbf{J} &= \int_0^T e^{\mathbf{A}v}\mathbf{B}\left\{u(n-1) + \left[\frac{u(n) - u(n-1)}{T}\right](T - v)\right\} dv \\ &= \int_0^T e^{\mathbf{A}v}\mathbf{B}\{u(n) + [u(n-1) - u(n)]v\}\, dv \\ &= (\mathbf{L}_1 - \mathbf{L}_2)\mathbf{B}u(n) + \mathbf{L}_2\mathbf{B}u(n-1)\end{aligned} \tag{5.177}$$

where $\mathbf{L}_1$ and $\mathbf{L}_2$ are as defined by (5.175).

The form of $\mathbf{J}$ in (5.177) is realizable in the sense that we estimate $\mathbf{x}(n+1)$ from $u(n)$ and $u(n-1)$; we find, however, that the resulting algorithm obtained by substituting (5.177) in (5.164) yields an output response with a full sample period delay. We can eliminate this delay if we use instead $u(n+1)$ and $u(n)$, respectively. We obtain the expression for method D:

$$\mathbf{J} = (\mathbf{L}_1 - \mathbf{L}_2)\mathbf{B}u(n+1) + \mathbf{L}_2\mathbf{B}u(n) \tag{5.178}$$

Substituting the above in (5.164) we now obtain an algorithm that is not real-time realizable without some digital processing delay; it is still nevertheless useful in non-real-time data-processing and simulation applications. Also, when the total digital processing time is small compared to T, then we can use method D to obtain an algorithm that has relatively little overall delay.

Other (Second-order) Methods. Liou,[1] Giese,[26] Cruickshank,[27] and Truxal[12] discuss methods for using higher-order approximations for the input; this writer has had little experience with the use of these approximations; based on limited studies, it appears that they will normally yield better phase response for large sample time, but often will not provide significant improvement in amplitude response. They may prove useful in some applications (see also the results of Baxter).[23]

Summary of Algorithms. All the above methods yield algorithms that are described in general by:

$$\mathbf{x}(n+1) = e^{\mathbf{A}T}\mathbf{x}(n) + \boldsymbol{\beta}_{-1}u(n-1) + \boldsymbol{\beta}_0 u(n) + \boldsymbol{\beta}_1 u(n+1) \quad (5.179a)$$

$$y(n) = \mathbf{C}\mathbf{x}(n) \quad (5.179b)$$

Table 5.8 summarizes the expression for the terms obtained from each

TABLE 5.8 EXPRESSIONS FOR TERMS IN EQ. (5.188a)

$$\mathbf{x}(n+1) = \tilde{\mathbf{A}}\mathbf{x}(n) + \boldsymbol{\beta}_{-1}u(n-1) + \boldsymbol{\beta}_0 u(n) + \boldsymbol{\beta}_1 u(n+1)$$

Method	$\boldsymbol{\beta}_{-1}$	$\boldsymbol{\beta}_0$	$\boldsymbol{\beta}_1$
O: Impulse invariant	0	0	$T\mathbf{B}$
A: Zero-order hold	0	$\mathbf{L}_1\mathbf{B}$	0
B: Linear extrapolative	$(\mathbf{L}_2 - \mathbf{L}_1)B$	$(2\mathbf{L}_1 - \mathbf{L}_2)\mathbf{B}$	0
C: Delayed linear interpolative	$\mathbf{L}_2\mathbf{B}$	$(\mathbf{L}_1 - \mathbf{L}_2)\mathbf{B}$	0
D: Undelayed linear interpolative	0	$\mathbf{L}_2\mathbf{B}$	$(\mathbf{L}_1 - \mathbf{L}_2)\mathbf{B}$

Where $\tilde{\mathbf{A}} = e^{\mathbf{A}T}$

$$\mathbf{L}_1 = \int_0^T e^{\mathbf{A}v}\,dv = \mathbf{A}^{-1}(e^{\mathbf{A}T} - \mathbf{I})$$

$$\mathbf{L}_2 = \frac{1}{T}\int_0^T e^{\mathbf{A}v}v\,dv = \frac{(\mathbf{A}^{-1})^2}{T}[\mathbf{I} - (\mathbf{I} - \mathbf{A}T)e^{\mathbf{A}T}]$$

of the methods. Appendix C discusses some of the computational aspects of evaluating the expressions (see also Ref. 37).

The Direct Input-Output Term D. In the general fixed linear system formulation of (5.20*b*), we include a constant term D relating the input and output, repeating here:

$$y(t) = \mathbf{C}\mathbf{x}(t) + Du(t)$$

Whenever D is nonzero this implies that the pole-zero excess (PZE) of the transfer function which is called $H(s)$ is zero (order of the denominator equals order of the numerator), and also that the system step response has a nonzero initial value. This would not be possible in a real-world system; however, the mathematical model of a system may neglect certain high-frequency effects, and may therefore have a nonzero value for D. Care must be taken in the design of the discretized filter resulting from such a system model. Where possible, if one is able to use $u(n)$ to calculate $y(n)$, then one can modify (5.179*b*) to

$$y(n) = \mathbf{C}\mathbf{x}(n) + Du(n) \tag{5.180}$$

If only $u(n - 1)$ and earlier values of u are to be used, then some form of prediction may be used; e.g., we may use

$$y(n) = \mathbf{C}\mathbf{x}(n) + D[2u(n - 1) - u(n - 2)] \tag{5.181}$$

which amounts to treating D via method B (linear extrapolation).

Equivalent Z-transform Methods. The state-space method presented above is convenient for computer-aided design of digital-filter algorithms; for hand calculation, equivalent Z-transform methods are easier to manipulate. Referring to Fig. 5.12, we can find the z-domain transfer function $H(z)$ as follows:

$$H(z) = \mathcal{Z}\{\mathcal{L}^{-1}[H_I(s)H(s)]\} \tag{5.182}$$

For each method, we use the appropriate input approximation transfer function.

Method O: Impulse Invariant. Here we use $H_I(s) = T$ to obtain

$$H_O(z) = \mathcal{Z}\{\mathcal{L}^{-1}[TH(s)]\} = T\mathcal{Z}[h(t)] \tag{5.183}$$

As mentioned earlier, this is equivalent to the impulse-invariant method discussed by Rader and Gold.[2]

Method A: Zero-order Hold. Here we use

$$H_I(s) = \frac{1 - e^{-sT}}{s} \tag{5.82b}$$

to obtain

$$\begin{aligned} H_A(z) &= \mathcal{Z}\left\{\mathfrak{L}^{-1}\left[\frac{1 - e^{-sT}}{s} H(s)\right]\right\} \\ &= (1 - z^{-1})\, \mathcal{Z}\left\{\mathfrak{L}^{-1}\left[\frac{H(s)}{s}\right]\right\} \\ &= \frac{z - 1}{z}\, \mathcal{Z}\left\{\mathfrak{L}^{-1}\left[\frac{H(s)}{s}\right]\right\} \end{aligned} \tag{5.184}$$

Method B: Linear Extrapolation. Here we use the transfer function of the conventional first-order hold

$$H_I(s) = (1 - e^{-sT})^2 \frac{1 + Ts}{Ts^2} \tag{5.82c}$$

to obtain

$$\begin{aligned} H_B(z) &= \mathcal{Z}\left\{\mathfrak{L}^{-1}\left[\frac{(1 - e^{-sT})^2(1 + Ts)}{Ts^2} H(s)\right]\right\} \\ &= \frac{(1 - z^{-1})^2}{T}\, \mathcal{Z}\left\{\mathfrak{L}^{-1}\left[\frac{H(s)}{s^2}(1 + Ts)\right]\right\} \\ &= \frac{(z - 1)^2}{Tz^2}\, \mathcal{Z}\left\{\mathfrak{L}^{-1}\left[\frac{H(s)(1 + Ts)}{s^2}\right]\right\} \end{aligned} \tag{5.185}$$

Method C: Delayed Linear Interpolation. At the expense of adding a full sample period of delay to the resulting discrete-time approximation to $h(t)$ we may use the more accurate realizable polygonal hold with

$$H_I(s) = \frac{(1 - e^{-sT})^2}{Ts^2} \tag{5.82d}$$

to obtain in a similar fashion

$$\begin{aligned} H_C(z) &= \frac{(1 - z^{-1})^2}{T}\, \mathcal{Z}\left\{\mathfrak{L}^{-1}\left[\frac{H(s)}{s^2}\right]\right\} \\ &= \frac{(z - 1)^2}{Tz^2}\, \mathcal{Z}\left\{\mathfrak{L}^{-1}\left[\frac{H(s)}{s^2}\right]\right\} \end{aligned} \tag{5.186}$$

Method D: Undelayed Linear Interpolation. If $u(n)$ is available at the time we are calculating $y(n)$, we may take advantage of this to implement an undelayed version of the above expression, which yields the transfer function of method D

$$H_D(z) = zH_C(z) = \frac{(z - 1)^2}{zT}\, \mathcal{Z}\left\{\mathfrak{L}^{-1}\left[\frac{H(s)}{s^2}\right]\right\} \tag{5.187}$$

Equivalence of State-variable and Z-transform Methods. It is possible to derive the numerator and denominator polynomial coefficients of $H(z)$ from the state-space algorithms of the form (5.179). The procedure is only summarized here, and is discussed in more detail in Appendix C. Since all of the algorithms discussed above are of the general form

$$\mathbf{x}(n+1) = \tilde{\mathbf{A}}\mathbf{x}(n) + \boldsymbol{\beta}_{-1}u(n-1) + \boldsymbol{\beta}_0 u(n) + \boldsymbol{\beta}_1 u(n+1)$$

and
$$Y(n) = \mathbf{C}\mathbf{x}(n)$$

where
$$\tilde{\mathbf{A}} = e^{\mathbf{A}T}$$

we can take the Z transform to obtain

$$z\mathbf{X}(z) - z\mathbf{x}(0_+) = \tilde{\mathbf{A}}\mathbf{X}(z) + (z^{-1}\boldsymbol{\beta}_{-1} + \boldsymbol{\beta}_0 + z\boldsymbol{\beta}_1)U(z) \qquad (5.188a)$$

and
$$Y(z) = \mathbf{C}\mathbf{X}(z) \qquad (5.188b)$$

We can manipulate (5.188a):

$$z\mathbf{X}(z) - \tilde{\mathbf{A}}\mathbf{X}(z) = z\mathbf{x}(0_+) + (z^{-1}\boldsymbol{\beta}_{-1} + \boldsymbol{\beta}_0 + z\boldsymbol{\beta}_1)U(z)$$

or

$$\mathbf{X}(z) = (z\mathbf{I} - \tilde{\mathbf{A}})^{-1}[z\mathbf{x}(0_+) + (z^{-1}\boldsymbol{\beta}_{-1} + \boldsymbol{\beta}_0 + z\boldsymbol{\beta}_1)U(z)] \qquad (5.189)$$

We can combine (5.189) and (5.188b) to obtain

$$Y(z) = z\mathbf{C}(z\mathbf{I} - \tilde{\mathbf{A}})^{-1}\mathbf{x}(0_+) + \mathbf{C}(z\mathbf{I} - \tilde{\mathbf{A}})^{-1}(z^{-1}\boldsymbol{\beta}_{-1} + \boldsymbol{\beta}_0 + z\boldsymbol{\beta}_1)U(z) \qquad (5.190)$$

If $\mathbf{x}(0_+) = 0$ then we obtain the transfer function

$$H(z) = \mathbf{C}(z\mathbf{I} - \tilde{\mathbf{A}})^{-1}(z^{-1}\boldsymbol{\beta}_{-1} + \boldsymbol{\beta}_0 + z\boldsymbol{\beta}_1) \qquad (5.191)$$

Note that

$$(z\mathbf{I} - \mathbf{A})^{-1} = z^{-1}\,\{Z \text{ transform of } (\tilde{\mathbf{A}})^n\}$$

and also that

$$\tilde{\mathbf{A}} = e^{\mathbf{A}T} \neq \mathbf{A}$$
$$= \sum_0^\infty \frac{(\mathbf{A}T)^n}{n!}$$
$$= \mathbf{I} + \mathbf{A}T + \cdots + \frac{(\mathbf{A}T)^n}{n!} + \cdots \qquad (5.192)$$

Equation (5.191) implies a rational function of z with a denominator polynomial identical to the denominator of $(z\mathbf{I} - \tilde{\mathbf{A}})^{-1}$. Appendix C discusses the computational aspects of finding $H(z)$ in rational-function form (5.111) from (5.191).

General Comments on the Methods. As pointed out by Kuo (Ref. 6, sec. 4.7), the direct Z-transform method can lead to misleading results if the pole-zero excess of the original continuous-time prototype $H(s)$ is less than two. Consider the transfer function

$$H(s) = \frac{a}{s+a} \qquad (5.193)$$

a simple unity-gain, first-order, low-pass filter. This $H(s)$ has a pole zero excess of only one. Using method O we obtain

$$H_O(z) = T\mathcal{Z}\{H(s)\} = \frac{aTz}{z - e^{-aT}} \tag{5.194}$$

We see that (5.194) is not real-time realizable; for example, we find the unit step response of the system has initial value aT, when in fact we know the continuous-time prototype has an initial response of zero. This is a considerable error, if aT is large.

Let us also look at the value of

$$H_O(e^{j\omega T})$$

when $\omega = 0$; we find

$$H_O(e^{j\omega T})\Big|_{\omega = 0} = H_O(1) = \frac{aT}{1 - e^{-aT}} \neq 1 \tag{5.195}$$

Thus the dc or low-frequency gain of $H_O(z)$ is in error.

If the prototype $H(s)$ has a pole-zero excess of at least two, or equivalently, in (5.20), if both $D = 0$ and $\mathbf{CB} = 0$, then the $H_O(z)$ obtained from method O will be real-time realizable; however, there will generally be an error in the overall filter gain at low frequencies. This error may not be serious, except in cases where the magnitude of important poles is a large fraction of $\omega_s/2$. We note again that in using method O for digital simulations $u(n + 1)$ must be available to calculate $\mathbf{x}(n + 1)$.

Methods A, B, and C lead to real-time realizable transfer functions if the pole-zero excess of $H(s)$ is at least one. Method D will always lead to a digital filter that is not real-time realizable. As we shall see in some forthcoming examples (Sec. 5.10), method O is generally quite accurate if the pole-zero excess of $H(s)$ is two or more, and is often the best choice when overall delay must be kept low, e.g., in a closed-loop digital simulation. Method B is generally to be preferred only for designing real-time realizable approximations whenever the prototype $H(s)$ has a pole-zero excess of less than two.

Method A usually provides the best amplitude accuracy, but does exhibit a delay of $T/2$. Method C would be used if a delay of T is more acceptable than a delay of only $T/2$. Method D, which is really only method C advanced one sample time, generally gives good overall accuracy. Of course, whenever method D is used to process real-time data, the output will be delayed by the time to calculate the effect of the current input value $u(n)$ on the current output $y(n)$.

Note also that method O leads to an $H(z)$ with the smallest number of numerator coefficients; method A has one more, and methods B, C, and D have two more than method O. This can offset the allowable value of T in a real-time processing situation.

The Folding Problem. A digital filter based upon a rational s-domain prototype $H(s)$ will have a frequency response that is periodic in ω with period $2\pi/T$. Thus as ω approaches $\omega_s/2$, the frequency response will differ from $H(s)$ due to the effect of terms of the form $H(s + jn\omega_s)$, $n \neq 0$; this is often called folding error. Unless $H(s)$ has a high pole-zero excess, this folding error can be large in a broadband filter (i.e., one designed to operate at frequencies approaching $\omega_s/2$). One way to eliminate the folding problem is to modify the prototype $H(s)$ to include a broadband, low-pass guard filter with flat amplitude and linear phase; this measure will, of course, raise the order of the resulting digital filter implementation.

Use of the Bilinear Z Transform (Ref. 1, sec. 7.5.2). The bilinear Z transformation

$$s = \frac{2(1 - z^{-1})}{T(1 + z^{-1})} \tag{5.196}$$

causes the entire left-hand side of the s plane to be mapped into the interior of the unit circle in the z plane, and folding errors are eliminated. The bilinear Z transformation is useful for designing digital low-pass, bandpass, and band-stop filters that have a piecewise constant magnitude function (passband and stop band). The effect of this transform may be more easily seen by considering it as two successive transforms.

Consider the transformation

$$s = \frac{2}{T} \tanh\left(\frac{\hat{s}T}{2}\right) \qquad \hat{s} = \hat{\sigma} + j\hat{\omega} \tag{5.197}$$

which maps the entire s plane into the horizontal strip in the $\hat{s}$ plane

$$-\frac{\omega_s}{2} \leq \text{Im}\,(\hat{s}) \leq \frac{\omega_s}{2} \tag{5.198}$$

This is thus a band-limiting transformation.

A second transform

$$z = e^{\hat{s}T} \tag{5.199}$$

maps the left half of the $\hat{s}$ plane into the unit circle in the z plane, and thus maps the left half of the s plane into the unit circle as well.

Note that the bilinear transform (5.196) is obtained by combining (5.197) and (5.199). Thus, to use the method, one finds the z-domain transfer function $H_{BI}(z)$ from

$$H_{BI}(z) = H(s)\Big|_{s=\frac{2(1-z^{-1})}{T(1+z^{-1})}} \tag{5.200}$$

Note from (5.197) that there is a frequency warping that accompanies this transformation, (5.200).

It is necessary to compensate for the warping by using the relationship

$$\omega_c = \frac{2}{T}\tan\frac{\omega_d T}{2} \tag{5.201}$$

where ω_c = cutoff or transition frequency of the s-domain prototype
ω_d = desired cutoff or transition frequency in final digital filter

Golden and Kaiser[38] also present additional low-pass to bandpass and band-stop transformations that may be used with (5.200) to facilitate the design of filters based on standard low-pass designs such as Butterworth, Chebyshev, etc.

Note that although the above technique is straightforward, and permits the design of broadband, high-order filters, it matches only the magnitude response of the digital filter and the continuous-time prototype. The bilinear transformation thus does not guarantee any preservation of phase response, and thus may not be suitable for applications where phase-shift and delay requirements must also be met, e.g., in a digital simulation.

5.10 Some Example Designs

The following examples are not intended to be comprehensive, but merely to illustrate some features of the design methods stressed previously. The reader is referred to the referenced articles for more details of special design techniques. In particular, the works of Rader and Gold[2] and Golden and Kaiser[38] provide more depth of coverage of the special problems associated with the design of high-order low-pass, bandpass, and band-stop filters. Our interest here has been primarily directed toward general design methods that may be incorporated into computer-aided analysis and simulation programs, although other applications have been kept in mind.

A 19-tap Nonrecursive Differentiator. The Fourier-series methods of Sec. 5.8 may be used to design a broadband differentiator; the desired transfer function is

$$H(s) = s \tag{5.202}$$

The associated upper Nyquist frequency limit is

$$\frac{\omega_s}{2} = \frac{\pi}{T}\text{ rad/sec}$$

hence the desired frequency-response curve is shown in Fig. 5.13a. We therefore want to find first a Fourier-series approximation to $H(s)$ over the range of radian frequencies $0 < \omega \leq \pi/T$. We note that

$$H(\omega) = U(\omega) + jV(\omega) \tag{5.203}$$

is odd and thus we choose

$$U(\omega) = 0 \qquad V(\omega) = \omega$$

Thus we have merely to find the Fourier series for the real, odd function $V = \omega$ over the interval $-1 \leq \omega \leq 1$. The resulting coefficients are given by

$$a_n = \frac{\beta_n}{2} = \frac{(-1)^{n+1}}{nT} \qquad n \neq 0 \tag{5.204}$$

Let us arbitrarily choose to use 19 taps [$J = 9$, (5.134)]. Table 5.9 lists the values of a_nT for $1 \leq n \leq 9$, hence we can design a nonrecursive filter of form (5.118) by dividing the table entries by T.

We note from Fig. 5.13*a*, however, that simple truncation of the Fourier series produces a filter with severe Gibb's-phenomenon oscillation of the frequency response. Table 5.9 lists coefficients for modified Fourier-series designs using both the Hamming weight [(5.141)] and the Kaiser weight [(5.143), $R = 6$]. Figure 5.13*b* shows the resulting frequency response of the latter two designs; the resulting passband ripple is greatly reduced. Figure 5.14 shows a more detailed comparison of the two weighting functions; here we see that the Kaiser-weighted filter yields less error at low frequencies, but in this case, the useful filter bandwidth is slightly reduced when compared to the Hamming-weighted filter.

Simple Second-order Tuned Filter (Recursive Design). In order to illustrate the properties of the recursive-filter-design methods discussed in Sec. 5.9 let us consider the simple high-Q second-order s-domain prototype filter ($Q = 10$, resonant at $\omega \approx 1$)

$$H(s) = \frac{1}{s^2 + 0.1s + 1} \tag{5.205}$$

TABLE 5.9 NORMALIZED COEFFICIENTS OF 19-TAP NONRECURSIVE DIFFERENTIATOR†

Tap index	*Direct*	*Hamming*	*Kaiser*
1	−1.000	−0.97225861	−0.96668240
2	0.50	0.44619022	0.43611705
3	−0.3333333	−0.25666667	−0.24396510
4	0.25	0.15496954	0.14179408
5	−0.20	−0.09202437	−0.08022842
6	0.1666667	0.05166667	0.04228436
7	−0.14285714	−0.02680279	−0.01973004
8	0.125	0.01346767	0.00742166
9	−0.1111111	−0.00888889	−0.00165259

† Entries are values of a_nT; $J = 9$.

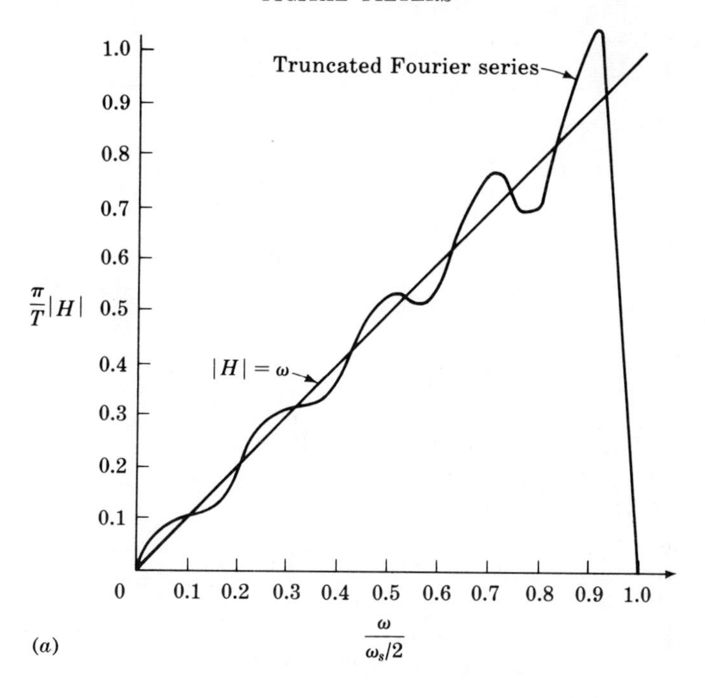

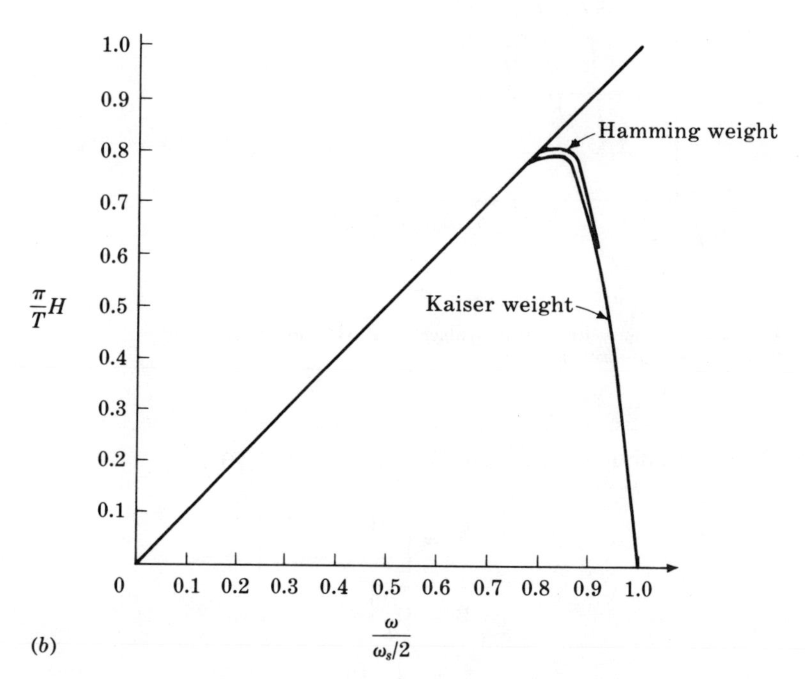

FIG. 5.13 Frequency response of 19-tap nonrecursive differentiator. (a) Truncated Fourier series; (b) Kaiser- and Hamming-weighted series.

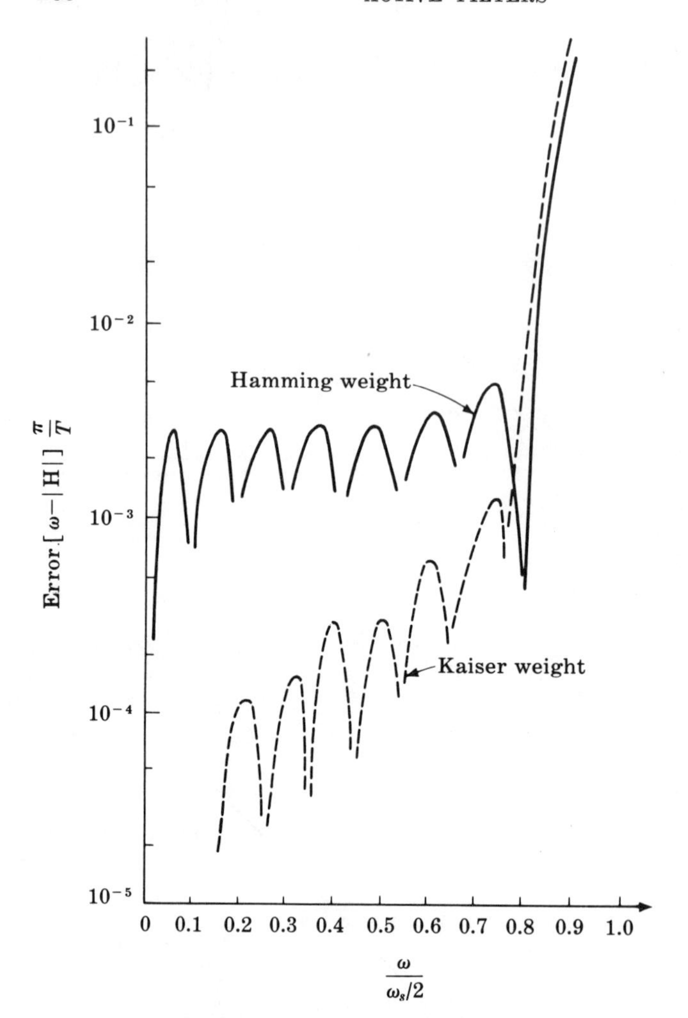

FIG. 5.14 Magnitude error, Kaiser- and Hamming-weighted nonrecursive differentiator.

the corresponding **A**, **B**, and **C** matrices [of (5.20)] are

$$\mathbf{A} = \begin{bmatrix} 0 & 1 \\ -1 & -0.1 \end{bmatrix} \tag{5.206}$$

$$\mathbf{B} = \begin{bmatrix} 0 \\ 1 \end{bmatrix}$$

$$\mathbf{C} = [1 \quad 0]$$

The methods A, B, C, and D applied to this system yield z-domain transfer functions of the form

$$\frac{a_0 + a_1 z^{-1} + a_2 z^{-2} + a_3 z^{-3}}{1 + b_1 a^{-1} + b_2 z^{-2}} \tag{5.207}$$

Table 5.10 shows values of the coefficients determined by each method for the case of $T = 0.1$ sec (that is about 20π samples per cycle at the resonant frequency). Figure 5.15 shows a plot of the error in magnitude (in db) vs. log ω for each method. Based on Fig. 5.15 it appears that method A is best; however, in many cases the phase-shift characteristic will be important, e.g., in digital simulation applications. A standard approach that is used in presenting the errors in analog computer simulation is to plot the magnitude of the error signal as a function of frequency. We will adapt this viewpoint here, and look at the magnitude of the output error expressed in percent (referred to the output level). For small errors, this can be estimated by

$$E \approx \sqrt{\left(\frac{dA}{A}\right)^2 + (d\theta)^2} \cdot 100\% \tag{5.208}$$

where A = desired output magnitude (ratio)
θ = desired output phase (radians)
dA = magnitude error
$d\theta$ = phase error

Figure 5.16 shows a plot of this *combined error* due to both gain and phase-shift errors; included is a plot of the error for method A if the error is compensated for the delay $T/2$. The plot for method C is not shown; typically the error is twice that of the uncompensated method A curve. From Fig. 5.16 we see that for broadband use method O generally is as acceptable as any method, although the low-frequency error is not as small as that of the other methods due to the dc gain error. Method A

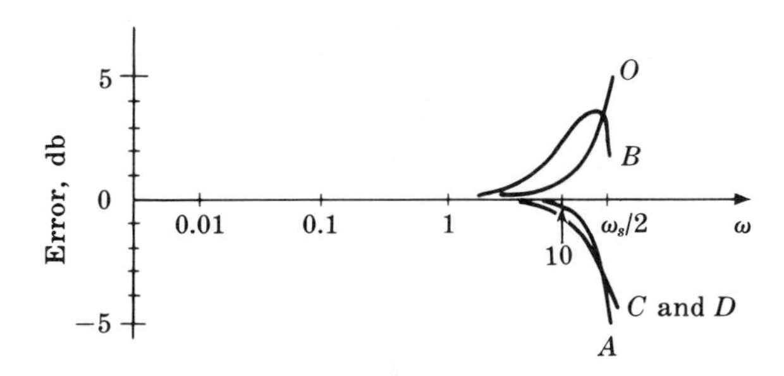

FIG. 5.15 Second-order tuned filter, magnitude error in db; $T = 0.1$; methods O, A, B, C, and D.

TABLE 5.10 COEFFICIENTS OF EQ. (5.207); $T = 0.1$; SECOND-ORDER TUNED FILTER

All methods: $b_1 = -1.980108$ $\quad b_2 = 0.99004983$

Numerator coefficients	*Method O*	*Method A*	*Method B*	*Method C*	*Method D*
a_0	0.0	0.0	0.0	0.0	1.661678×10^{-3}
a_1	9.933591×10^{-3}	4.9792263×10^{-3}	6.6409042×10^{-3}	1.661678×10^{-3}	6.6268103×10^{-3}
a_2	0.0	4.9626510×10^{-3}	6.6102350×10^{-3}	6.626810×10^{-3}	1.6533889×10^{-3}
a_3	0.0	0.0	$-3.3092620 \times 10^{-3}$	1.6533889×10^{-3}	0.0

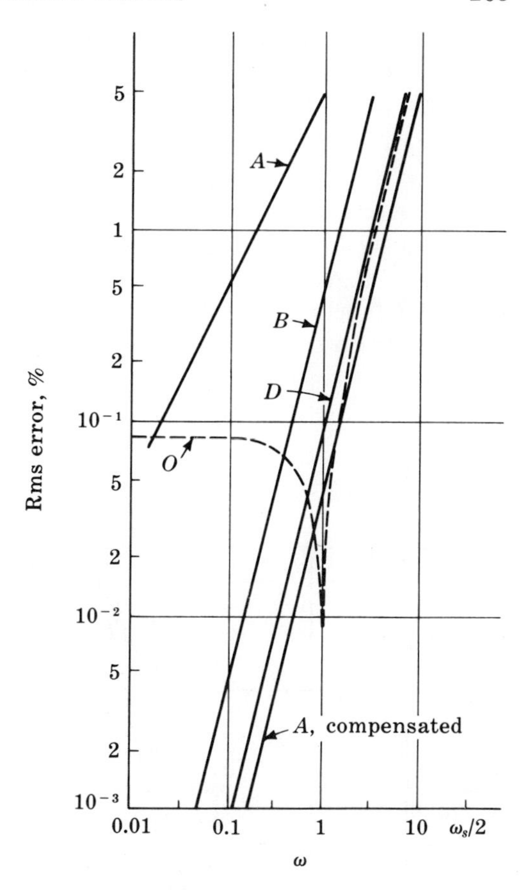

FIG. 5.16 Second-order tuned filter, rms error in percent; $T = 0.1$; methods O, A, B, D, and A compensated for $T/2$ delay. Note filter is tuned to $\omega = 1$.

would be best, neglecting the $T/2$ delay. Method D has less error than method O at low frequencies, but of course, the resulting $H(z)$ has a nonzero value for a_0 (not real-time realizable without some processing delay). Method C would have the same error as method D, except for the phase error caused by the delay T. Comparisons made with larger and smaller values of T yield similar results. Comparisons made for different values of Q are also similar.

Since method O does lead to a real-time realizable digital filter with the lowest combined error, and an overall gain accuracy comparable to method A, it seems that this basic "impulse-invariant" method is perhaps the best to use. Indeed, this judgment seems to be generally valid, for relatively small values of T, if the pole-zero excess of the prototype $H(s)$ is two or more.

Simple First-order Low-pass Filter. Consider the prototype low-pass filter

$$H(s) = \frac{1}{s+1} \tag{5.209}$$

This transfer function has a pole-zero excess of only one, and thus we can expect that method O will not be the best for approximating this particular function. Table 5.11 shows values of the coefficients in the z-domain transfer function for each method. Figure 5.17 shows a plot

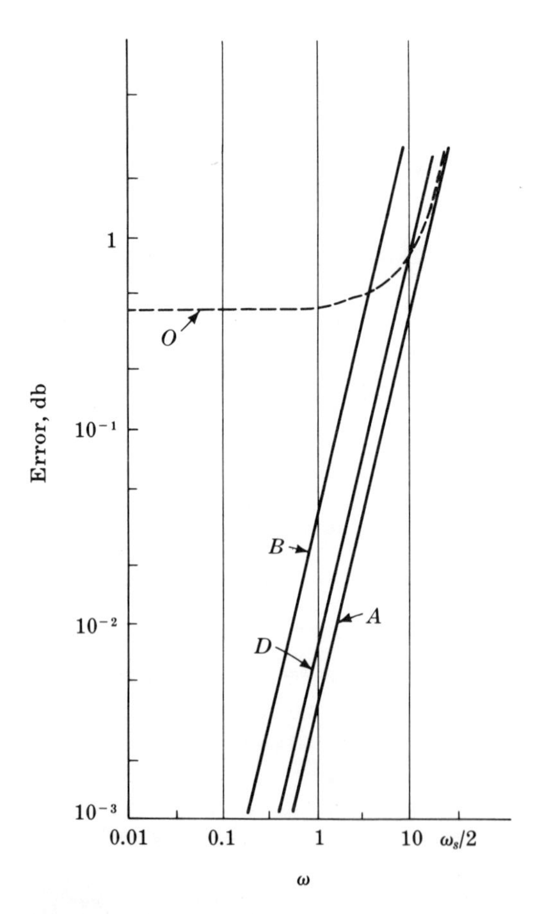

FIG. 5.17 First-order low-pass filter, magnitude error in db; $T = 0.1$; methods O, A, B, and D; note error in method D is negative, others are positive.

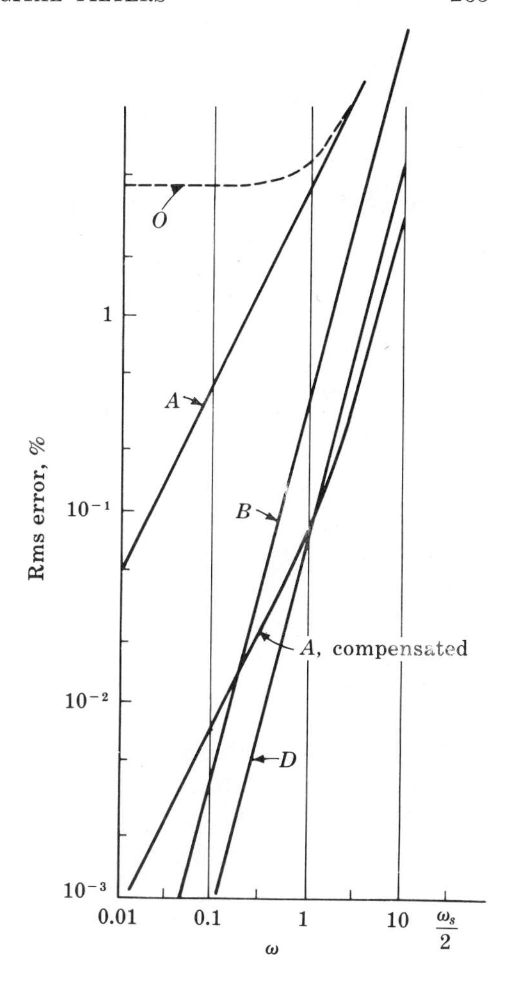

FIG. 5.18 First-order low-pass filter, rms error in percent; $T = 0.1$; methods O, A, B, D, and A compensated for $T/2$ delay. Note filter roll off is at $\omega = 1$.

of the magnitude error in db vs. frequency for the case $T = 0.1$ (that is, we are using 20π samples per cycle at the roll-off frequency). Figure 5.18 shows the combined error, including both amplitude and phase errors. From these plots we can conclude that method O is generally inferior to all the other methods, and further, that method O yields an $H(z)$ that is not real-time realizable. Here then we have a simple example where method B is probably best for digital simulation of this particular transfer function where total error is important, and we require

Table 5.11 Coefficients of Eq. (5.207); First-order Low-pass Filter; $T = 0.1$

All methods: $b_1 = -0.9048374$					
Numerator coefficients	*Method O*	*Method A*	*Method B*	*Method C*	*Method D*
a_0	0.1	0.0	0.0	0.0	4.83742×10^{-2}
a_1	0.0	9.516258×10^{-2}	1.4353676×10^{-1}	4.83742×10^{-2}	4.678838×10^{-2}
a_2	0.0	0.0	-4.837418×10^{-2}	4.678838×10^{-2}	0.0

real-time realizability. Note, however, that except for the delay of $T/2$, method A yields a fairly low error, and it shows the best amplitude response. Method D yields the lowest combined error, but, of course, is not strictly real-time realizable. The error of method C is the same as for method D except for a delay of T.

Although the above simple examples cannot present a complete picture of the relative merits of the various input approximation methods, they are representative of the behavior of recursive-digital-filter designs based upon matching a continuous-time prototype. Generally these methods lead to more efficient filter form (fewer coefficients) than the nonrecursive designs.

A Third-order Butterworth Filter: Use of the Bilinear Transform. Consider next the third-order Butterworth filter with s-domain transfer function

$$H(s) = \frac{1}{s^3 + 2s^2 + 2s + 1} \tag{5.210}$$

which has a -3-db cutoff frequency $\omega \approx 1$ rad/sec. Suppose we are interested in designing an approximating digital filter with $T = 1$; only amplitude response is of interest.

Here is a situation where the cutoff frequency is fairly close to $\omega_s/2 \approx \pi$. Figure 5.19 shows the magnitude response curves obtained by using methods O and A; for comparison, the response of the original $H(s)$ is plotted, along with the response of a filter designed by means of the bilinear transform. Table 5.12 lists the actual magnitude errors associated with each method. We see that although the bilinear transform provides a sharper cutoff, the actual magnitude curve does not match a third-order Butterworth response as well as do the methods O and A curves. Thus, if the digital filter is being designed strictly for low-pass filtering, then the bilinear-transform response better meets the actual requirements. However, if one is attempting to *simulate* the given $H(s)$ response, the state-variable methods are more accurate.

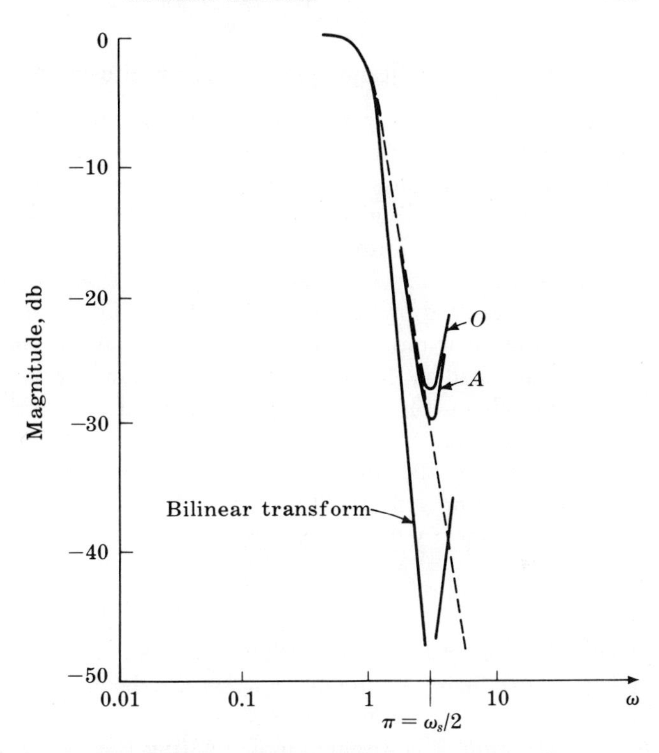

FIG. 5.19 Third-order Butterworth filter, magnitude in db; $T = 1$; methods O, A, and bilinear transform; prototype response is shown as a dotted line.

TABLE 5.12 MAGNITUDE ERRORS (db), THIRD-ORDER BUTTERWORTH; $T = 1.0$

Angular frequency	*Method A*	*Method O*	*Bilinear transform*
1	−0.38	0.1	−0.01
2	−1.43	0.7	−9
2.5	−1.8	−0.05	−21
π	0.45	2.0	†

† The bilinear transform produces a transmission zero at $\omega = \pi = \omega_s/2$.

5.11 Quantization (Round-off) Errors: Finite-word-length Effects

Thus far we have been considering the behavior of digital filters assuming that they are ideal discrete-time operators. We have made no reference to the possible performance degradations associated with implementing the filter algorithms with actual digital hardware, which of necessity uses a finite number of digits for representing all variables and coefficients. Precise prediction of these effects is often difficult, and perhaps the best approach is to simulate the filter behavior on a general-purpose digital computer and study the effects of shortening the word length empirically. It is nevertheless possible for the designer to place useful bounds on these effects, particularly if the variable quantization is not too coarse.

In a binary fixed-point digital system, all quantities X are represented by a corresponding M-bit digital machine variable X_D (a many-to-one-mapping), typically represented by

$$X \approx FX_D = F(-D_0 + \sum_{i=1}^{M-1} D_i 2^{-i}) \tag{5.211}$$

where $D_i = 0$ or 1 for all i. (The above represents quantities in two's complement form; a similar analysis can be made based upon signed-magnitude or one's complement forms.)

The constant F is a scale factor relating the variable X to the binary machine variable X_D. Our choice of notation implies that

$$-1 \leq X_D \leq 1 - 2^{1-M} \tag{5.212}$$

which for reasonably large M can be considered to be approximately the range $-1 \leq X_D < +1$. The mapping of a continuous-valued real quantity into its corresponding digital counterpart FX_D is shown in Fig. 5.20a. Figure 5.20b shows the error between X and FX_D. Since the finest step of X_D has value 2^{1-M} then X changes in quantization steps of size

$$q = F2^{1-M} \tag{5.213}$$

Thus if proper rounding off is done the maximum error in X is

$$|X - X_D|_{\max} = F2^{1-M} = \frac{q}{2} \tag{5.214}$$

Using the notation of (5.214) we see that increasing the word length M reduces the quantization error correspondingly. Note also that the fractional precision of representation is

$$\frac{q/2}{|X|_{\max}} = 2^{-M} \tag{5.215}$$

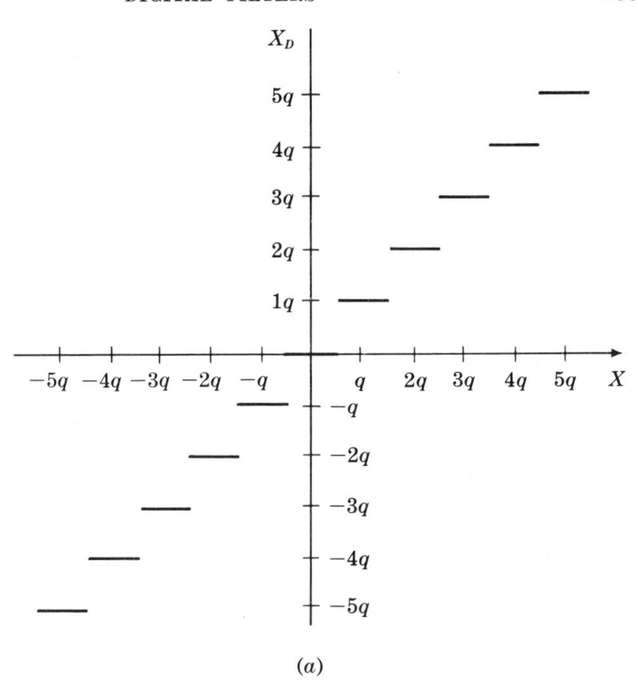

(a)

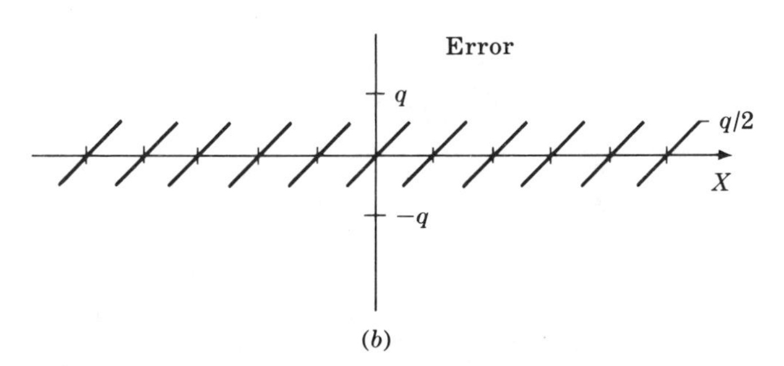

(b)

FIG. 5.20 Quantization with round off. (a) Input-output mapping; (b) quantization error.

Quantization or round-off errors cause the performance of the actual hardware implementation of a digital filter to differ from the ideal discrete-time operator designed with the assumption of infinite precision.

The obvious way to avoid finite-word-length effects is to use a large number of bits in the hardware system. This of course increases the final implementation cost and in serial-arithmetic systems may lower the achievable computing speed. Thus it is important to use only as many bits as are required. If the digital filtering is being done by a general-purpose digital computer, it is also important to be able to determine whether a particular machine will do the job.

There are a number of subtle ramifications of the round-off error problem which affect the entire filter design: for example, increasing filter order or decreasing the sampling time makes the filter behavior more sensitive to round-off errors. Some implementations (notably the direct and canonical forms of Fig. 5.10*a* and *b*) are more sensitive to round-off errors than are others (the cascade and parallel forms of Fig. 5.10*c* and *d*); unfortunately, the less-sensitive implementations require more computer operations per time step. Generally there are a number of design trade offs of a fairly subtle and hard-to-estimate nature that enter into the proper choice of word length, implementation algorithm, and filter order to realize an optimum design.

It is our intention here to point out some of the pitfalls and to provide reference to current useful work on this subject; the interested reader is left to pursue this somewhat tortuous subject in the references.

We note that digital-filter implementations all can be expressed in the general form of (5.43), repeated here

$$\mathbf{x}(n+1) = \tilde{\mathbf{A}}\mathbf{x}(n) + \tilde{\mathbf{B}}u(n)$$
$$y(n) = \tilde{\mathbf{C}}\mathbf{x}(n) + \tilde{D}u(n)$$

Three major sources of error arise from a finite-word-length implementation of these algorithms:

1. If the input values $u(n)$ are derived from the output of an analog/digital converter, then there will be an *input quantization error*.
2. Each multiplication operation will in general produce a *multiplicative round-off error* (except multiplication by 0 or ± 1). Normally in a fixed-point binary system the other operations (addition, subtraction, transfer) do not introduce additional error sources.
3. Each coefficient used in the algorithms is represented by a precision determined by the number of bits used in its representation (5.211); hence there is an additional error made due to the *limited coefficient precision*.

Bennett[28] and Widrow[29] have established a basis for a statistical description of quantization error. Although quantization error is actually deterministically related to the input-signal source, nevertheless the nature of the error lends itself to a statistical description; i.e., we find that if the quantization is fine enough, then round-off error effects may be treated as an additive quasi-random noise. An inspection of Fig. 5.20 lends justification to the following description of fine quantization noise:

1. The noise is uniformly amplitude distributed with mean-value zero and variance $q^2/12$ (rms value $= q/\sqrt{12}$). This assumes that round off to the nearest digital is always performed.
2. The bandwidth of the quantization noise is quite broad compared to that of the parent unquantized signal (many zero crossings per cycle of the original signal); hence the quantization noise is essentially uncorrelated with the parent signal; moreover we can usually approximate the autocorrelation function of the noise as an impulse function, i.e.,

$$\phi(n) \approx \frac{q^2}{12}\,\delta(n) \tag{5.216}$$

Widrow[29] has established the conditions under which the above assumptions hold.

Knowles and Edwards,[31,32] Knowles and Olcayto,[33] Gold and Rader,[34] and others have successfully used this quasi-random model to estimate effects of all the above types in finely quantized systems. The technique involves finding the total rms error produced at the filter output from the various quantization "noise" sources; this may be done either using direct simulation or via evaluation of contour integrals of the noise spectral density. Rather than elaborate on their results, the following summary is given here.

1. The quasi-random quantization noise model is useful for estimating finite-word-length effects in low-pass, bandpass, and band-stop filters with reasonably fine quantization (e.g., 15 or more bits), especially if the errors are small.
2. Of considerable importance, it seems consistently true that parallel-programmed implementations (Fig. 5.10*d*) are least sensitive to round-off error effects, the cascade is next most sensitive, and the direct and canonic implementations are most sensitive. Knowles and Olcayto[33] cite an example of a 22d-order band-stop filter that demonstrated comparable performance with a 15-bit parallel implementation, a 24-bit cascade form, and a 45-bit direct form (viz., a round-off noise of 10^{-3}, Figs. 6, 8, and 10 of the referenced paper). These results are of considerable significance to the digital-filter designer (note that Knowles and Olcayto used direct simulation to obtain many of their results).

Kuo and Kaiser (Ref. 1, sec. 7.6.1) have made an analysis of the type-3 (coefficient-accuracy) errors that also provides insight into the nature of the problem. Their results show that for an nth-order low-pass filter of given bandwidth and a sampling rate $1/T$, an absolute lower bound on the number of binary digits required to represent the coefficients b_j in (5.111) is

$$n_d \geq 1 + \log_2 \left[\frac{2^{n+2}}{5\sqrt{n} \prod_{k=1}^{n} (s_k T)} \right] \tag{5.217}$$

$$\geq n \left[1 + \log_2 \left(\frac{1}{T} \right) \right] - \sum_{k=1}^{n} \log_2 s_k \tag{5.218}$$

where the s_k are the poles of the original s-domain prototype filter. Note that this bound really only ensures *stability* of the filter, not acceptable accuracy of pole location, which may take several more bits.

This result indicates that the number of bits required is approximately proportional to $n \log_2 (1/T)$. Thus doubling the filter order doubles the number of bits; doubling the sampling rate requires n more bits.

Consider the third-order Butterworth filter with unity-radian bandwidth discussed in the previous section.

Here
$$\prod_{k=1}^{n} s_k \approx 1$$

and we obtain the word length estimates of 13 bits for $T = 0.1$, 23 bits for $T = 0.01$. To ensure acceptable accuracy, the actual word length might actually have to be 5 to 10 bits more than shown here. As a corroboration of this, it was actually found that for $T = 0.01$ it was not possible to accurately calculate the coefficients of the filter with floating-point arithmetic that used only a 27-bit mantissa.

We should also note that it is considerably more difficult to estimate the word-length requirements in a floating-point arithmetic system; however, most implementations for real-time data processing would be fixed-point for reasons of speed and cost. The problem of analyzing floating-point systems has not been as thoroughly treated; see, however, Sandberg.[35]

This section has been intended primarily to present some insight into the nature of the word-length problem; obviously it is not a trivial task to estimate word-length requirements. In this writer's opinion, the safest, and perhaps in the long run the quickest, way of analyzing quantization error effects is by means of digital computer simulation in a manner similar to Knowles' and Olcayto's work.[33]

5.12 Summary

As indicated in Sec. 5.1, we have made no attempt to cover every topic in the rapidly growing field of digital-filter design. We hope the presentation will permit the reader to go further in his study of particular areas of interest. For example, Rader and Gold[2] present a discussion of frequency-sampling techniques that are useful for the efficient design of a bank of adjacent bandpass filters (comb filters). Many of the mean-square optimization techniques of sampled-data control systems[6,7] can certainly be applied to the optimal design of discrete-time filters for wide-sense stationary random signals. Certainly one must mention in passing the great recent interest in fast-Fourier-transform techniques for digital spectral analysis (see Ref. 41 for a tutorial review of this topic at an introductory level).

5.13 Acknowledgments

The author is indebted to members of the faculty and staff of the University of Arizona for their assistance in programming the examples in Sec. 5.10; included are Prof. L. P. Huelsman, and Messrs. R. Burkhardt, A. Trevor, and R. White. These empirical studies were supported in part by the National Science Foundation under Grant GK-1860 and by the National Aeronautics and Space Administration Grant NsG-646.

Appendix A Digital Filter Steady-state Frequency Response

Although many writers (see, e.g., Refs. 1 and 2) make use of the operator $H(e^{j\omega T})$ to describe the magnitude and phase response of digital filters, this writer has not seen any formal justification of this in terms of a z-domain demonstration, except by Hurewicz (Ref. 4, p. 238). Certainly, the technique is straightforward and intuitively acceptable.

Consider a scalar digital filter with transfer function $H(z)$ and consider the complex sequence

$$u(n) = \sum_{n=1}^{\infty} e^{jn\omega T} = \sum_{n=1}^{\infty} \cos(n\omega T) + j \sin(n\omega T) \tag{5A.1}$$

which has Z transform

$$\frac{z}{z - e^{j\omega T}} \tag{5A.2}$$

Consider also the real part of $u(n)$, viz.,

$$\operatorname{Re}\{u(n)\} = \operatorname{Re}\left\{ \sum_{n=1}^{\infty} e^{jn\omega T} \right\} = \sum_{n=1}^{\infty} \cos(n\omega T) \tag{5A.3}$$

with transform

$$\text{Re}\left\{\frac{z}{z - e^{j\omega T}}\right\} = \frac{z(z - \cos \omega T)}{z^2 - 2z \cos (\omega T) + 1} \tag{5A.4}$$

If the sequence $u(n)$ is applied to a system with transfer function $H(z)$, then the output sequence will have the Z transform

$$Y(z) = X(z)H(z) \qquad \text{(see, e.g., Ref. 5, p. 247)}$$

$$= \frac{z}{z - e^{j\omega T}} H(z) \tag{5A.5}$$

$Y(z)$ is assumed to be expandable in partial-fraction form to yield

$$Y(z) = \left[\sum \frac{C_i}{z - \alpha_i}\right] + \frac{H(e^{j\omega T})e^{j\omega T}}{z - e^{j\omega T}} \tag{5A.6}$$

where the α_i are the poles of $H(z)$. We assume that the bracketed expression corresponds to a transient response. The remaining term corresponds to a steady-state solution $Y_{ss}(z)$, with

$$Y_{ss}(z) = \frac{H(e^{j\omega T})e^{j\omega T}}{z - e^{j\omega T}} \tag{5A.7}$$

which may be rewritten

$$Y_{ss}(z) = H(e^{j\omega T})e^{j\omega T}z^{-1} \sum_{n=0}^{\infty} e^{jn\omega T}z^{-n} \tag{5A.8}$$

By the change of index to $m = n + 1$, we have

$$Y_{ss}(z) = H(e^{j\omega T}) \sum_{m=1}^{\infty} e^{jm\omega T}z^{-m} \tag{5A.9}$$

which for $m \geq 1$ is the transform of a sinusoidally varying sequence [as expected, the response at $m = 0$ is 0, as is the case for any physically realizable $H(z)$].

If, for example, we want the steady-state response due to a sequence of cosine-wave samples, i.e.,

$$u(n) = \cos n\omega T \tag{5A.10}$$

we may find the response from (5A.9) as follows:

$$Y_{\cos}(z) = \text{response to } \cos(n\omega T)$$

$$= \text{Re}\{Y_{ss}(z)\}$$

Let

$$H(e^{j\omega T}) = U + jV \tag{5A.11}$$

then

$$Y_{\cos}(z) = \text{Re}\left\{(U + jV) \sum_{n=1}^{\infty} (\cos n\omega T + j \sin n\omega T)\, z^{-n}\right\} \tag{5A.12}$$

$$\begin{aligned} y(nT) &= \text{Re}\{(U + jV)(\cos n\omega T + j \sin n\omega T)\} \qquad n \geq 1 \qquad (5A.13) \\ &= U \cos n\omega T - V \sin n\omega T \\ &= \sqrt{U^2 + V^2}\left(\frac{U}{\sqrt{U^2 + V^2}} \cos n\omega T - \frac{V}{\sqrt{U^2 + V^2}} \sin n\omega T\right) \\ &= |H(e^{j\omega T})|\,[\cos\theta \cos(n\omega T) - \sin\theta \sin(n\omega T)] \\ &= |H(e^{j\omega T})| \cos(n\omega T + \theta) \qquad (5A.14) \end{aligned}$$

where
$$\theta = \tan^{-1}\frac{V}{U} = \arg H(e^{j\omega T}) \tag{5A.15}$$

Appendix B Short Table of Z Transforms

The following short table of Z transforms is provided for quick reference; more extensive tables appear in a number of books on sampled-data systems (see, e.g., Refs. 6, 7, 8, or 10).

TABLE 5B.1 A SHORT TABLE OF Z TRANSFORMS

Laplace transform	*Time function*	*Z transform*
$\frac{1}{s}$	$U(t)$	$\frac{z}{z-1}$
$\frac{1}{s^2}$	$t\,U(t)$	$\frac{Tz}{(z-1)^2}$
$\frac{1}{s+a}$	$e^{-at}\,U(t)$	$\frac{z}{z-e^{-aT}}$
$\frac{1}{(s+a)^2+b^2}$	$\frac{e^{-at}\sin bt}{b}\,U(t)$	$\frac{ze^{-aT}\sin bT}{b(z^2 - 2ze^{-aT}\cos bT + e^{-2aT})}$
$\frac{1}{s(s+a)}$	$\frac{1-e^{-at}}{a}\,U(t)$	$\frac{(1-e^{-aT})z}{a(z-1)(z-e^{-aT})}$
$\frac{1}{s[(s+a)^2+b^2]}$	$\frac{1 - e^{at}\sec\phi\cos(bt+\phi)}{a^2+b^2}\,U(t)$ where $\phi = \tan^{-1}\frac{-a}{b}$	$\frac{1}{a^2+b^2}\left[\frac{z}{z-1} - \frac{z^2 - ze^{-aT}\sec\phi\cos(bT-\phi)}{z^2 - 2ze^{-aT}\cos bT + e^{-2aT}}\right]$

Appendix C Computational Aspects of Using State-variable Methods

Transforming a Rational $H(s)$ to State-variable Form. In Sec. 5.9, several recursive design methods are presented that involve state-variable representations of both the prototype continuous-time transfer function $H(s)$ and the discrete-time digital approximating algorithms. Briefly we must begin with a given $H(s)$ which may be given in rational form; assume that the form is

$$H(s) = \frac{c_m s^m + c_{m-1}s^{m-1} + \cdots + c_1 s + c_0}{s^k + d_{k-1}s^{k-1} + \cdots + d_1 s + d_0} \tag{5C.1}$$

From this we first must obtain a state-variable form

$$s\mathbf{X}(s) = \mathbf{AX}(s) + \mathbf{B}U(s) \tag{5C.2a}$$

$$Y(s) = \mathbf{CX}(s) + DU(s) \tag{5C.2b}$$

The method for this is straightforward (see Ref. 37, sec. 2.5). Assuming that a phase-variable representation of the system will suffice it is possible to find **A, B, C,** and D from (5C.1) directly by the following relationships:

$$\mathbf{A} = \begin{bmatrix} 0 & 1 & 0 & \cdots & 0 & 0 \\ 0 & 0 & 1 & \cdots & 0 & 0 \\ \cdots & \cdots & \cdots & \cdots & \cdots & \cdots \\ 0 & 0 & 0 & \cdots & 0 & 1 \\ -d_0 & -d_1 & -d_2 & \cdots & -d_{k-2} & -d_{k-1} \end{bmatrix}$$

$$\mathbf{B} = \begin{bmatrix} 0 \\ 0 \\ \cdot \\ \cdot \\ \cdot \\ 0 \\ 1 \end{bmatrix} \qquad \mathbf{C} = [c_0 \quad c_1 \quad \cdots \quad c_m \quad 0 \quad \cdots \quad 0] \qquad m < k$$

$$D = \begin{Bmatrix} 0 & m < k \\ c_m & m = k \end{Bmatrix}$$

Evaluating the Matrices of the Discrete-time Approximation. The evaluation of the matrices $\tilde{\mathbf{A}}$, $\boldsymbol{\beta}_{-1}$, $\boldsymbol{\beta}_0$, and $\boldsymbol{\beta}_1$ in (5.179) is relatively straightforward, once the **A**, **B**, and **C** matrices are known. This writer has found the Taylor-series expansion method of Liou (Ref. 1, pp. 103–106) useful. Using this approach, one finds $\tilde{\mathbf{A}}$ from

$$e^{\mathbf{A}T} = \tilde{\mathbf{A}} \approx \mathbf{I} + \sum_{n=1}^{N} \frac{T^n(\mathbf{A})^n}{n!} \tag{5C.3}$$

where n is chosen to reduce the truncation error to some desired value. (See alternatively Ref. 36.) Similarly, this writer has had success in evaluating the matrices $\mathbf{L}_1$ and $\mathbf{L}_2$ in (5.175) by a similar series approach. Here we use the same value of N used in (5C.3).

The series for $\mathbf{L}_1$ and $\mathbf{L}_2$ are

$$\mathbf{L}_1 \approx \sum_{n=1}^{N} \frac{T^n}{n!} (\mathbf{A})^{n-1} \tag{5C.4}$$

and

$$\mathbf{L}_2 \approx \mathbf{L}_1 - \sum_{n=2}^{N} \frac{T^{n-1}}{n!} \mathbf{A}^{n-2} \tag{5C.5}$$

Obtaining $H(z)$ in Rational-function Form. As indicated in Sec. 5.9, the state-variable algorithms of (5.179) imply a z-domain transfer function $H(z)$, which may be expressed by

$$H(z) = \mathbf{C}(z\mathbf{I} - \mathbf{A})^{-1}(z^{-1}\boldsymbol{\beta}_{-1} + \boldsymbol{\beta}_0 + z\boldsymbol{\beta}_1)$$

This form is not directly usable for finding the implementations described in Sec. 5.7; we require $H(z)$ in the form

$$H(z) = \frac{a_0 + a_1 z^{-1} + \cdots + a_m z^{-m}}{1 + b_1 z^{-1} + \cdots + b_k z^{-k}}$$

This form can be found using the Souriau-Frame algorithm (Ref. 5, sec. 7.3) or the algorithm described by Tuel et al.[40] This writer has successfully used the latter method, as implemented by Melsa[39] for obtaining $H(z)$ in rational form. The reader is referred to these references for details.

References

1. Kuo, F. F., and J. F. Kaiser: "System Analysis by Digital Computer," John Wiley & Sons, Inc., New York, 1966.
2. Rader, C. M., and B. Gold: Digital Filter Design Techniques in the Frequency Domain, *Proc. IEEE*, vol. 55, no. 2, pp. 149–171, February, 1967.
3. Monroe, A. J.: "Digital Processes for Sampled-data Systems," John Wiley & Sons, Inc., New York, 1962.
4. Hurewicz, W.: Filters and Servo Systems with Pulsed Data, in H. M. James, N. B. Nichols, and R. S. Phillips, "Theory of Servomechanisms," chap. 5, McGraw-Hill Book Company, New York, 1947.
5. Schwarz, R. J., and B. Friedland: "Linear Systems," McGraw-Hill Book Company, New York, 1965.
6. Kuo, B. C.: "Analysis and Synthesis of Sampled-data Control Systems," Prentice-Hall, Inc., Englewood Cliffs, N.J., 1963.
7. Tou, J. T.: "Digital and Sampled-data Control Systems," McGraw-Hill Book Company, New York, 1959.
8. Jury, E. I.: "Theory and Application of the Z-transform Method," John Wiley & Sons, Inc., New York, 1964.
9. Papoulis, A.: "The Fourier Integral and Its Applications," McGraw-Hill Book Company, New York, 1962.
10. Ragazzini, J. R., and G. F. Franklin: "Sampled-data Control Systems," John Wiley & Sons, Inc., New York, 1958.

11. Wilts, C. H.: "Principles of Feedback Control," Addison-Wesley Publishing Company, Inc., Reading, Mass., 1960.
12. Truxal, J. G.: Numerical Analysis for Network Design, *Trans. IRE/PGCT*, vol. 1, no. 3, pp. 49–60, September, 1954.
13. Karplus, W. J.: Error Analysis of Hybrid Computer Systems, *Simulation*, pp. 120–136, February, 1966.
14. Wait, J. V.: Data Reconstruction Errors: Fractional and Polygonal Holds, *University of Arizona, Department of Electrical Engineering, ACL Memo* 140, July, 1967.
15. Leneman, O. A. Z.: Random Sampling of Random Processes: Optimum Linear Interpolation, *J. Franklin Inst.*, vol. 281, no. 4, pp. 302–313, April, 1966.
16. Leneman, O. A. Z., and J. B. Lewis: Random Sampling of Random Processes: Mean-square Comparison of Various Interpolators, *Trans. IEEE/PGAC*, vol. AC-11, no. 3, pp. 396–403, July, 1966.
17. PDP-9 User's Handbook, *Digital Equipment Corporation Document F*-95, Maynard, Mass.
18. Brubaker, T. A., and D. R. Stevens: Time Domain Design of Finite Memory Digital Filters, *J. Comp. Phys.*, June, 1968.
19. Brubaker, T. A., and J. B. Peterson: Time Domain Synthesis of Finite Memory Digital Filters Using Orthogonal Polynomials, *Proc. Hawaii Intern. Conf. Syst. Sci.*, January, 1968.
20. Taylo, J. T.: Digital Filters for Non-Real-Time Data Processing, NASA CR-880, Northeast Louisiana State College, October, 1967, contract NAS 8-11492, CFSTI, Springfield, Va.
21. Fleischer, P. E.: Digital Realization of Complex Transfer Functions, *Simulation*, vol. 6, no. 3, pp. 1–25, March, 1966.
22. Wait, J. V.: State-space Methods for Designing Digital Simulations of Continuous Fixed Linear Systems, *Trans. IEEE/PGEC*, vol. EC-16, no. 3, pp. 351–354, June, 1967.
23. Baxter, D. C.: Digital Simulation Using Approximate Methods, *National Research Council, Division of Mechanical Engineering Analysis Section, Rept. MK*-15, July, 1965.
24. Mantey, P. E., and G. F. Franklin: Comment on "Digital Filter Design Techniques in the Frequency Domain," *Proc. IEEE*, vol. 55, no. 12, pp. 2196–2197, December, 1967.
25. Numerical Techniques for Real Time Digital Flight Simulation, *IBM Corp., Technical Publications Dept., Application Note E*20-0029-0, White Plains, N.Y.
26. Giese, C.: State Variable Difference Methods for Digital Simulation, *Simulation*, pp. 263–271, May, 1967.
27. Cruickshank, A. J. O.: Time Series and Z Transform Methods of Analysis of Linear and Non-linear Control Systems, *Proc. Intern. Cong. IFAC, 1st.*, vol. 1, pp. 277–285, published by Butterworth & Co. (Publishers), Ltd., London, 1961.
28. Bennett, W. R.: Spectra of Quantized Signals, *Bell Syst. Tech. J.*, vol. 27, pp. 446–472, July, 1948.
29. Widrow, B.: Analysis of Amplitude-quantized Sampled-data Systems, *AIEE Trans.-Pt. II, Appl. and Industry*, vol. 79, pp. 555–567, January, 1961.
30. Bertram, J. E.: The Effect of Quantization in Sampled-feedback Systems, *AIEE Trans.-Pt. II, Appl. and Industry*, vol. 77, pp. 177–182, September, 1958.
31. Knowles, J. B., and R. Edwards: Finite Word-length Effects in Multirate Direct Digital Control Systems, *Proc. IEEE*, vol. 112, no. 12, pp. 2376–2384, December, 1965.

32. Knowles, J. B., and R. Edwards: Effect of a Finite-word-length Computer in a Sampled-data Feedback System, *Proc. IEEE*, vol. 112, no. 6, pp. 1197–1207, June, 1965.
33. Knowles, J. B., and E. M. Olcayto: Coefficient Accuracy and Digital Filter Response, *Trans. IEEE/PGCT*, vol. CT-15, no. 1, pp. 31–41, March, 1968.
34. Gold, B., and C. M. Rader: Effects of Quantization Noise in Digital Filters, *Proc. Spring Joint Computer Conf.*, pp. 213–219, 1966.
35. Sandberg, I. W.: Floating-point-roundoff Accumulation in Digital-filter realizations, *Bell Syst. Tech. J.*, vol. 46, pp. 1775–1791, October, 1967.
36. Liou, M. L.: A Novel Method of Evaluating Transient Response, *Proc. IEEE*, vol. 54, pp. 20–23, January, 1966.
37. Schultz, D. G., and J. L. Melsa: "State Functions and Linear Control Systems," McGraw-Hill Book Company, New York, 1968.
38. Golden, R. M., and J. F. Kaiser: Design of Wideband Sampled-data Filters, *Bell Syst. Tech. J.*, vol. 43, pp. 1533–1546, July, 1964.
39. Melsa, J. L.: A Digital Computer Program for the Analysis and Design of State Variable Feedback Systems, *University of Arizona Engineering Experiment Station, Tucson, NASA CR*-850, March, 1967 (NASA date August, 1967).
40. Tuel, W. G., Jr.: On the Transformation to (Phase Variable) Canonical Form, *Trans. IEEE/PGAC*, vol. AC-11, p. 607, July, 1967; see also accompanying notes in same volume by M. R. Chidambara (pp. 607–608), D. S. Rane (p. 608), and C. D. Johnson and W. M. Wonham (pp. 609–610).
41. Brigham, E. O., and R. E. Morrow: The Fast Fourier Transform, *IEEE Spectrum*, vol. 4, no. 12, pp. 63–70, December, 1967.

CHAPTER 6

PARAMETRIC FREQUENCY CONVERTERS

B. J. Leon
Purdue University
Lafayette, Ind.

6.1 Energy Conversion in Nonlinear Reactances

Electronic amplifiers use energy from a basic electrical source to increase the power available in a desired signal wave form. In the amplifiers of the previous chapters the basic electrical source was the bias battery or dc power supply. The class of devices considered in this chapter are ones whose basic electrical energy comes from a steady-state sinusoidal power source. Techniques for generating such sinusoidal power are discussed in the chapters on oscillators in most electronics texts. In this chapter the power source is assumed to be available. The problem is the conversion of energy from the source to a useful signal whose wave form is different from the wave form of the source. Two problems are considered. The first, which is treated only briefly, is the conversion to a steady-state sinusoid of a different frequency. The second, which is the primary topic of this chapter, is the use of the source energy to amplify a comparatively low-power input signal.

A. Basic Analysis Method. The basic idea of frequency conversion in the class of nonlinear systems that includes parametric frequency converters is most easily seen by reference to the feedback diagram shown in Fig. 6.1. For a first heuristic discussion of the operation of this system as a frequency converter let us assume that $u(t)$ is a steady-state sinusoid. That is

$$u(t) = A \cos (\omega t + \phi) \tag{6.1}$$

If the feedback loop is open then $x(t)$ is also a steady-state sinusoid of the same frequency. That is

$$x(t) = B \cos (\omega t + \theta) \tag{6.2}$$

The constants B and θ of the output are related to the corresponding

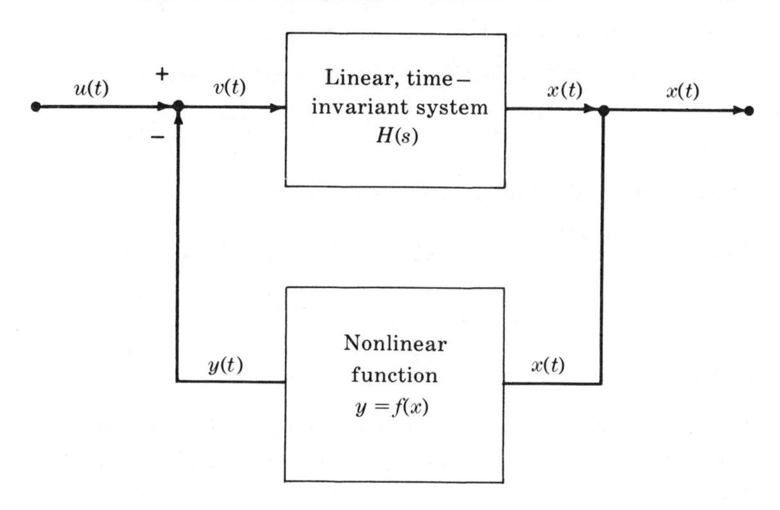

Fig. 6.1 Basic nonlinear system.

constants A and ϕ of the input through the transfer function $H(s)$ by

$$\begin{aligned} B &= A|H(j\omega)| \\ \theta &= \phi + \underline{/H(j\omega)} \end{aligned} \tag{6.3}$$

In the open-loop system there is no frequency conversion.

When the feedback loop of Fig. 6.1 is closed, the system is nonlinear. The exact mathematical characterization is a nonlinear differential equation or a nonlinear integral equation. For all except the most special cases there is no method for getting a closed-form solution to these equations. By a heuristic argument that simultaneously uses steady-state and transient reasoning, it is possible to get an idea of how the system works. For this argument it is assumed that the nonlinear function in the feedback loop is a polynomial. That is

$$f(x) = \alpha_1 x + \alpha_2 x^2 + \cdots + \alpha_n x^n \tag{6.4}$$

For the initial discussion, we only need consider the case where $n = 2$.

The reasoning is by successive approximation to the output of Fig. 6.1. For the first approximation we assume there is no feedback and the output is given by (6.2). For the second approximation, we assume that the first approximant is the input of the nonlinearity. The resulting output of the nonlinearity subtracted from the system input is the input to the linear time-invariant system. The resulting output of the linear system is the second approximant.

For the special case where n in (6.4) is 2 we have

$$y_1 = f(x_1) = \alpha_1 B \cos(\omega t + \theta) + \alpha_2 B^2 \cos^2(\omega t + \theta)$$
$$= \frac{\alpha_2 B^2}{2} + \alpha_1 B \cos(\omega t + \theta) + \frac{\alpha_2 B^2}{2} \cos(2\omega t + 2\theta) \qquad (6.5)$$

Then the second approximation to the output is

$$x_2(t) = -\frac{\alpha_2 \beta^2 H(0)}{2} + B \cos(\omega t + \theta) - \alpha_1 B |H(j\omega)| \cos(\omega t + 2\theta - \phi)$$
$$- \frac{\alpha_2 B^2 |H(j2\omega)|}{2} \cos(2\omega t + \psi) \qquad (6.6)$$

where $\psi = 2\theta + \underline{/H(j2\omega)}$

The second approximant, (6.6), has a dc term (if the linear system passes direct current), a term at the input frequency, and a term at the second harmonic.

The third approximation to the output is obtained by feeding the second approximant back through the nonlinearity, subtracting the result from the input, and passing this difference through the linear, time-invariant transfer function. This third approximant will contain terms at direct current, the input frequency, and the second, third, and fourth harmonics of the input frequency. As successively higher approximations are made by the same procedure, they will have all harmonics of the input up to some maximum. In the case of the square-law nonlinearity, this maximum for the nth approximant is 2^{n-1}. For higher-order polynomial nonlinearities, the maximum will be higher. In any event, as $n \to \infty$ all harmonics eventually appear in successive approximants.

If the procedure described above does give successively better approximations to the output, then we see that the device is a harmonic generator. If the desired output is the Kth harmonic, then as a single frequency circuit at that harmonic frequency any circuit described by the block diagram of Fig. 6.1 is active. The energy, of course, comes from the input which is at a different frequency and thus not directly a part of the single frequency system whose basic frequency is the Kth harmonic.

When the input to the system of Fig. 6.1 consists of two sinusoids at two different frequencies, then the procedure described above leads to a set of approximants that contain all harmonics of both frequencies and all sums and differences of all these harmonics. Specifically, the input is

$$u(t) = A \cos(\omega_0 t + \phi) + B \cos(\omega_1 t + \theta) \qquad (6.7)$$

The output as predicted by the procedure is of the form

$$x(t) = \sum_{n=0}^{\infty} \sum_{m=-\infty}^{\infty} C_{mn} \cos(|n\omega_0 + m\omega_1| t + \psi_{mn}) \qquad (6.8)$$

The extension of this procedure to n distinct input signals at n noncommensurate frequencies is straightforward. The result is like (6.8) with an n-fold summation, rather than a double summation, over all combinations of sums and differences of the input frequencies and their harmonics. The response so obtained is correct for systems wherein the input-signal amplitudes are not too large. For large input signals it can be shown[1,2,3] that it is possible to generate additional frequency components not included in the various summations. Thus this iterative procedure is not a general method of analysis for nonlinear circuits with sinusoidal excitations. For many problems the method is adequate. The exact conditions that guarantee the success of the method are not known.

As an illustrative example, let us consider the nonlinear circuit of Fig. 6.2. By proper choice of parameters we shall see that this circuit can convert energy from the source at ω_b to the frequency ω_a. Furthermore the response at ω_a can be nearly linear with respect to the amplitude of the signal source at ω_a. To apply the procedure to this circuit we successively approximate the charge $q(t)$, which is the integral of the current $i(t)$ with the constant of integration chosen so that $q(t)$ is a steady-state quantity. For the first approximant, we linearize the nonlinear capacitance by neglecting the q^2 term. Thus in complex exponential notation

$$q_1(t) = \frac{\mathcal{V}_a Y(j\omega_a)}{j\omega_a} e^{j\omega_a t} + \frac{\mathcal{V}_b Y(j\omega_b)}{j\omega_b} e^{j\omega_b t} + \text{complex-conjugate terms} \tag{6.9}$$

where

$$\mathcal{V}_a = \tfrac{1}{2} V_a e^{j\phi_a}$$

$$\mathcal{V}_b = \tfrac{1}{2} V_b e^{j\phi_b}$$

$$Y(j\omega) = \frac{j\omega}{(j\omega)^2 + 0.1(j\omega) + 1}$$

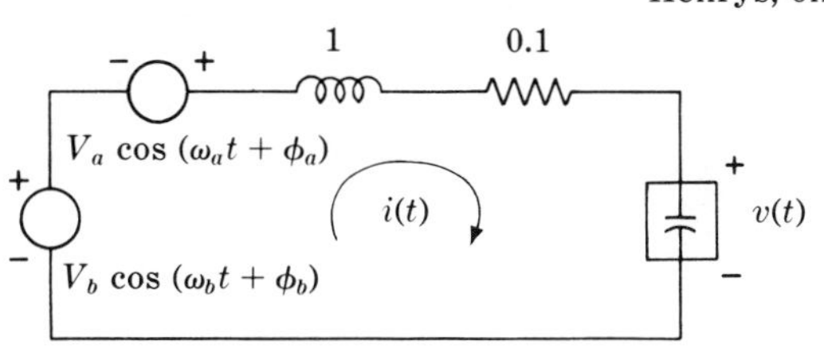

FIG. 6.2 A simple nonlinear circuit.

To get the second approximant, we pass $q_1(t)$ through the nonlinear part of the capacitor and add the resulting voltage to the source voltage. Relative to Fig. 6.1, the nonlinear function in the feedback loop is

$$y(t) = 0.1q^2(t) \tag{6.10}$$

The feedback is negative since the voltage across the capacitor and the source voltage have opposite polarity. The forward transfer function for this example relates the charge on the linear part of the capacitor to the combined source and nonlinear capacitor voltage. Thus

$$H(s) = \frac{Y(s)}{s} = \frac{1}{s^2 + 0.1s + 1} \tag{6.11}$$

For the second approximant, the voltage across the nonlinear capacitor due to the first approximant must be computed. It is

$$\begin{aligned} y_1(t) = 0.1q_1(t) = {} & 0.1\mathcal{V}_a{}^2H^2(j\omega_a)e^{j2\omega_a t} + CC \\ & + 0.1\mathcal{V}_b{}^2H^2(j\omega_b)e^{j2\omega_b t} + CC \\ & + 0.2\mathcal{V}_a\mathcal{V}_bH(j\omega_a)H(j\omega_b)e^{j(\omega_a+\omega_b)t} + CC \\ & + 0.2\mathcal{V}_a\mathcal{V}_b{}^*H(j\omega_a)H^*(j\omega_b)e^{j(\omega_a-\omega_b)t} + CC \\ & + 0.2[|\mathcal{V}_a|^2|H(j\omega_a)|^2 + |\mathcal{V}_b|^2|H(j\omega_b)|^2] \end{aligned} \tag{6.12}$$

where CC indicates an added complex-conjugate term. Now the second approximant is

$$\begin{aligned} q_2(t) = q_1(t) - 0.1\{ & \mathcal{V}_a{}^2H^2(j\omega_a)H(2j\omega_a)e^{2j\omega_a t} + CC \\ & + \mathcal{V}_b{}^2H^2(j\omega_b)H(2j\omega_b)e^{2j\omega_b t} + CC \\ & + 2\mathcal{V}_a\mathcal{V}_bH(j\omega_a)H(j\omega_b)H(j(\omega_a + \omega_b))e^{j(\omega_a+\omega_b)t} + CC \\ & + 2\mathcal{V}_a\mathcal{V}_b{}^*H(j\omega_a)H^*(j\omega_b)H(j(\omega_a - \omega_b))e^{j(\omega_a-\omega_b)t} + CC \\ & + 2H(0)[|\mathcal{V}_a|^2|H(j\omega_a)|^2 + |\mathcal{V}_b|^2|H(j\omega_b)|^2]\} \end{aligned} \tag{6.13}$$

Note that the second approximant is the first approximant plus terms with a 0.1 factor. Squaring q_2 and multiplying by 0.1 to get y_2 will produce all the terms of y_1 plus terms with a $(0.1)^2$ factor plus terms with a $(0.1)^3$ factor. Thus the third approximant q_3 will be q_2 plus new terms with $(0.1)^2$ and $(0.1)^3$ factors. Continuing through y_3 to q_4 reprodûces q_3 plus terms with higher powers of (0.1) multipliers, etc. Thus as one anticipates computing each higher approximant, he can estimate how much change there will be in the results. For the present example we choose numbers so that the third approximant gives suitably accurate results.

Proceeding to the next approximant gives a very large number of terms. We do not need to compute all these to see how the circuit can be used as an energy converter. To begin, let us compute the component of $q_3(t)$ at ω_a. For this term, we need the component of y_2 at ω_a, call it y_{2a}.

$$\begin{aligned} y_{2a} = -0.02[V_a|V_a|^2|H(j\omega_a)|^2H(2j\omega_a)H(j\omega_a) \\ + 2V_a|V_b|^2H(j\omega_a)|H(j\omega_b)|^2H(j(\omega_a + \omega_b)) \\ + 2V_a|V_b|^2H(j\omega_a)|H(j\omega_b)|^2H(j(\omega_a - \omega_b)) \\ + 2V_a|V_a|^2|H(j\omega_a)|^2H(j\omega_a)H(0) \\ + 2V_a|V_b|^2|H(j\omega_b)|^2H(j\omega_a)H(0)] \end{aligned} \quad (6.14)$$

Then the complex amplitude of the third approximation to the charge on the linear capacitor at the frequency ω_a is

$$\begin{aligned} Q_{3a} = \mathcal{V}_aH(j\omega_a)\{1 + 0.02H(j\omega_a)[|\mathcal{V}_a|^2|H(j\omega_a)|^2(2H(0) + H(2j\omega_a)) \\ + 2|\mathcal{V}_b|^2|H(j\omega_b)|^2(H(0) + H(j(\omega_a - \omega_b)) + H(j(\omega_a + \omega_b)))]\} \end{aligned} \quad (6.15)$$

The complex amplitude Q_{3a} contains terms that are linear in the signal amplitude $\mathcal{V}_a$ and terms that are cubic. If we want essentially linear signal processing, the parameters must be chosen so that the cubic terms are negligible. On the other hand if we want some energy conversion due to the nonlinearity we cannot make the entire bracket multiplying $0.02H(j\omega_a)$ negligible, for this would make the result that of a linear circuit. The desirable linear terms have the factor $|\mathcal{V}_b|^2|H(j\omega_b)|^2$. Correspondingly the undesired terms have the factor $|\mathcal{V}_a|^2|H(j\omega_a)|^2$. By making the power-source amplitude $\mathcal{V}_b$ large enough compared to the signal-source amplitude $\mathcal{V}_a$, the desired almost-linear result is obtained.

With the simple circuit of Fig. 6.2 we can increase the effect of the desired terms most by making $\omega_a \ll \omega_b$ and making the resonance at $\omega_b - \omega_a$. Then $|H(j(\omega_a - \omega_b))|$ is the largest of the various transfer functions. The frequency ω_a should not be too far from $\omega_b - \omega_a$ because $H(j\omega_a)$ must also have an appreciable value. With this choice of frequencies $H(2j\omega_a)$ will be fairly small as it should be. To be specific, let $\omega_a = 1.1$ and $\omega_b = 2.1$. Then $\omega_b - \omega_a$ is 1, the resonant frequency of the linear circuit. With these frequencies

$$\begin{aligned} H(j\omega_a) &= 4.22\underline{/-152.4^\circ} \\ |H(j\omega_a)|^2 &= 17.8 \\ H(j(\omega_a - \omega_b)) &= j10 \\ |H(j\omega_b)|^2 &= 0.086 \\ H(j2\omega_a) &= 0.260\underline{/176.9^\circ} \\ H(j(\omega_a + \omega_b)) &= -0.108 \\ H(0) &= 1.0 \end{aligned} \quad (6.16)$$

To keep the signal small we let

$$\mathcal{V}_a = 0.1$$

To make the power-source amplitude large enough compared to the signal we choose $\mathcal{V}_b$ so that $|\mathcal{V}_b|^2|H(j\omega_b)|^2$ is 10 times the corresponding ω_a terms. Thus

$$|\mathcal{V}_b|^2 = \frac{10|\mathcal{V}_a|^2|H(j\omega_a)|^2}{|H(j\omega_b)|^2} = 20.7$$

With these numbers (6.15) becomes

$$\begin{aligned} Q_{3a} &= 0.422\angle{-152.4°}\ \{1 + 0.0844\angle{-152.4°}\ [0.178(2 \\ &\qquad + 0.260\angle{176.9°}) + 1.78(1 + j10 - 0.108)]\} \\ &= 0.422\angle{-152.4°}\ (1 + 1.50\angle{-67.5°}) \\ &= 0.885\angle{166.1°} \end{aligned} \tag{6.17}$$

From the first line of (6.17) we see that the term that is nonlinear in $\mathcal{V}_a$ is 1/50 the linear term. Thus the circuit will perform an essentially linear operation on the signal at frequency ω_a. The second line shows that the term that does not also appear in the first approximant is 1.5 times the first approximant. Thus we are on dangerous ground if we take this iteration as the last. The next iteration will produce a term that our heuristic reasoning implies to be about 0.15. This may make a noticeable difference in the answer. Rather than carry out the very messy computation to get the next approximant exactly, let us wait until the end of Sec. 6.2 wherein another computation method for circuits satisfying the linear signal-processing conditions is presented. Then we can compare the present answer with the answer by another method.

For the present let us stay with the unjustified assumption that Q_{3a} is the complex amplitude of the charge at frequency ω_a when $\mathcal{V}_a = 0.1$ and $|\mathcal{V}_b| = 4.53$. In order to have a better justified result let us also compute the charge when the excitations are smaller by a factor of 4. Then the new term in the third approximant is smaller by a factor of 16. Let a caret over the letter represent these new conditions. Now

$$\begin{aligned} \hat{Q}_{3a} &= 0.1055\angle{-152.4°}\ (1 + 0.094\angle{-67.5°}) \\ &= 0.109\angle{-157.2°} \end{aligned} \tag{6.18}$$

To see that in both cases this third approximation to the circuit response indicates that energy is converted from the source at ω_b to frequency ω_a let us compute the power flow at ω_a for both conditions. The currents in the circuit are

$$\begin{aligned} I_{3a} &= j\omega_a Q_{3a} = 0.974\angle{-103.9°} = -0.234 - j0.944 \\ \hat{I}_{3a} &= 0.120\angle{-67.2°} = 0.0465 - j0.110 \end{aligned} \tag{6.19}$$

The power delivered by the voltage source at frequency ω_a in each case is

$$P_{s3a} = 2 \text{ Re } (\mathcal{V}_a I_{3a}^*) = -0.0468 \text{ watt}$$
$$\hat{P}_{s3a} = 2 \text{ Re } [1/40 \hat{I}_{3a}^*] = 0.00233 \text{ watt}$$

The power delivered to the resistor in each case is

$$P_{R3a} = 2|I_{3a}|^2 R = 0.189 \text{ watt}$$
$$\hat{P}_{R3a} = 0.00288 \text{ watt}$$

In the first case this third approximant says that all the load power plus 0.0468 watt that the source absorbs is coming from the source at frequency ω_b. In the second case 0.00055 watt of power is converted, this being excess load power over that delivered by the source. In both cases it seems that energy is converted in the nonlinear capacitance. If we drew a single frequency equivalent circuit to show the power flow at ω_a, the nonlinear capacitance would be replaced by a capacitor plus a negative resistor. This negative resistor can provide the required energy.

B. The Manley-Rowe Formulas. The above example demonstrates that a nonlinear capacitance can be used to convert power from one frequency to another in such a way as to amplify a small signal. In order to design the linear, passive, time-invariant circuit characterized by $H(s)$ in the analysis method above one needs a set of power-flow formulas that show how energy is converted in the nonlinear capacitance without detailed information about $H(s)$. The Manley-Rowe formulas[4] give such information.

When a nonlinear reactance (capacitance or inductance) is driven by two noncommensurate steady-state sources at ω_a and ω_b through a linear, passive, time-invariant circuit such as the example above, then all frequencies $(m\omega_a + n\omega_b)$ appear in the resulting voltages and currents. Here m and n take on all positive and negative integer values as well as zero. We define P_{mn} as the power at the frequency $|m\omega_a + n\omega_b|$ that the reactance absorbs. Since a reactance is lossless and absorbs no power at direct current† conservation of energy gives

$$\sum_{m=1}^{\infty} \sum_{n=-\infty}^{\infty} P_{mn} = \sum_{m=-\infty}^{\infty} \sum_{n=1}^{\infty} P_{mn} = 0 \tag{6.20}$$

The Manley-Rowe formulas state

$$\begin{aligned} \sum_{m=1}^{\infty} \sum_{n=-\infty}^{\infty} \frac{mP_{mn}}{m\omega_a + n\omega_b} &= 0 \\ \sum_{m=-\infty}^{\infty} \sum_{n=1}^{\infty} \frac{nP_{mn}}{m\omega_a + n\omega_b} &= 0 \end{aligned} \tag{6.21}$$

† For a capacitor there is no dc current and for an inductor the dc voltage is zero.

When the two formulas (6.21) are added the result is the conservation-of-energy formula (6.20). Thus the Manley-Rowe formulas contain a statement of conservation of energy plus the information that each double sum in (6.21) must be zero.

The derivation of (6.21) is given for a simple nonlinear inductor or capacitor in Ref. 4. Penfield[5] showed that the formulas apply to any lossless time-invariant nonlinear system if the system has a well-defined hamiltonian. He also proved the logical extension of the two formulas to k formulas when there are k noncommensurate frequencies driving the system. We shall not repeat the proofs.

The application of the Manley-Rowe formulas to a circuit such as that of Fig. 6.2 begins by noting that except at frequencies near the resonant frequency of the series-tuned circuit, there is no way that the nonlinear capacitor can deliver much power to the circuit. Thus the only appreciable terms in the formulas (6.21) with ω_a and ω_b as used in the example above are the terms at frequencies ω_a, $\omega_b - \omega_a$, and ω_b. The last term is not near resonance, but there is a source to provide the power at this frequency. With only these three terms, the formulas become

$$\begin{aligned} \frac{P_{1.0}}{\omega_a} + \frac{P_{1,-1}}{\omega_a - \omega_b} &= 0 \\ \frac{P_{0,1}}{\omega_b} + \frac{P_{-1,1}}{-\omega_a + \omega_b} &= 0 \end{aligned} \tag{6.22}$$

The power at the difference frequency must be negative since there is no source to supply power. Thus the power at ω_b is positive to satisfy the second equation when $\omega_b > \omega_a$.

For the first of equations (6.22) we have $P_{1,-1} = P_{-1,1} < 0$. Since $\omega_b > \omega_a$ the term $P_{1,-1}/(\omega_a - \omega_b)$ is positive. Thus $P_{1,0}$, the power delivered to the capacitor at ω_a, is negative. The capacitor is supplying power to the circuit at frequency ω_a. This is in agreement with the numbers in the example above.

The formulas (6.22) contain quantitative information about the conversion efficiency of the circuit in question. If the first equation is rewritten it shows that the ratio of power out of the capacitor at ω_a and $\omega_b - \omega_a$ is proportional to the frequencies. That is

$$\frac{P_{1,0}}{P_{1,-1}} = \frac{\omega_a}{\omega_b - \omega_a}$$

Thus if $\omega_a > \omega_b - \omega_a$, more power is converted to frequency ω_a than to frequency $\omega_b - \omega_a$.

The use of the Manley-Rowe formulas for quantitative information about power flow must be exercised with caution. There is little difficulty in choosing a circuit so that the power flow at each of the unimportant frequencies is small. The problem arises because in each formula there are an infinite number of these frequencies with small power flow. Consequently, even though each neglected term is small the infinite sum of neglected terms may be appreciable. Thus the Manley-Rowe formulas give design guidelines, but they are not reliable for quantitative design of frequency converters.

C. Spurious Responses and Subharmonics. The analysis method of Sec. 6.1A generated a solution with specific frequency terms. These terms included all exciting frequencies, all their harmonics, and all sums and differences of exciting frequencies and their harmonics. The Manley-Rowe formulas of Sec. 6.1B are based on the assumption that these frequencies make up the total response. In all except degenerate cases, the solution generated by the process of Sec. 6.1A is a steady-state solution to the system equations. Problems arise because in many interesting engineering situations this solution is not unique. There are other solutions to the system equations. Often in such cases, the physical system response will be one of the other solutions because the other solutions may be more stable under small perturbations of the excitation such as the perturbations caused by physical noise.

Lumped, linear, time-invariant† amplifiers will usually oscillate if the gain is too large. The amplifier example above will oscillate also if the power source, the voltage source at ω_b, is of too large an amplitude. The analysis method presented above will not reveal such oscillatory conditions. Nonlinear circuits that can amplify by the nonlinear mixing process can almost always be made to oscillate if the power source is of large enough amplitude. The oscillations will initially start at the frequency where maximum gain occurs. As the power-source amplitude is increased, the oscillating frequency may shift considerably.

The easiest type of oscillations to explain are the true subharmonic oscillations. For some simple equations such as the circuit equation of Fig. 6.2, Hale,[1] Hayashi,[2] Minorsky,[3] and many others show a way of predicting such oscillations. In order to predict such oscillations one must first know the frequency at which they will occur. If the circuit of Fig. 6.2 is driven at a frequency very nearly twice the resonant frequency of the linear circuit, then for voltage-source amplitudes above some threshold amplitude, the response will suddenly change so that there is a component at one-half the driving frequency. The analysis method described above does not allow such a subharmonic.

The basic procedure for predicting subharmonic oscillations is to assume a response that contains these extra frequency terms with unknown coefficients. These extra frequency terms mix with all other

† Of course all amplifiers are basically nonlinear, but the results apply even in the small-signal case.

frequencies in the nonlinear element. Thus the amplitude of the response at each frequency depends on the amplitudes of the extra terms. By equating the final response at the driving frequency to the source amplitude and the responses at each of the other frequencies to zero, there results a set of nonlinear algebraic equations in the various amplitudes. When these equations have a solution, then the assumed form is one possible response. Even for the simplest examples, the procedure is quite complicated. In most practical amplifier design, the oscillation thresholds are found empirically.

6.2 Exact Formalism for Nonlinear Frequency Converters

In the previous section, the analysis of nonlinear frequency converters was presented in a fairly heuristic manner. The analysis method presented does not always give the response of a particular circuit. Furthermore, when the method does lead to the response, only an approximation is obtained with a reasonable amount of calculation. In this section, we examine the equations of the circuits of the previous section in more detail. We proceed by stating those properties that can be proved, and pointing out where gaps in the present knowledge of the equations lie. Unfortunately for a circuit sufficiently complex to be practical, exact methods are not available. The methods of the previous section are those that must be used initially. The additional techniques presented in this section are necessary for getting precise analytical results for the signal amplifying properties of these circuits.

A. Conversion of Circuit Equations to Standard Form. The basic one-varactor-diode nonlinear frequency converter is shown schematically in Fig. 6.3. The circuit consists of a source representing the signal or

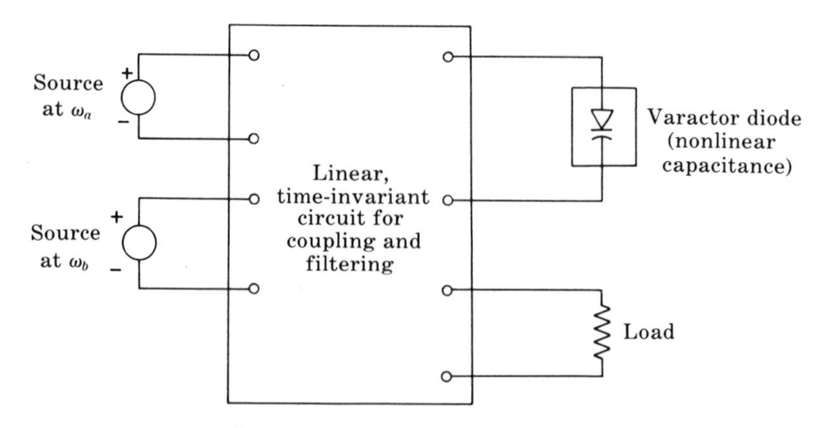

FIG. 6.3 Basic parametric amplifier.

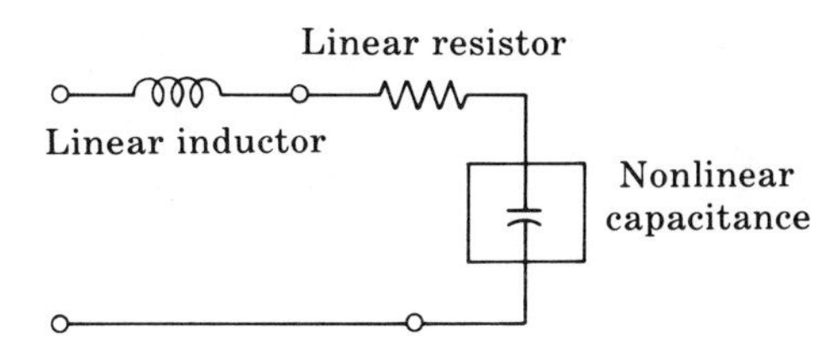

FIG. 6.4 Varactor-diode model.

signals to be amplified, the basic steady-state power source, the varactor diode which contains the nonlinear capacitance necessary for energy conversion, a load, and a network of linear, time-invariant elements for coupling and filtering. A good circuit model for the varactor diode is that of Fig. 6.4. This model consists of a linear resistor and inductor in series with a nonlinear capacitance. The details of the model are not particularly important for much of the mathematical development that follows.

The available mathematics for nonlinear circuits is based on two standard equation forms. One form is a system of first-order differential equations—one equation for each energy-storage element in the system. The other is a system of nonlinear integral equations—one equation for each nonlinear element in the system. For the circuit of Fig. 6.3, the integral equation form is the most convenient. The basic equation for a circuit with one nonlinear element is of the form

$$x(t) = g(t) + \int_{\alpha}^{t} h(t - \tau)N(x(\tau))\, d\tau \tag{6.23}$$

where $x(t)$ is the unknown
$g(t)$ is known in terms of the excitations and known circuit parameters
$h(t)$ is a circuit impulse response in terms of known circuit parameters
$N(x)$ is the nonlinear function of the nonlinear element

To get an equation in the form (6.23) from the circuit of Fig. 6.3 requires several steps. A nonlinear capacitance is characterized by a nonlinear charge-voltage relationship. Specifically, the charge is

$$q(t) = \bar{q}(v(t)) \tag{6.24}$$

or the voltage is

$$v(t) = \bar{v}(q(t)) \tag{6.25}$$

where the overbarred symbol represents a functional relationship. If the charge-voltage relationship is monotonic, then both (6.24) and (6.25)

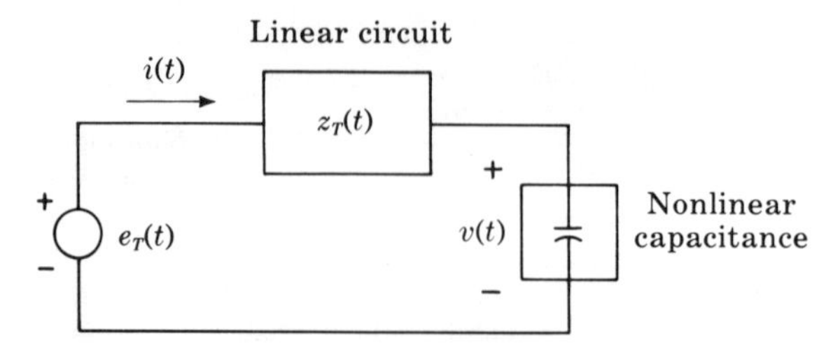

FIG. 6.5 Thévenin circuit.

apply and $\bar{q}$ is the function inverse to $\bar{v}$. That is

$$\bar{q} = \bar{v}^{-1}$$
$$\bar{v} = \bar{q}^{-1}$$

The incremental capacitance is the slope of the charge as a function of voltage. Thus

$$C(v) = \frac{d}{dv}\bar{q}(v) \tag{6.26}$$

Similarly, the incremental elastance is the slope of the voltage as a function of charge. Thus

$$S(q) = \frac{d}{dq}\bar{v}(q) \tag{6.27}$$

If $C(v) \neq 0$ and $S(q) \neq 0$ then

$$S(q) = \frac{1}{C(\bar{v}(q))}$$

or

$$C(v) = \frac{1}{S(\bar{q}(v))} \tag{6.28}$$

The charge-voltage relationship for most capacitive diodes used in frequency converters is monotonic. Thus all the formulas above apply. The first step in setting up an equation in the form (6.23) for the circuit of Fig. 6.3 is to make a Thévenin equivalent of the linear, time-invariant network to the left of the nonlinear capacitance within the varactor diode. The result is the circuit shown in Fig. 6.5. For this circuit Kirchhoff's voltage law gives

$$e_T(t) = \int_{\alpha}^{t} z_T(t - \tau)i(\tau)\, d\tau + v(t) \tag{6.29}$$

where $e_T(t)$ is the Thévenin (open-circuit) voltage of the linear, time-invariant circuit including the effect of energy storage at $t = \alpha$.

$z_T(t)$ is the impulse response of the linear, time-invariant circuit as seen from the capacitor terminals. The Thévenin impedance is the Laplace transform of $z_T(t)$.

α is an initial time when all energy-storage effects are known. For steady-state solutions $\alpha \to -\infty$.

There are two unknowns, $v(t)$ and $i(t)$, in (6.29). These can be related by using the nonlinear charge-voltage relation of the capacitor. Integrating the convolution in (6.29) by parts makes the charge $q(t)$ rather than the current $i(t)$ the unknown. Then

$$e_T(t) = z_T(0)q(t) - z_T(t-\alpha)q(\alpha) + \int_\alpha^t \dot{z}_T(t-\tau)q(\tau)\,d\tau + v(t) \quad (6.30)$$

where $\dot{z}(t)$ is the derivative of the impulse-response function. Equation (6.30) is correct only if the impulse response z_T is a function in the usual sense. If z_T contains an impulse or the derivative of an impulse, then the equation is not valid. When the incremental capacitance is positive, the case with virtually all semiconductor varactor diodes, there is no difficulty setting up the circuit equations so that z_T is a function. If we consider the nonlinear capacitor as two capacitors in parallel as shown in Fig. 6.6, then the linear part C_0 can be included with the linear circuit. The Thévenin equivalent can be taken looking back from terminal pair $x - x$. As long as $C_0 < \min C(v)$, with $C(v)$ the original nonlinear capacitance, then $C_1(v)$ in Fig. 6.6 is still positive and the remainder of the required computations can be executed. Now z_T is a function and $z_T(0)$ is $1/C_0$.

In terms of the charge as unknown (6.30) becomes

$$e_T(t) + z_T(t-\alpha)q(\alpha) = \bar{v}(q(t)) + z_T(0)q(t) + \int_\alpha^t \dot{z}_T(t-\tau)q(\tau)\,d\tau \quad (6.31)$$

where $\bar{v}$ is the nonlinear voltage-charge relationship for the nonlinear capacitor in the final analysis. To convert (6.31) to the desired form (6.23) we define $\bar{v}_1(q)$ by

$$\bar{v}_1(q) = \bar{v}(q) + z_T(0)q \quad (6.32)$$

If $\bar{v}(q)$ is monotonic increasing (a positive capacitance) so is $\bar{v}_1(q)$. Thus $\bar{v}_1$ has an inverse function $\bar{q}_1(v_1(t))$, where $v_1(t)$ is $\bar{v}_1(q(t))$. In terms of

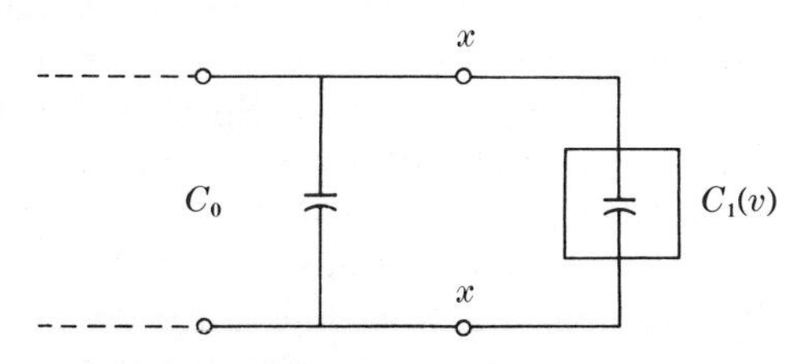

FIG. 6.6 Separation of a linear part from a nonlinear capacitance.

$v_1(t)$, (6.31) becomes

$$e(t) + z_T(t - \alpha)q(\alpha) = v_1(t) + \int_\alpha^t \dot{z}_T(t - \tau)\bar{q}_1(v_1(\tau))\, d\tau \tag{6.33}$$

The left side of (6.33) is a known function of time, call it $g(t)$. Thus (6.33) can be rewritten

$$v_1(t) = g(t) - \int_\alpha^t \dot{z}(t - \tau)\bar{q}_1(v_1(\tau))\, d\tau \tag{6.34}$$

This is exactly the desired form (6.23).

B. The Physical Significance of the Volterra Equation Form. Equations (6.34) and (6.23) are in a form known as a Volterra integral equation. In Tricomi[6] it is shown that such an equation has a unique solution for any $t > \alpha$. That is, if $g(t)$ and $\dot{z}(t)$ and $q_1(v_1)$ are all known and have the properties discussed above, then there is a unique $v_1(t)$ for all $t \geq \alpha$. From this $v_1(t)$ the charge q can be found through the formula (6.32), and the voltage across the nonlinear capacitor can be found from the known formula $\bar{v}(q(t))$.

The theory of Volterra integral equations says that the voltage on the nonlinear capacitor is unique when the input $g(t)$ in (6.34) is specified. At first glance this does not seem to agree with the statements made in the previous sections concerning subharmonics and spurious responses. The apparent discrepancy is resolved when we realize that a knowledge of $g(t)$ requires much more than the knowledge of the steady-state forcing function required in the previous section. The Thévenin voltage $e_T(t)$ depends on the condition of the energy-storage elements at $t = \alpha$. Also $g(t)$ defined by the left side of (6.33) depends on $q(\alpha)$, the condition of the nonlinear capacitor at $t = \alpha$. Thus if at any fixed time $t = \alpha$, the condition of the energy-storage elements both linear and nonlinear have particular values,† and if the input from $t = \alpha$ to some other t is fixed, then the response is uniquely determined at that value of t.

The steady-state responses discussed in Sec. 6.1 are solutions to (6.34) when $\alpha = -\infty$. The uniqueness proof for Volterra equations is not valid in the limit as α approaches $-\infty$. Thus this theory says nothing about uniqueness of steady-state solutions.

C. The Proper Conditions for Amplification. Let us examine the basic circuit of Fig. 6.3 and see what should be true if we are to obtain amplification of a signal. Specifically suppose the source at ω_a in the figure is an information-bearing signal that requires a band of frequencies centered at ω_a. The source at ω_b is a steady-state sinusoidal source whose purpose is to supply the power for amplification of the information-bearing signal. Even though the circuit is nonlinear, we should like the processing of the information-bearing signal to be essentially linear.

† In "modern control" terms we would say the state of the system at $t = \alpha$ is fixed.

In terms of (6.34) the requirement of linear signal processing can be stated as follows. The forcing term $g(t)$ in (6.34) is linear in the inputs and the terms due to energy storage. That is

$$g(t) = g_a(t) + g_b(t) + g_i(t) \tag{6.35}$$

where $g_a(t)$ is linear in terms of the source at ω_a
$g_b(t)$ is linear in terms of the source at ω_b
$g_i(t)$ is linear in terms of the initial conditions at $t = \alpha$

The effect of the source at ω_a on the response can be ascertained by examining the system first with this source set to zero and then with the source nonzero. The difference between the responses under these two input conditions is the signal-processing function of the circuit.

Mathematically the response to the signal at ω_a is constructed as follows: Let v_{1b} be the solution to (6.34) with $g_a(t) = 0$. That is

$$v_{1b}(t) = g_b(t) + g_i(t) - \int_\alpha^t \dot{z}(t - \tau)\bar{q}_1(v_{1b}(\tau))\, d\tau \tag{6.36}$$

Then the quantity $v_{1a}(t)$ due to the source at ω_a is the difference between $v_{1b}(t)$ of (6.36) and $v_1(t)$ from (6.34). Subtracting (6.36) from (6.34) gives $v_{1a}(t)$ as the solution to the equation

$$v_{1a}(t) = g_a(t) - \int_\alpha^t \dot{z}(t - \tau)[\bar{q}_1(v_{1b}(\tau) + v_{1a}(\tau)) - \bar{q}_1(v_{1b}(\tau))]\, d\tau \tag{6.37}$$

The response (capacitor voltage) due to the source at ω_a can be found by first using v_{1a} and v_{1b} in the formula (6.32) that relates the charge on the nonlinear capacitor to the quantity v_1. Then this charge relates to the voltage through the charge-voltage function of the nonlinear capacitor. Specifically we first define $q_b(t)$ by the equation†

$$v_{1b}(t) = \bar{v}(q_b(t)) + z_T(0)q_b(t) \tag{6.38}$$

Then $q_a(t)$ is defined by the equation

$$v_{1a}(t) + v_{1b}(t) = \bar{v}(q_a(t) + q_b(t)) + z_T(0)q_a(t) + z_T(0)q_b(t) \tag{6.39}$$

With this q_a and q_b the response due to the source at ω_a is

$$v_a(t) = \bar{v}(q_a(t) + q_b(t)) - \bar{v}(q_b(t)) \tag{6.40}$$

The overall relation between the response $v_a(t)$ and the signal input is the linear equation that gives $g(t)$ in terms of the input plus the composite effect of the three nonlinear equations (6.37), (6.39), and (6.40). To see how well the circuit performs as a linear amplifier we must ascertain the departure from linearity of the composite effect of the three equations. Such an exact expression for the nonlinear distortion is extremely difficult

† Recall that $\bar{v}$ is the nonlinear voltage-charge function of the nonlinear capacitor.

to ascertain. An estimate of the distortion can be obtained by first finding the departure from linearity for each equation separately. Then the total distortion is less than or equal to the sum of the absolute values of the three terms from the three equations.

To estimate the nonlinear effects in each of the equations we apply Taylor's theorem to the respective nonlinear functions. Thus for (6.37) we estimate the nonlinearity in $\bar{q}_1$ by

$$q_1(v_{1b}(\tau) + v_{1a}(\tau)) = \bar{q}_1(v_{1b}(\tau)) + \bar{q}_1'(v_{1b}(\tau))v_{1a}(\tau) + R_{q1}(v_{1a}(\tau)) \quad (6.41)$$

where $\bar{q}_1'$ is the derivative of the function $\bar{q}_1$

R_{q1} is the Taylor remainder characterized by

$$\lim_{x \to 0} \frac{R_{q1}(x)}{x} = 0$$

If $v_{1a}(t)$ is sufficiently small then the remainder $R_{q1}(v_{1a}(t))$ is smaller and, hopefully, negligible. Neglecting the remainder reduces (6.37) to the linear equation

$$v_{1a}(t) = g_a(t) - \int_\alpha^t z(t-\tau)\bar{q}_1'(v_{1b}(\tau))v_{1a}(\tau)\,d\tau \quad (6.42)$$

For (6.39) we apply Taylor's theorem to $\bar{v}$. Thus

$$\bar{v}(q_b(t) + q_a(t)) = \bar{v}(q_b(t)) + \bar{v}'(q_b(t))q_a(t) + R_{\bar{v}}(q_a(t)) \quad (6.43)$$

Using (6.43) and (6.38) in (6.39) gives

$$v_{1a} = \bar{v}'(q_b(t))q_a(t) + z_T(0)q_a(t) + R_{\bar{v}}(q_a(t)) \quad (6.44)$$

Again if the input is small, $q_a(t)$ is small and $R_v(q_a(t))$ is negligible. Finally substituting (6.43) in (6.40) gives

$$v_a(t) = \bar{v}'(q_b(t))q_a(t) + R_{\bar{v}}(q_a(t)) \quad (6.45)$$

Again the hope is that $R_{\bar{v}}(q_a(t))$ is negligible.

There is no good technique for estimating the nonlinear distortion in the various equations above precisely. All the theory shows is that if the signals are small compared to the power-source amplitude, then the distortion is small. For examples where detailed estimates have been made,[7] the results are very conservative. For engineering purposes one normally assumes that the signals are small enough and proceeds to use (6.42) or a similar linear equation to see how the system processes small signals.

When the signals are sufficiently small so that the linearizations described above are valid, then the processing of the signals by the circuit is essentially linear. This is one desired feature of a parametric amplifier. There is another required condition that is revealed from a study of (6.34) with $g(t)$ given by (6.35). If the circuit is turned on at $t = \alpha$ and we wait a reasonable length of time before applying the signal to be amplified, it is important that the signal processing be independent of α and specific initial conditions at that time. This requirement is satisfied if (6.36), the equation of the circuit without a small-signal term, has a unique solution as $\alpha \to -\infty$ and this solution is independent of $g_i(t)$. Efforts have been made[7,8] at developing conditions that guarantee the uniqueness of the steady-state response. To date these efforts have produced results that are too conservative for practical amplifier design. Thus for engineering applications the conditions are usually found empirically.

D. The Linear Parametric Amplifier Model. When all the conditions of the previous section are met, then (6.42) or a similar equation can be used to analyze the small-signal processing of the circuit. The function $\bar{q}_1'$ is the derivative of a charge as a function of voltage. As pointed out by (6.26) such a function is just an incremental capacitance. In fact the composite effect of the linearized equations (6.42) plus (6.44) and (6.45) with the remainders neglected is to replace the nonlinear capacitance in the circuits of Fig. 6.5 with a linear, time-variant capacitor. This linear capacitance $C(t)$ is constructed from the nonlinear capacitor of Fig. 6.5 by computing

$$C[\bar{v}(q_b(t))] = C(t) \tag{6.46}$$

where $C(v)$ is $d/dv\ (\bar{q}(v))$ and $q_b(t)$, defined by (6.38), is the charge on the nonlinear capacitance due to the power source alone.

When the conditions of the previous section are met, then $q_b(t)$ is a periodic function whose fundamental frequency is the frequency of the power source. It can be found by the methods of Sec. 6.1A. The time-variant capacitance $C(t)$ is also periodic with the same fundamental frequency. The circuit that must be analyzed is shown in Fig. 6.7.

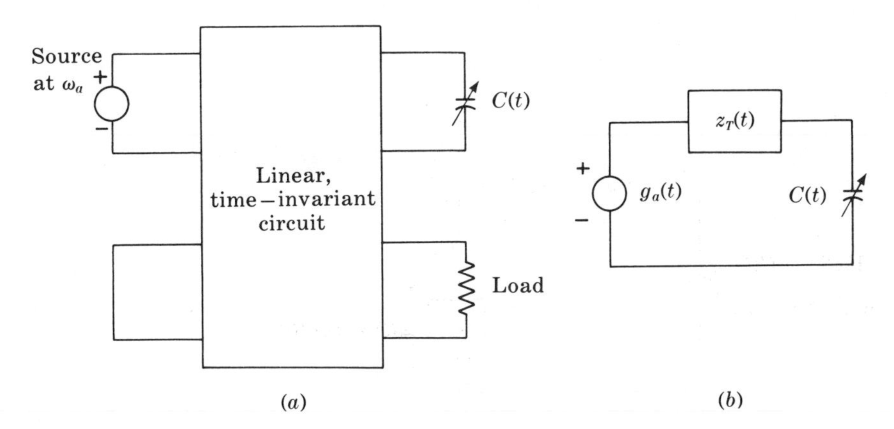

FIG. 6.7 Linearized circuits. (a) Linearized Fig. 6.3; (b) linear Thévenin circuit.

The first form, Fig. 6.7*a*, is derived from the circuit of Fig. 6.3 by replacing the nonlinear capacitance with $C(t)$ and by setting the source at ω_b to zero. The second form, Fig. 6.7*b*, is derived from Fig. 6.5 by replacing the nonlinear capacitance with $C(t)$ and by replacing the total Thévenin voltage source with only that part due to the small signal.

To demonstrate the two-step procedure for analyzing the small-signal processing properties of nonlinear frequency converters let us analyze the same circuit that we considered in Sec. 6.1A. That is, the circuit of Fig. 6.2 with $\mathcal{V}_a = 0.1$, $|\mathcal{V}_b| = 4.53$, and again with $\hat{\mathcal{V}}_a = 0.025$ and $|\hat{\mathcal{V}}_b| = 1.13$. The first step is to analyze the circuit of Fig. 6.8. For this circuit the first approximation to the charge on the linear part of the capacitor is

$$q_1(t) = H(j\omega_b)\mathcal{V}_b e^{j\omega_b t} + CC$$

Then, the first approximation to the voltage across the nonlinearity is

$$y_1(t) = 0.1[H^2(j\omega_b)\mathcal{V}_b{}^2 e^{j\omega_b t} + CC + 2|H(j\omega_b)|^2|\mathcal{V}_b|^2] \tag{6.47}$$

The second approximation to the charge is

$$\begin{aligned} q_2(t) = &- 0.2H(0)|H(j\omega_b)|^2|\mathcal{V}_b|^2 \\ &+ H(j\omega_b)\mathcal{V}_b e^{j\omega_b t} + CC \\ &- [0.1H^2(j\omega_b)H(2j\omega_b)\mathcal{V}_b{}^2 e^{j2\omega_b t} + CC] \end{aligned} \tag{6.48}$$

Proceeding to the third approximant with either value of $\mathcal{V}_b$ or $\hat{\mathcal{V}}_b$ will make no significant change in the results. Thus we terminate with this second approximant.

Since the circuit under consideration is a series circuit, the most convenient linearization is to use the elastance as the linear, time-variant element. Now

$$v(q) = (q + 0.1q^2)$$

Then

$$S(q) = v'(q) = 1 + 0.2q$$

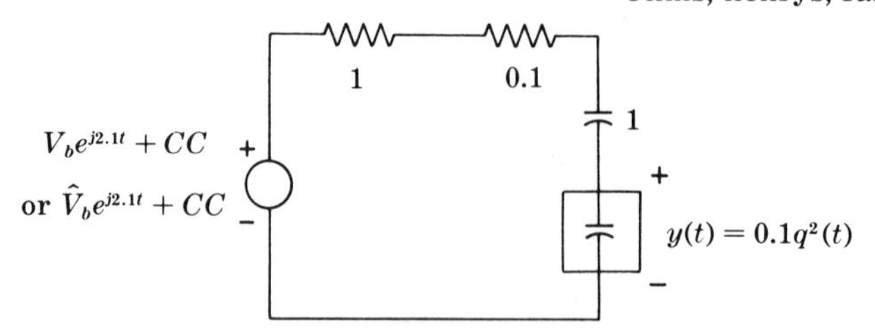

FIG. 6.8 Nonlinear circuit with power source only.

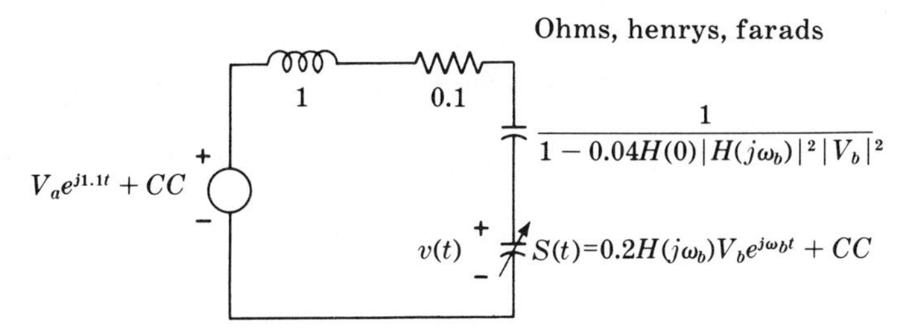

FIG. 6.9 Linearized simple circuit.

Using q_2 for q and neglecting the second harmonic term [since $H(j2\omega_0)$ is small] gives the circuit of Fig. 6.9.

To analyze the circuit of Fig. 6.9 we assume that the only significant frequency components of the current and charge are those at 1.1 rad/sec and at 1.0 rad/sec. That is, we assume

$$q(t) = Q_a e^{j1.1t} + CC + Q_- e^{j1.0t} + CC \tag{6.49}$$

The complex amplitudes Q_a and Q_- are to be determined. This charge on the time-variant elastance produces a voltage

$$\begin{aligned} v(t) &= 0.2H(j\omega_b)\mathcal{V}_b Q_-^* e^{j1.1t} + CC \\ &\quad + 0.2H(j\omega_b)\mathcal{V}_b Q_a^* e^{j1.0t} + CC \end{aligned} \tag{6.50}$$

In this expression only the assumed significant terms are included.

To find the complex amplitudes Q_a and Q_- we apply Kirchhoff's voltage law around the loop of Fig. 6.9. Since sinusoids at different frequencies are linearly independent, the complex amplitudes of the various voltages at each of the two frequencies must satisfy Kirchhoff's voltage law separately. Thus

$$\mathcal{V}_a = \left[\frac{1}{H(j\omega_a)} - 0.04H(0)|H(j\omega_b)|^2|\mathcal{V}_b|^2\right] Q_a + 0.2H(j\omega_b)\mathcal{V}_b Q_-^* \tag{6.51}$$

$$0 = 0.2H^*(j\omega_b)\mathcal{V}_b^* Q_a + \left[\frac{1}{H(j(\omega_a - \omega_b))} - 0.04H(0)|H(j\omega_b)|^2|\mathcal{V}_b|^2\right] Q_-^*$$

The second equation is the complex amplitudes of the voltages at frequency $\omega_a - \omega_b$, a negative frequency, so that the results are two equations in two unknowns. The solution is

$$Q_a = \frac{\mathcal{V}_a H(j\omega_a)[1 - 0.04H(0)H(j(\omega_a - \omega_b))|H(j\omega_b)|^2|\mathcal{V}_b|^2]}{\begin{aligned}&[1 - 0.04H(j\omega_a)H(0)|H(j\omega_b)|^2|\mathcal{V}_b|^2] \\ &\quad \times [1 - 0.04H(0)H(j(\omega_a - \omega_b))|H(j\omega_b)|^2|\mathcal{V}_b|^2] \\ &\quad - 0.04H(j(\omega_a - \omega_b))H(j\omega_a)|H(j\omega_b)|^2|\mathcal{V}_b|^2\end{aligned}} \tag{6.52}$$

A comparison of (6.52) with (6.15) above can be made when the terms preceded by 0.04 are all small compared to 1. Then the usual expansion of the denominator and neglecting of terms involving $(0.04)^2$ leaves exactly the linear terms in (6.15). When these higher-order terms are not negligible, then (6.52) contains terms involving $H(0)H(j\omega_a - \omega_b)$. Such terms do not appear in Q_{3a} in (6.15), but they will appear when Q_{4a} is computed. Thus the linearization (6.52) may be more accurate than Q_{3a}.

With the values of parameters used above we have

$$Q_a = \mathcal{V}_a \frac{(4.22\underline{/-152.4^\circ})(1 - j0.712)}{(1 - 0.30\underline{/-152.4^\circ})(1 - j0.712) - 3.0\underline{/-62.4^\circ}}$$

$$= 0.273\underline{/82.2^\circ} \tag{6.53}$$

$$\hat{Q}_a = \hat{\mathcal{V}}_a \frac{(4.22\underline{/-152.4^\circ})(1 - j0.0445)}{(1 - 0.01875\underline{/152.4^\circ})(1 - j0.0445) - 0.1875\underline{/-62.4^\circ}}$$

$$= 0.116\underline{/-161.4^\circ} \tag{6.54}$$

Comparison of Q_a with Q_{3a}, (6.17) above, shows considerable discrepancy as anticipated. The terms in (6.53) that must be small compared to 1 if the two methods are to give the same answers are respectively 0.712, 0.30, and 3.0. Because of the last of these we certainly do not expect Q_a and Q_{3a} to agree. On the other hand $\hat{Q}_a$ of (6.54) and $\hat{Q}_{3a}$ of (6.18) are virtually identical considering that the computations were made by slide rule. Continuing the computations to show power flow will show that both cases appear to convert energy to the frequency ω_a. There is no particular value to the power calculations since in the first case the result is not reliable and in the second case the result is the same as the one obtained before.

The purpose of the above examples was to show that the analysis that engineers are forced to use for these complex circuits contains many unjustified assumptions. Analysis by different methods may give very different answers if one is not careful to see that the assumptions of each method are justified. The nonlinear part of the analysis can be improved by computing higher-order approximations. In the next section we take a more detailed look at the analysis of linear circuits with periodic, time-variant parameters. Such circuits can amplify a signal. The amplifying power comes from the time-variant parameters. The term *parametric amplifier* is used for those amplifying circuits wherein the amplifying power comes from the variation of a parameter as in the case of the circuit of Fig. 6.9.

6.3 Linear Circuits with Periodic Parameters

In the previous section, we saw that if a nonlinear frequency-converting circuit acts as a linear signal processor then the signal-processing operation could be ascertained by analyzing a linear circuit with periodic parameter variation. In this section, we discuss the analysis of this class of circuits. The basic analysis method applies to a circuit with any type of periodic time-variant parameter—resistor, inductor, or capacitor. To be specific, the discussion is carried out for a circuit with one time-variant capacitor.

A. The Basic Circuit Equation. The circuit under consideration in this section is that of Fig. 6.10. The current $i(t)$ along with the impulse-response function $y(t)$ represents the Norton equivalent of the small-signal source and the linear, time-invariant network as seen from the terminals of the time-variant capacitor. The time-variant capacitance is a uniformly continuous function of time and thus can be represented by a Fourier series. It can be shown[7,9] that the response computation is not too inaccurate if the infinite Fourier series is replaced by a finite trigonometric polynomial. In fact, most analyses truncate the polynomial after one time-dependent term. Then the capacitance variation is sinusoidal. The methods of Ref. 7 can be used to see the error in this drastic truncation. For most practical amplifiers, the error is not serious.

For this section, we discuss the circuit of Fig. 6.10 with

$$C(t) = C_0 + C_1 e^{j\omega_b t} + C_1^* e^{-j\omega_b t} \tag{6.55}$$

where

$$|C_1| \leq \frac{C_0}{2}$$

The inequality relating C_0 and C_1 guarantees that $C(t)$ is always positive. Furthermore, choosing C_1 real is equivalent to a specific selection of the time origin. This simplifies the notation without loss of generality. The notation is further simplified if the time scale is normalized so that $\omega_b = 1$; that is, we use $\tau = \omega_b t$ as the time variable. A final normalization to simplify notation is the selection of an impedance level factor so that $C_1 = 1$. Then, (6.55) becomes

$$C(\tau) = C_0 + e^{j\tau} + e^{-j\tau} = C_0 + 2\cos\tau \tag{6.56}$$

where $C_0 > 2$.

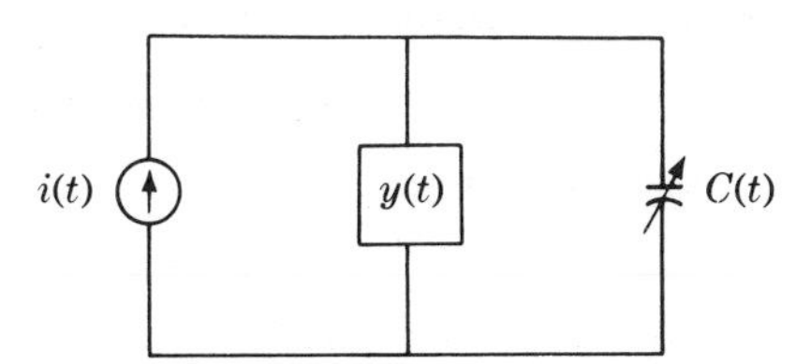

Fig. 6.10 Basic linear circuit.

With the above normalizations, the Kirchhoff-current-law equation for the circuit of Fig. 6.10 is

$$i(\tau) = \int_{\alpha}^{\tau} y(\tau - \lambda)v(\lambda)\, d\lambda + \frac{d}{d\tau} c(\tau)v(\tau) \tag{6.57}$$

If the system is stable and the linear, time-invariant system is damped, then the initial time α can be taken as $-\infty$. These two properties apply if the derivation of the linear model in the previous section of this chapter is valid. With α set to $-\infty$ and the formula (6.56) used for $C(\tau)$, (6.57) becomes

$$i(\tau) = \int_{-\infty}^{\tau} y(\tau - \lambda)v(\lambda)\, d\lambda + C_0 \frac{dv(\tau)}{d\tau} + \frac{d}{d\tau}[(e^{j\tau} + e^{-j\tau})v(\tau)] \tag{6.58}$$

The primary application of parametric amplifiers is at UHF and microwave frequencies. For such amplifiers the basic design criteria—gain, bandwidth, phase shift—are given in the frequency domain. Consequently, for design purposes, it is most convenient to Laplace-Fourier transform (6.58). The result of this transformation is a difference equation in the frequency domain. The notation is simplified if ω is taken as the frequency variable rather than the usual s of the conventional Laplace-Fourier transform.† With this transform the individual terms in (6.58) have the following representations in the frequency domain.

The current $i(\tau)$ transforms to $I(\omega)$. The convolution transforms to the product of the admittance $Y(\omega)$ and the frequency-domain voltage $V(\omega)$. Since differentiation in the time domain is equivalent to multiplication by $j\omega$ in the frequency domain, the next term transforms to $j\omega C_0 V(\omega)$. By the shifting theorem for Laplace transforms the time-domain term $e^{j\tau}v(\tau)$ transforms to $V(\omega - 1)$. Then differentiation gives a factor $j\omega$ in the frequency domain. The final result is

$$I(\omega) = Y(\omega)V(\omega) + j\omega C_0 V(\omega) + j\omega V(\omega + 1) + j\omega V(\omega - 1) \tag{6.59}$$

B. The Steady-state Frequency Response. In principle (6.59) can be solved for $V(\omega)$. This result can be inverse transformed to find $v(\tau)$, the solution to (6.58). Although the difference equation (6.59) has infinitely many solutions, only one will have an inverse transform. The transform of this unique solution is the unique $v(\tau)$. For amplifier design, the important considerations occur for real frequencies, that is, when the input $i(\tau)$ is a steady-state sinusoid at frequency ω_a. In the frequency domain this means

$$I(\omega) = \mathscr{I}\delta(\omega - \omega_a) + \mathscr{I}^*\delta(\omega - \omega_a) \tag{6.60}$$

† Specifically given $x(\tau)$ then

$$X(\omega) = \int_{-\infty}^{\infty} x(\tau)e^{-j\omega\tau}\, dt$$

where ω is a complex variable.

where δ is the Dirac delta function and $\mathcal{I}$ is the complex amplitude of the sine wave relative to the previously selected time origin.

With this current the resulting voltage is an infinite series of steady-state sinusoids at frequencies $|k + \omega_a|$ with k taking on all integer values from $-\infty$ to ∞. Since the system is linear, the complex amplitude of each of the frequencies of the response is proportional to the complex amplitude of the input. Thus, the time-domain response is

$$v(\tau) = Z_0(\omega_a)\mathcal{I}e^{j\omega_a\tau} + Z_0^*(\omega_a)\mathcal{I}^*e^{-j\omega_a\tau} + \sum_{k=1}^{\infty} [Z_k(\omega_a)\mathcal{I}e^{j(k+\omega_a)\tau} + Z_k^*(\omega_a)\mathcal{I}^*e^{-j(k+\omega_a)\tau} + Z_{-k}(\omega_a)\mathcal{I}e^{j(\omega_a-k)\tau} + Z_{-k}^*(\omega_a)\mathcal{I}^*e^{-j(\omega_a-k)\tau}] \quad (6.61)$$

where the Z_k are computable in terms of the parameters of the circuit. The dimension of the Z_k is impedance. Each is the ratio of a voltage amplitude to a current amplitude.

Although there are no closed-form formulas for the Z_k in (6.61), there are known formulas in the form of convergent continued fractions.[10,†] They are

$$Z_0(\omega) = \frac{1}{j\omega[A(\omega) + L(\omega) + L^*(-\omega^*)]} \quad (6.62)$$

where

$$A(\omega) = \frac{Y(\omega)}{j\omega} + C_0$$

$$L(\omega) = \cfrac{-1}{A(\omega - 1) + \cfrac{1}{A(\omega - 2) - \cfrac{1}{A(\omega - 3) + \cfrac{1}{A(\omega - 4) - \cfrac{1}{A(\omega - 5) + \cfrac{1}{\ddots}}}}}}$$

For $k > 0$

$$\begin{aligned} Z_k(\omega) &= Z_0L^*(-\omega^*)L^*(-\omega^* - 1) \cdots L^*(-\omega^* - k + 1) \\ Z_{-k}(\omega) &= Z_0L(\omega)L(\omega - 1) \cdots L(\omega - k + 1) \end{aligned} \quad (6.63)$$

The formulas for the Z_k are all fairly simple once the continued fraction $L(\omega)$ is computed for the various values of ω required. When $Y(\omega)$ is the admittance of a stable *RLC* circuit, then the continued fraction converges. The value of $L(\omega)$ can be computed to better and better

† These formulas are all defined for complex frequencies. Thus ω can be complex. In the present discussion we are interested in real frequencies only.

approximations by terminating the computation after more and more terms. For most engineering problems only one or two terms are required. In fact, virtually all parametric amplifiers are designed by terminating after only one term. Then the computation is extremely simple.

In Ref. 10 an upper-bound estimate of the error introduced by truncating the continued fraction after any given number of terms is presented. In addition, a formula for adding an extra term to the finite continued fraction without having to repeat all the computations of the original fraction is given. Although these formulas are important from a theoretical point of view, they are rarely used in amplifier design. The usual design formulas are based on the physical assumption that the time-invariant circuit characterized by the admittance $Y(\omega)$ can be built so that at all frequencies above the pump (power-source) frequency the circuit presents essentially a short circuit across the capacitor terminals. Then $Y(\omega)$ and $A(\omega)$ defined after (6.62) are both very large for $\omega > 1$. For these amplifiers the signal-source frequency ω_a is less than 1. The only important voltages are those at frequencies ω_a and $1 - \omega_a$. Thus, only Z_0 and Z_{-1} have significance in (6.61). The formulas (6.62) and (6.63) give

$$Z_0(\omega_a) = \frac{1}{j\omega_a[A(\omega_a) + L(\omega_a) + L^*(-\omega_a)]} \tag{6.64}$$

$$Z_{-1}(\omega_a) = L(\omega_a)Z_0(\omega_a) \tag{6.65}$$

with

$$A(\omega_a) = \frac{Y(\omega_a)}{j\omega_a} + C_0$$

$$L(\omega_a) = \frac{-1}{Y(\omega - 1)a/[j(\omega_a - 1)] + C_0}$$

$$L(-\omega_a) = \frac{-1}{A(-\omega_a - 1)} = 0$$

This last term is zero because the frequency $-\omega_a - 1$ is larger than 1 in absolute values. Thus, $A(-\omega_a - 1)$ is assumed to be essentially infinite.

With the above assumptions the formulas for Z_0 and Z_1 become

$$\begin{aligned} Z_0(\omega_a) &= \frac{Y(\omega_a - 1) + j(\omega_a - 1)C_0}{[Y(\omega_a - 1) + j(\omega_a - 1)C_0][Y(\omega_a) + j\omega_a C_0] + \omega_a(\omega_a - 1)} \\ &= \frac{Y^*(1 - \omega_a) - j(1 - \omega_a)C_0}{[Y^*(1 - \omega_a) - j(1 - \omega_a)C_0][Y(\omega_a) + j\omega_a C_0] - \omega_a(1 - \omega_a)} \end{aligned} \tag{6.66}$$

where the last expression has all frequencies positive.

$$Z_1(\omega_a) = \frac{j(1 - \omega_a)}{[Y^*(1 - \omega_a) - j(1 - \omega_a)C_0][Y(\omega_a) + j\omega_a C_0] - \omega_a(1 - \omega_a)} \tag{6.67}$$

The two formulas (6.66) and (6.67) are the ones used by virtually all engineers for the design of two-frequency parametric amplifiers.

A common design for a two-frequency parametric amplifier is that of Fig. 6.11. The tank circuit on the left consisting of L_a, C_a, and G_a is tuned to the signal frequency ω_a. The tank circuit on the right is tuned to the idler (difference) frequency $1 - \omega_a$. The method of selecting the values of the circuit parameters is best illustrated by picking an appropriate set of numbers and then saying why they are good after the analysis is completed. In the next section, methods of designing parametric amplifiers will be given in more detail.

The signal frequency is chosen as 0.2. Then the idler is at 0.8 rad/sec. The circuit parameters chosen are

$$\begin{aligned} C_0 &= 10 \\ C_a &= 90 \\ L_a &= 0.25 \\ C_i &= 90 \\ L_i &= 0.0158 \\ G_a &= 0.25 \\ G_i &= 1.0 \end{aligned} \tag{6.68}$$

With these numbers

$$\begin{aligned} Y(0.2) &= 0.264 - j1.98 \\ Y(0.8) &= 1.21 - j7.75 \end{aligned} \tag{6.69}$$

Then the formulas (6.66) and (6.67) give

$$\begin{aligned} Z_0(0.2) &= \frac{1.21 + j7.75 - j8.0}{(1.21 - j0.26)(0.26 - j2 + j2) - 0.2(0.8)} \\ &= \frac{1.21 - j0.25}{0.14 - j0.0625} = 7.95\underline{/-12°} \end{aligned} \tag{6.70}$$

$$Z_1(0.2) = \frac{j0.8}{0.153\underline{/-24°}} = 5.23\underline{/114°} \tag{6.71}$$

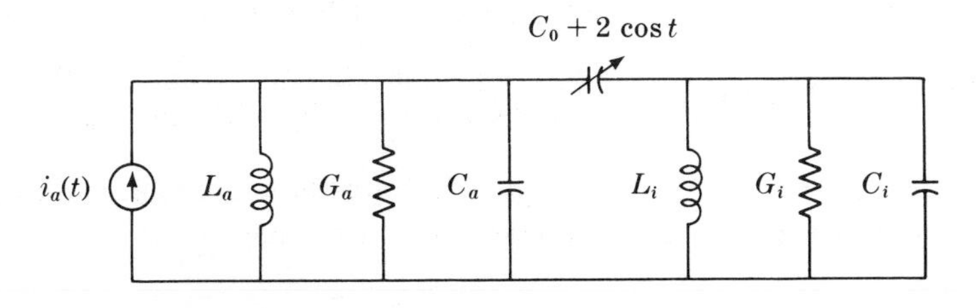

Fig. 6.11 A two-frequency amplifier.

To see that this circuit converts energy from the time-variant capacitance to the equivalent circuit loads we compute the dissipated power and the source power for a 1-amp input. At frequency ω_a the dissipated power is

$$P_a = |Z_0(0.2)|^2 \operatorname{Re}[Y(0.2)] = 15.8 \text{ watts} \tag{6.72}$$

At frequency $1 - \omega_a$ the dissipated power is

$$P_{-1} = |Z_{-1}(0.8)|^2 \operatorname{Re}[Y(0.8)] = 33.2 \text{ watts} \tag{6.73}$$

The power delivered by the source can be found by computing the voltage across the source at ω_a. Since the idler tank is virtually a short circuit at ω_a, this voltage is equal to $Z_0(0.2)$. Thus

$$P_s = \operatorname{Re}[7.95\underline{/-12^\circ}] = 7.8 \text{ watts} \tag{6.74}$$

Thus 8.0 watts at ω_a and 33.2 watts at $1 - \omega_a$ are supplied by the time-variant capacitance.

The above numbers check quite well with the Manley-Rowe equations. The total power delivered by the capacitance is 41.2 watts, the sum of the signal and idler power. This is $P_{0,1}$ in (6.22). The signal power of 8.0 watts is minus $P_{1,0}$; and the idler power of 33.2 watts is minus $P_{1,-1}$. The specific terms for (6.22) are

$$\frac{P_{0,1}}{\omega_b} = \frac{41.2}{1} = 41.2$$

$$\frac{P_{-1,1}}{-\omega_a + \omega_b} = \frac{-33.2}{-0.2 + 1} = -41.5$$

$$\frac{P_{1,0}}{\omega_a} = \frac{-8}{0.2} = -40.0$$

The three numbers are not exactly equal in magnitude due to the slide-rule accuracy and the fact that we neglected power flow at other frequencies.

In the derivation of the formulas (6.66) and (6.67) it was assumed that for $\omega > 1$, $Y(\omega)$ was essentially infinite. This permitted truncation of $L(\omega_a)$ as defined by (6.62) after only one term and the assumption that $L^*(-\omega_a)$ is zero. For the example of Fig. 6.11

$$Y(\omega) = \frac{(1 - 22.5\omega^2 + j0.0625\omega)(1 - 1.42\omega^2 + j0.0158\omega)}{-0.0014\omega^2 - j(0.72\omega^3 - 0.266\omega)} \tag{6.75}$$

For $\omega > 1$ the numerator is real and the denominator imaginary for all practical purposes. Thus, for $\omega > 1$ we may write

$$Y(\omega) \approx j\,\frac{32\omega^4 - 23.9\omega^2 + 1}{\omega(0.72\omega^2 - 0.226)} \tag{6.76}$$

The important frequencies are $\omega = 1.2, 1.8, 2.2, 2.8$, etc. At these frequencies the formulas (6.62) and (6.76) yield

$$\begin{aligned} A(1.2) &= 38.4 \\ A(1.8) &= 48.5 \\ A(2.2) &= 50 \\ A(2.8) &= 52.2 \end{aligned} \tag{6.77}$$

As $\omega \to$ H, $A(\omega) \to 55$. For larger values of ω we may as well take $A(\omega) = 55$.

With the above numbers for $A(\omega)$ for $\omega > 1$ and $Y(1 - \omega_a)$ from (6.69) the formula for $L(\omega)$ is

$$L(0.2) = \cfrac{-1}{\cfrac{1.21 - j7.75}{-j0.8} + 10 + \cfrac{1}{48.5 - \cfrac{1}{52.2 + \cfrac{55 - \sqrt{55^2 - 4}}{2}}}} \tag{6.78}$$

where $\dfrac{-K + \sqrt{K^2 - 4}}{2}$ is the result of working out the continued fraction

$$\cfrac{-1}{K + \cfrac{1}{K - \cfrac{1}{K \cdot \ddots}}}$$

An examination of (6.78) shows that the error in $L(\omega)$ by using only the first term is negligible. For $L^*(-\omega)$ we have

$$L^*(-0.2) = \cfrac{-1}{A^*(1.2) + \cfrac{1}{A^*(2.2) - \cfrac{1}{A^*(3.2) + \cfrac{1}{\ddots}}}}$$

$$= \cfrac{-1}{38.4 - \cfrac{1}{50 - \cfrac{55 - \sqrt{55^2 - 4}}{2}}} \tag{6.79}$$

To slide-rule accuracy we may as well take

$$L^*(-0.2) = \frac{1}{38.4} = 0.026$$

Thus neglecting $L^*(-0.2)$ is not a bad approximation.

For the example of Fig. 6.11 the approximate formulas (6.66) and (6.67) are well within slide-rule accuracy of the exact expression. The reason is that for $\omega > 1$, $A(\omega)$ is quite large. Thus the effect of the neglected terms in the continued fraction is negligible. If we had chosen C_a and C_i to be much smaller than C_0, the effect of the neglected terms would not be negligible. In this case a better approximation is to assume that for $\omega > 1$,

$$\frac{Y(\omega)}{j\omega} = \lim_{\omega \to \infty} \frac{Y(\omega)}{j\omega} = C_p \tag{6.80}$$

where C_p is the parasitic shunt capacitance of the time-invariant network. In this case for $\omega_a < 1$ we have

$$L(\omega_a) = \frac{-1}{A^*(1 - \omega_a) - [C_0 + C_p - \sqrt{(C_0 + C_p)^2 - 4}]/2}$$

$$L^*(-\omega) = -\frac{C_0 + C_p - \sqrt{(C_0 + C_p)^2 - 4}}{2} \tag{6.81}$$

Inclusion of the extra terms in (6.81) in the formula (6.62) for $Z_0(\omega_a)$ gives

$$Z_0(\omega_a) = \frac{1}{j\omega_a\{A(\omega_a) - 1/[A^*(1 - \omega_a) - \mu] - \mu\}} \tag{6.82}$$

where $$\mu = -\frac{C_0 + C_p - \sqrt{(C_0 + C_p)^2 - 4}}{2}$$

Similar formulas for the Z_k can also be constructed when μ is not negligible.

6.4 Design Considerations for Linear, Two-frequency, Parametric Amplifiers

The analysis method of the previous sections of this chapter resulted in a pair of comparatively simple formulas for the voltage at the signal and idler (difference) frequencies in a two-frequency parametric amplifier. These formulas can be used to design amplifiers for good gain, bandwidth, and noise figure. Of course, when the formulas are used for a design, the engineer must be sure the design includes the circuit constraints required to make the formulas valid. In the formulas (6.66) and (6.67) the admittance $Y(\omega)$ of the linear, time-invariant network always appears in conjunction with certain other terms. Thus the important design quantity is not this admittance by itself but rather

$$A(\omega) = \frac{Y(\omega) + j\omega C_0}{j\omega} \tag{6.83}$$

The numerator of (6.83) is the admittance of $Y(\omega)$ in parallel with the capacitance C_0. Since there is much literature on the synthesis of lumped, linear, time-invariant networks with a parallel capacitance at one or more ports, it is convenient to consider $Y(\omega) + j\omega C_0$ as the admittance to be synthesized. Since the time-variant capacitance is $C_0 + 2\cos t$, it can be considered as C_0 in parallel with a capacitor $2\cos t$. These associations result in the circuit of Fig. 6.12.

A. The Equivalent Admittance Concept. The basic design philosophy for selecting the circuit in the dotted box of Fig. 6.12 is to synthesize a matching network to take maximum power over a band of frequencies from the time-variant capacitance on the right. In order to synthesize a matching network we first need to find the admittance to be matched. That is, we need an equivalent admittance for the time-variant capacitance of Fig. 6.12. At frequency ω_a the complex amplitude of the voltage across the time-variant capacitance is obtained from (6.61) and (6.62). It is

$$V_a = \mathcal{I} Z_0(\omega_a) = \frac{\mathcal{I}}{j\omega[L(\omega_a) + A(\omega_a) + L^*(-\omega_a)]} \tag{6.84}$$

The complex amplitude of the current through the time-variant capacitor at frequency ω_a is the difference between the source current and the current that flows through the admittance in the dotted box of Fig. 6.12. Thus,

$$\begin{aligned} \mathcal{I}_{C_a} &= \mathcal{I} - V_a[Y(\omega_a) + j\omega_a C_0] \\ &= \mathcal{I}\left[1 - \frac{A(\omega_a)}{L(\omega_a) + A(\omega_a) + L^*(-\omega_a)}\right] \\ &= \mathcal{I}\left[\frac{L(\omega_a) + L^*(-\omega_a)}{L(\omega_a) + A(\omega_a) + L^*(-\omega_a)}\right] \end{aligned} \tag{6.85}$$

Dividing (6.85) by (6.84) gives the equivalent admittance for the time-variant capacitor at frequency ω_a. It is

$$Y_C(\omega_a) = [L(\omega_a) + L^*(-\omega_a)]j\omega_a \tag{6.86}$$

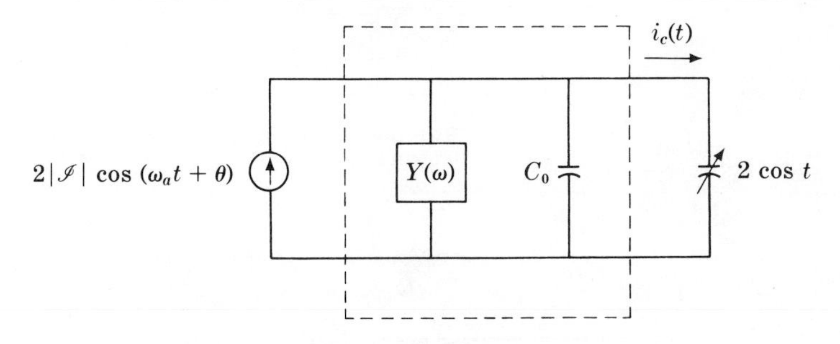

FIG. 6.12 Separation of the variable capacitance.

The formula (6.86) is exact for the circuit of Fig. 6.12. If the approximation that only frequencies below the pump frequency are significant is used, then the formula simplifies still further. The result is

$$Y_C(\omega_a) = \frac{\omega_a(\omega_a - 1)}{Y(\omega_a - 1) + j(\omega_a - 1)C_0}$$

$$= -\frac{\omega_a(1 - \omega_a)}{Y^*(1 - \omega_a) - j(1 - \omega_a)C_0} \tag{6.87}$$

This last formula shows the conditions that should be met for an amplifier. If the admittance of the dotted box in Fig. 6.12 is real and positive at frequency $1 - \omega_a$, then $Y_C(\omega_a)$ is real and negative. If we design a negative-resistance amplifier for frequencies ω_a over some band and maintain the admittance real over the band $1 - \omega_a$, then we have designed a parametric amplifier.

B. The Design of Negative-resistance Amplifiers.† Before discussing the design of a negative-resistance parametric amplifier that has a constraint at the idler frequency, let us go over the design of a simple negative-resistance amplifier. There are many ways to utilize negative resistances to make an amplifier. At UHF and microwave frequencies, the frequency range where parametric amplifiers are most useful, a common configuration is the circulator—the negative-resistance amplifier of Fig. 6.13. With this configuration the problem is the design of the

† For a more complete discussion of negative-resistance amplifiers see Ref. 11.

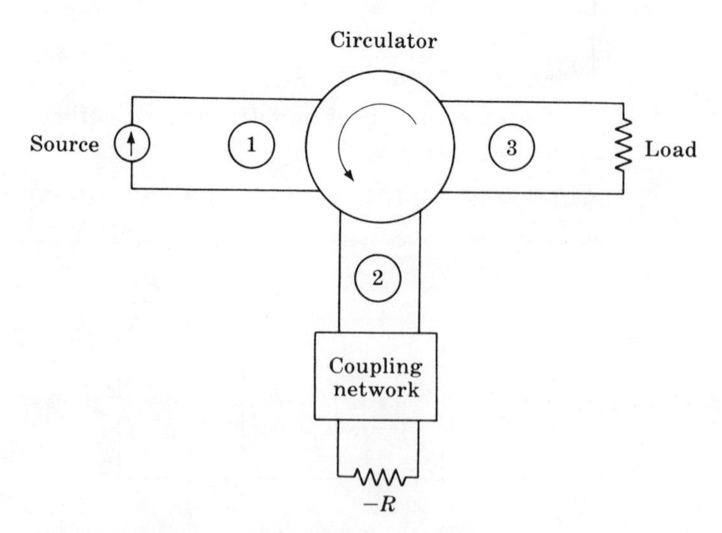

FIG. 6.13 Basic negative-resistance amplifier.

coupling network to give the desired gain and bandwidth properties while accommodating the parasitic shunt capacitance of the negative resistor.

While our ultimate objective is a high-frequency bandpass amplifier, we start by designing a low-pass amplifier. This result is then converted to meet the center frequency and bandwidth requirements by a low-pass to bandpass transformation.

An amplifier that is made to connect a source with an internal impedance to a finite impedance load is normally designed on the basis of available power gain. That is, the gain is the power delivered to the load divided by the power available from the source. For the circuit of Fig. 6.13, when the source and load are both matched to the characteristic impedance of the circulator at their respective ports, the available power gain is equal to the squared magnitude of the reflection coefficient seen looking into the coupling network from the circulator. The characteristic impedance used for computing this reflection coefficient is the circulator characteristic impedance at port 2.

The problem of design of a negative-resistance amplifier with the configuration of Fig. 6.13 reduces to the problem of synthesizing the coupling network of Fig. 6.14 so that the reflection coefficient ρ has the desired gain-bandwidth properties. In the present discussion, we shall consider only the problem of synthesizing a lossless coupling network. This is adequate for the circuit model of Fig. 6.10. If a varactor-diode model including a spreading resistance such as that of Fig. 6.4 is to be included, then more involved design procedures are needed.[12]

Any two-port network, such as the coupling network of Fig. 6.14, can be characterized by a 2×2 scattering matrix. This matrix contains two reflection coefficients and two transmission coefficients. A one-port, such as the negative resistance in Fig. 6.14, is characterized by a reflection coefficient. For the present discussion, we consider only real characteristic impedances. For a one-port load the relation between the reflection coefficient and the load impedance is

$$\rho_L = \frac{Z_0 - Z_L}{Z_0 + Z_L} \tag{6.88}$$

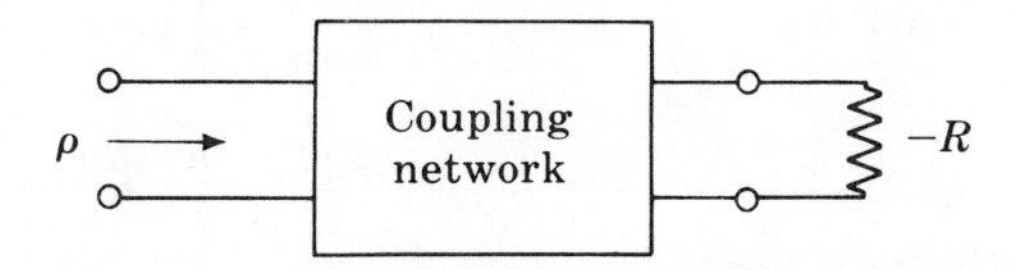

FIG. 6.14 Transformation of the negative resistance.

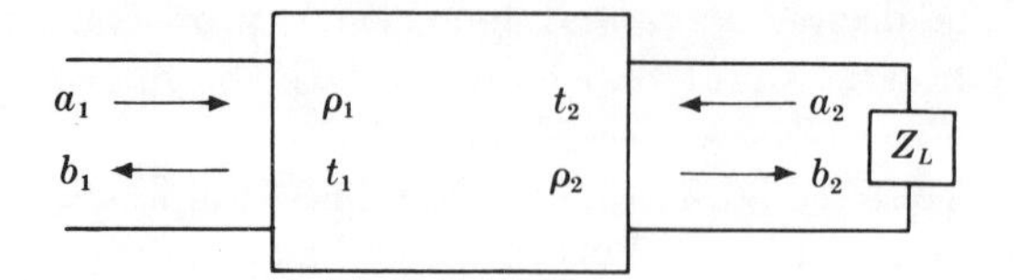

FIG. 6.15 Scattering variables.

Since we are interested in getting all the power from the negative resistor to the load in Fig. 6.14 we want losses in the coupling network minimized. For the first pencil-and-paper design the best we can do is to make the coupling network lossless. A lossless network of L's, C's, and ideal transformers is reciprocal. Thus, the two transmission coefficients, the off-diagonal scattering matrix elements, are equal. The lossless property relates the transmission and reflection coefficients by

$$1 - |\rho_1|^2 = |t_1|^2 = 1 - |\rho_2|^2 = |t_2|^2 \tag{6.89}$$

and

$$\rho_1 t_2^* + t_1^* \rho_2 = 0$$

Furthermore since $t_1 = t_2$, $\rho_1 = \rho_2^*$.

To get the interconnection rules and the requirements on the lossless networks for a good amplifier, let us refer to Fig. 6.15. Let a_1 and a_2 be the complex amplitudes of the waves incident on the two-port. Let b_1 and b_2 be the two waves leaving the two-port as shown. Then the scattering matrix description of the two-port is

$$\begin{bmatrix} b_1 \\ b_2 \end{bmatrix} = \begin{bmatrix} \rho_1 & t_2 \\ t_1 & \rho_2 \end{bmatrix} \begin{bmatrix} a_1 \\ a_2 \end{bmatrix} \tag{6.90}$$

The one-port load is described by

$$a_2 = \rho_L b_2 \tag{6.91}$$

The desired reflection coefficient ρ is the ratio of b_1 to a_1. Combining (6.90) and (6.91) gives

$$b_1 = \left(\rho_1 + \frac{t_1 t_2 \rho_L}{1 - \rho_2 \rho_L}\right) a_1 \tag{6.92}$$

Therefore

$$\rho = \frac{b_1}{a_1} = \left(\rho_1 + \frac{t_1 t_2 \rho_L}{1 - \rho_2 \rho_L}\right)$$

$$= \frac{\rho_1 - \rho_L(\rho_1 \rho_2 - t_1 t_2)}{1 - \rho_2 \rho_L} \tag{6.93}$$

When the two-port in Fig. 6.15 is lossless and the load is a negative resistance as in Fig. 6.14, then there are some interesting simplifications of (6.93). The losslessness formula (6.89) reduces (6.93) to

$$\rho = \frac{\rho_1 - \rho_L}{1 - \rho_2 \rho_L} \tag{6.94}$$

When the load is a negative resistance the formula (6.94) relates in an interesting way to the same lossless network with a positive-resistance load. If we use a prime to designate the reflection coefficient of a positive-resistance load of the same absolute value we have both ρ_L and ρ'_L real and

$$\rho'_L = \frac{1}{\rho_L}$$

The overall reflection coefficient of Fig. 6.14 with the negative resistor replaced by a positive resistor is

$$\rho' = \frac{\rho_1 - \rho'_L}{1 - \rho_2 \rho'_L} = \frac{\rho_1 \rho_L - 1}{\rho_L - \rho_2} \tag{6.95}$$

Since ρ_L is real and $\rho_1 = \rho_2^*$

we have
$$\rho' = \frac{\rho_L \operatorname{Re}(\rho_1) - 1 + j\rho_L \mathcal{I}_m \rho_1}{\rho_L - \operatorname{Re}(\rho_1) - j\mathcal{I}_m \rho_1} \tag{6.96}$$

and
$$\rho = \frac{\operatorname{Re}(\rho_1) - \rho_L + j\mathcal{I}_m \rho_1}{1 - \rho_L \operatorname{Re}(\rho_1) - j\rho_L \mathcal{I}_m \rho_1} \tag{6.97}$$

A comparison of ρ and ρ' shows that

$$\rho = \frac{1}{\rho'} \tag{6.98}$$

Because of the relationship (6.98) the synthesis of a lossless network terminated in a negative resistor reduces to the well-documented problem of the synthesis of a lossless network terminated in a positive resistor. In fact the standard filter networks with properly chosen constants lead to pretty good negative-resistance amplifiers. First let us consider a Butterworth filter designed to work between a 1-ohm internal resistance source and slightly mismatched load. The transmission coefficient† for this filter is

$$|t_1(j\omega)|^2 = \frac{K^2}{1 + \omega^{2n}} \tag{6.99}$$

with $K < 1$. Since the filter is lossless, the magnitude of the reflection coefficients are both

$$|\rho_2|^2 = |\rho_1|^2 = 1 - |t(j\omega)|^2 = \frac{1 - K^2 + \omega^{2n}}{1 + \omega^{2n}} \tag{6.100}$$

† Since the circuit is reciprocal, $t_1 = t_2$ and there is only one transmission coefficient.

The formula (6.99) is the available power gain of the filter when the terminating resistor is the characteristic impedance at port 2. When the filter is so terminated, (6.100) is not only the reflection coefficient of the lossless two-port but also the input reflection coefficient $|\rho'|^2$. When the same lossless two-port is terminated in a negative resistor of equal magnitude, the input reflection coefficient has magnitude

$$|\rho(j\omega)|^2 = \frac{1}{|\rho'(j\omega)|^2} = \frac{1 + \omega^{2n}}{1 - K^2 + \omega^{2n}} \tag{6.101}$$

This last expression has a maximum value of $1/(1 - K^2)$ at direct current and decreases monotonically to 1 at ∞.

As an example suppose we choose $K^2 = 0.96$ so that the dc power gain is 25. To make the computations easy but nontrivial, we choose a second-order Butterworth filter characteristic. Thus with positive termination the transmission function is

$$|t(j\omega)|^2 = \frac{0.96}{1 + \omega^4}$$

The reflection coefficient is

$$|\rho_1(j\omega)|^2 = \frac{0.04 + \omega^4}{1 + \omega^4} \tag{6.102}$$

The realization follows the standard procedure. (See Ref. 13, Sec. 12-4.)

When $|\rho(j\omega)|^2$ is given by (6.102) the reflection coefficient for complex s is given by

$$\begin{aligned} \rho_1(s)\rho_1(-s) &= \frac{s^4 + 0.04}{s^4 + 1} \\ &= \frac{(s^2 + 0.6325s + 0.2)(s^2 - 0.6325s + 0.2)}{(s^2 + \sqrt{2}\, s + 1)(s^2 - \sqrt{2}\, s + 1)} \end{aligned} \tag{6.103}$$

The denominator factor with all positive terms belongs to $\rho_1(s)$ since left-half-plane poles are required. Either numerator factor can be assigned to $\rho_1(s)$ and the other to $\rho_1(-s)$. Let us assign the first to $\rho_1(s)$. Then

$$\rho_1(s) = \frac{s^2 + 0.6325s + 0.2}{s^2 + \sqrt{2}\, s + 1} \tag{6.104}$$

The input impedance for the terminated filter is

$$Z_{\text{in}} = \frac{1 - \rho(s)}{1 + \rho(s)}$$

Thus

$$\begin{aligned} Z_{\text{in}}(s) &= \frac{0.7815s + 0.8}{2s^2 + 2.046s + 1.2} \\ &= \frac{1}{2.56s + 1/(0.650s + 0.667)} \end{aligned} \tag{6.105}$$

The realization of (6.105) is shown in Fig. 6.16.

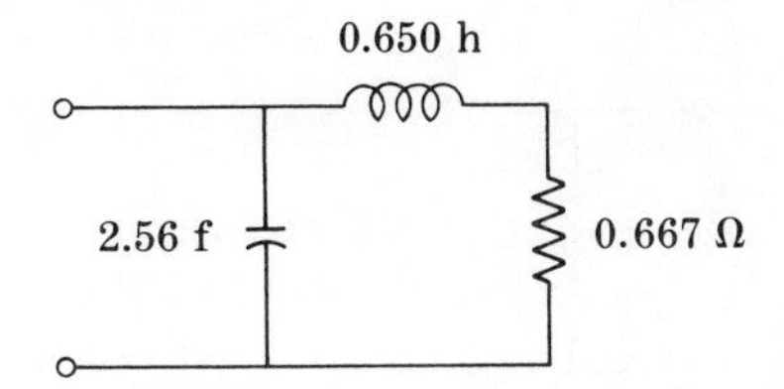

FIG. 6.16 Basic prototype.

The prototype element values in Fig. 6.16 can now be used in a low-pass to bandpass transformation to give the parameters for a high-frequency bandpass amplifier. Then the impedance level can be scaled to match the particular negative-resistance device that is to be used for the amplifier. As a filter the prototype of Fig. 6.16 has a bandwidth of 1 rad/sec. When the positive resistor is replaced by a negative resistor, the resulting amplifier does not have the same bandwidth. Before we can perform the necessary frequency transformation we must know the bandwidth of the prototype negative-resistance amplifier.

The formula for the available power gain of the prototype amplifier as a function of frequency is given by (6.101). With the values of Fig. 6.16 the formula is

$$|\rho(j\omega)|^2 = \frac{1 + \omega^4}{0.04 + \omega^4} \tag{6.106}$$

Since the dc gain is 25, the half-power point occurs when

$$\frac{1 + \omega_c{}^4}{0.04 + \omega_c{}^4} = 12.5 \tag{6.107}$$

where ω_c is the cutoff frequency. Solving (6.107) for ω_c gives

$$\omega_c = 0.4566 \tag{6.108}$$

The prototype negative-resistance amplifier with available power gain of 25 and bandwidth 1 rad/sec operating out of a 1-ohm source can be found by frequency scaling the prototype of Fig. 6.16. The scaling factor is given by (6.108). Such scaling changes the inductor and capacitor to the values in Fig. 6.17.

Since the magnitude of the reflection coefficient is all that matters for the gain-bandwidth considerations and since for the lossless network $|\rho_2| = |\rho_1|$, we could design the negative-resistance amplifier going the other way. That is, we could operate out of a 0.667-ohm positive-resistance source into a negative 1-ohm load. The gain and bandwidth would be the same. Since the available power gain is independent of overall impedance-level change, we could then scale all impedances up by a factor of $3/2$. The result would be an amplifier operating from a 1-ohm source to a $-3/2$-ohm load. The element values for this prototype

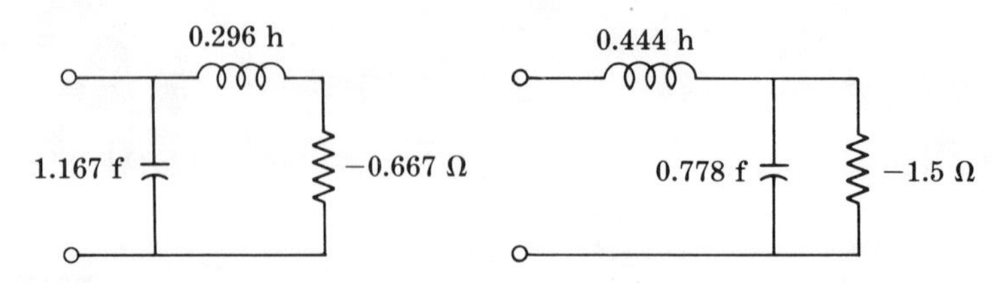

FIG. 6.17 Negative-resistance prototype.

FIG. 6.18 Reversed prototype.

are shown in Fig. 6.18. This prototype is more convenient when a tunnel diode is to be used for the negative resistor because there is a capacitor to accommodate the parasitic shunt capacitance of the diode.

For the final design let us begin by specifying an amplifier with center frequency of 1 gigaradian/sec with a 10 percent bandwidth. Furthermore, let the negative resistor be a tunnel diode with negative resistance of −10 ohms shunted by 10 pf. The source and load are 50 ohms and the circulator has a characteristic impedance of 50 ohms. In this case, the low-pass to bandpass transformation is

$$\omega_{LP} = \frac{\omega_{BP}^2 - 10^{18}}{10^8 \omega_{BP}} \tag{6.109}$$

With the transformation (6.109) and overall impedance-level scaling to 50 ohms the bandpass amplifier is that of Fig. 6.19. Then converting the ell of inductors to a transformer and reducing the impedance level on the right by a factor of 7.5 gives the final amplifier of Fig. 6.20.

There are, of course, many other ways to choose the prototype to meet the specifications. Others will certainly give more practical values in the realization than those of Fig. 6.20. The computations and procedures are similar for other amplifiers. The practical problems of coming out with a buildable final design are not the main objectives of this chapter.

C. The Design of Negative-resistance Parametric Amplifiers. In Sec. 6.3A we saw that a time-variant capacitance $C(t) = 2 \cos t$ could be

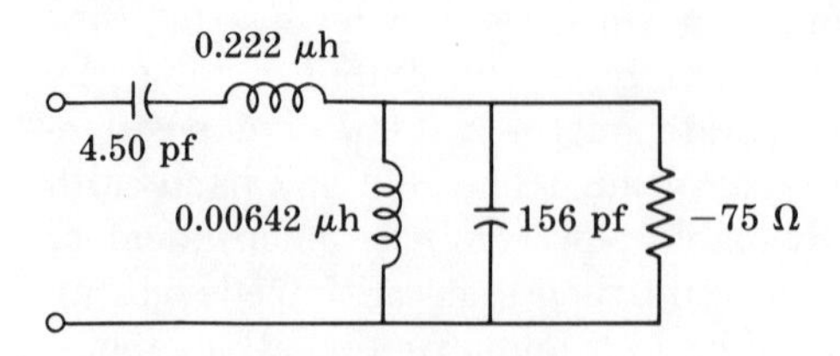

FIG. 6.19 Unnormalized filter.

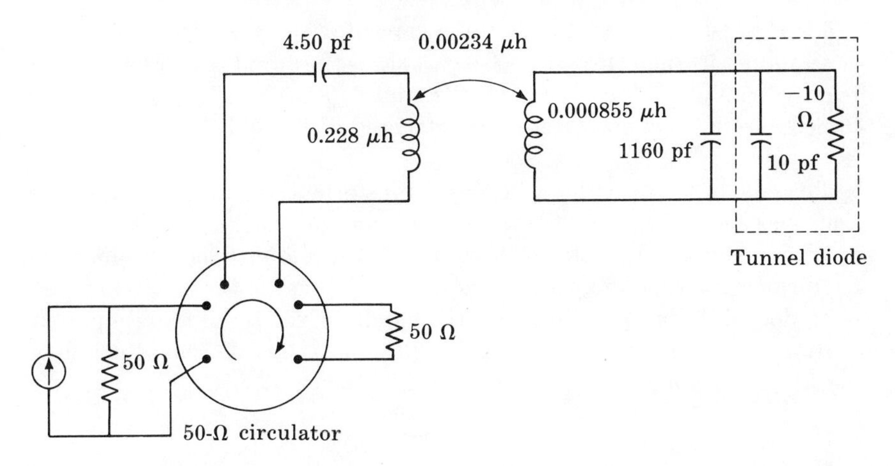

FIG. 6.20 Tunnel-diode amplifier.

made to look like a negative resistance at a frequency ω_a provided the admittance seen by the capacitor at frequency $1 - \omega_a$ was real. In Sec. 6.3B we saw how to utilize a negative resistance to make a bandpass amplifier. To make a parametric amplifier we can modify the technique of the previous section to satisfy the impedance constraint at the idler frequency. At the center frequency of the bandpass amplifier of Fig. 6.20 the impedance looking back from the negative resistor is real and equal to 50 ohms. Thus if we make a double-bandpass amplifier with a passband at the signal frequency and another at the idler frequency we can satisfy conditions for the negative-resistance effect and for a bandpass amplifier.

The problem can be solved by using a low-pass to double-bandpass transformation. This transformation uses the same philosophy as the low-pass to bandpass transformation of (6.109). Of course it is a bit more complex. The transformation (6.109) satisfies five conditions:

1. When $\omega_{LP} = 0$, $\omega_{BP} = \omega_0$, the center frequency of the passband.
2. When $\omega_{LP} = 1$, $\omega_{BP} = \omega_{c_1}$ or ω_{c_2} where ω_{c_1} and ω_{c_2} are the upper and lower cutoff frequencies of the bandpass amplifier. Since the bandwidth was specified $\omega_{c_1} - \omega_{c_2}$ is the bandwidth.
3. When $\omega_{LP} \to \infty$, $\omega_{BP} \to \infty$ or $\omega_{BP} \to 0$. Since the low-pass filter does not pass infinite frequencies, the bandpass filter does not pass dc or infinite frequencies.
4. The transformation is monotone for $0 \leq \omega_{BP} < \omega_0$, and for $\omega_0 < \omega_{BP} < \infty$. Thus the monotone nature of the Butterworth characteristic is preserved.

5. If $j\omega_{LP}C$ is the admittance of a capacitor and $j\omega_{LP}L$ the impedance of an inductor, then $[j(\omega_{BP}^2 - \omega_0^2)/b\omega_{BP}]C$ is the admittance of a network readily constructed of L's and C's and $[j(\omega_{BP}^2 - \omega_0^2)/b\omega_{BP}]L$ is the impedance of a network readily constructed of L's and C's.

The low-pass to double-bandpass transformation for the parametric amplifier should satisfy conditions corresponding to (1), (3), (4), and (5) as given below. For the bandwidth of the parametric amplifier the condition corresponding to (2) is not enough because the negative resistance does not stay constant over the band. Even so, as a first approximation we make a transformation from ω_{LP} to ω_{DBP} satisfying the following five conditions:

1′. When $\omega_{LP} = 0$, $\omega_{DBP} = \omega_a$ and $1 - \omega_a$.

2′. When $\omega_{LP} = 1$, $\omega_{DBP} = \omega_{c_1}$ and ω_{c_2}, the upper and lower cutoff frequencies of the amplifier centered at ω_a and $\omega_{c_3} = 1 - \omega_{c_2}$ and $\omega_{c_4} = 1 - \omega_{c_1}$.

3′. When $\omega_{LP} \to \infty$, $\omega_{DBP} \to \infty$ or 0 or some other frequency not in the passband or the idler band.

4′. The transformation is monotone between its poles and zeros.

5′. The admittance and impedance $j\omega_{LP}C$ and $j\omega_{LP}L$ transform to an admittance and impedance that are readily constructed of L's and C's.

In order to satisfy requirement (1′) the transformation must have zeros at ω_a and $1 - \omega_a$. To satisfy requirement (5′) the transformation must be a reactance function. Since a reactance function has poles and zeros that alternate along the $j\omega$ axis, a possibility is

$$\omega_{LP} = \frac{(\omega_{DBP}^2 - \omega_n^2)(\omega_{DBP}^2 - \omega_i^2)}{K\omega_{DBP}(\omega_{DBP}^2 - \omega_X^2)} \tag{6.110}$$

where K is constant and ω_X is a number between ω_a and ω_i. The choices of K and ω_X set the bandwidth requirements.

To see specifically how ω_X and K determine the width of the two bands we set the right side of (6.109) to plus and minus one. That is, we find ω_{c_1} and ω_{c_2} such that

$$(\omega_{c_1}^2 - \omega_a^2)(\omega_{c_1}^2 - \omega_i^2) = K\omega_{c_1}(\omega_{c_1}^2 - \omega_X^2) \tag{6.111a}$$

and

$$(\omega_{c_2}^2 - \omega_a^2)(\omega_{c_2}^2 - \omega_i^2) = -K\omega_{c_2}(\omega_{c_2}^2 - \omega_X^2) \tag{6.111b}$$

Multiplying (6.111a) out gives the quartic

$$\omega_{c_1}^4 - K\omega_{c_1}^3 - (\omega_a^2 + \omega_i^2)\omega_{c_1}^2 + K\omega_X^2\omega_{c_1} + \omega_a^2\omega_i^2 = 0 \tag{6.112}$$

This equation has four roots, $\omega_\alpha, \omega_\beta, \omega_\gamma, \omega_\delta$. We must select K and ω_X so that the roots are real.

Going from factored form to (6.112) shows that

$$K = \omega_\alpha + \omega_\beta + \omega_\gamma + \omega_\delta \tag{6.113}$$

and

$$\omega_a{}^2\omega_i{}^2 = \omega_\alpha\omega_\beta\omega_\gamma\omega_\delta \tag{6.114}$$

Then either all roots are positive or two are positive and two negative. The other signs in (6.112) show that the latter is the case. In fact the roots of (6.111*b*) have the same four absolute values but opposite signs.

To be specific let us order the roots so that

$$|\omega_\alpha| > |\omega_\beta| > |\omega_\gamma| > |\omega_\delta| \tag{6.115}$$

If we have two bands as desired, ω_α is the upper cutoff frequency of the higher frequency band and $-\omega_\beta$ is the lower cutoff frequency of that band. Similarly ω_γ is the upper cutoff frequency of the lower frequency band and $-\omega_\delta$ is the lower cutoff frequency. Since we want the two bandwidths to be equal we must make the bandwidth

$$b = \omega_\alpha + \omega_\beta = \omega_\gamma + \omega_\delta = \frac{K}{2} \tag{6.116}$$

Thus K is set in terms of the desired bandwidth b. We still must find ω_X.

The remaining two coefficients in (6.112) are related to the roots by

$$\begin{aligned} -(\omega_a{}^2 + \omega_i{}^2) &= \omega_\alpha\omega_\beta + (\omega_\alpha + \omega_\beta)(\omega_\gamma + \omega_\delta) + \omega_\gamma\omega_\delta \\ &= \omega_\alpha\omega_\beta + \omega_\gamma\omega_\delta + b^2 \end{aligned} \tag{6.117}$$

and

$$-K\omega_X{}^2 = \omega_\alpha\omega_\beta(\omega_\gamma + \omega_\delta) + \omega_\gamma\omega_\delta(\omega_\alpha + \omega_\beta) = \frac{K}{2}(\omega_\alpha\omega_\beta + \omega_\gamma\omega_\delta) \tag{6.118}$$

Then (6.118) can be rewritten

$$-\omega_X{}^2 = \frac{\omega_\alpha\omega_\beta + \omega_\gamma\omega_\delta}{2} \tag{6.119}$$

In the low-pass to bandpass transformation (6.109) the center frequency of the band is the geometric mean of the two cutoff frequencies. Such a condition satisfies (6.114), but it satisfies (6.117) only for bandwidth zero. If the bandwidth is small compared to the center frequencies in both bands, then the condition is still almost satisfied. If it is, then by (6.119)

$$\omega_X{}^2 \approx \frac{\omega_i{}^2 + \omega_a{}^2}{2} \tag{6.120}$$

This is the other required formula.

With the design problems of the low-pass to double-bandpass transformation taken care of, let us return to the problem of the parametric amplifier. When such a filter is used, then at the center frequency the admittance seen looking back from the time-variant capacitor at the idler frequency is the source conductance multiplied by whatever internal impedance-level transformation there is in the filter.† Call this conductance G_{SL} for source conductance seen from load. Now by (6.87) the negative conductance presented by the time-variant capacitance at the signal frequency is

$$G = \frac{\omega_a(1 - \omega_a)}{G_{SL}} \tag{6.121}$$

Let us carry out a specific design. Let the desired center-frequency gain be 25 so that we can use the prototype of the previous section. Let the signal frequency be 0.2 and then the idler frequency is 0.8. To maintain the narrowband conditions needed for (6.120) let the bandwidth be 0.02. These numbers will give us a prototype parametric amplifier. Then by frequency and magnitude scaling we can see what values of the parameters are required for various useful frequency ranges.

For a gain of 25 we can use the prototype of Fig. 6.18. This circuit requires $1.5G = G_{SL}$. With this fact and the desired frequencies (6.121) yields

$$G_{SL} = \sqrt{1.5 \times 0.2 \times 0.8} = 0.491 \text{ ohm} \tag{6.122}$$

Thus the appropriate prototype to which we apply the low-pass to double-bandpass transformation is that of Fig. 6.21. This prototype was obtained from that of Fig. 6.18 by raising the impedance level by 1/0.491.

We should note that we would use exactly the same prototype for the situation wherein the signal and idler frequencies were interchanged since the formula (6.121) is unchanged by this frequency interchange.

With the center frequencies and bandwidths given, the low-pass to double-bandpass transformation (6.110) becomes

$$\omega_{LP} = \frac{(\omega_{DBP}{}^2 - 0.04)(\omega_{DBP}{}^2 - 0.64)}{0.04(\omega_{DBP}{}^2 - 0.25)\omega_{DBP}} \tag{6.123}$$

There are many ways to synthesize an admittance or impedance corresponding to (6.123). For the admittance that replaces the capacitance in the prototype of Fig. 6.21, we choose a realization that has the shunt capacitance separate. This way the capacitance C_0 of the diode can be

† In the filter of Fig. 6.20, the tunnel diode's negative resistance sees 50/7.5 ohms looking back.

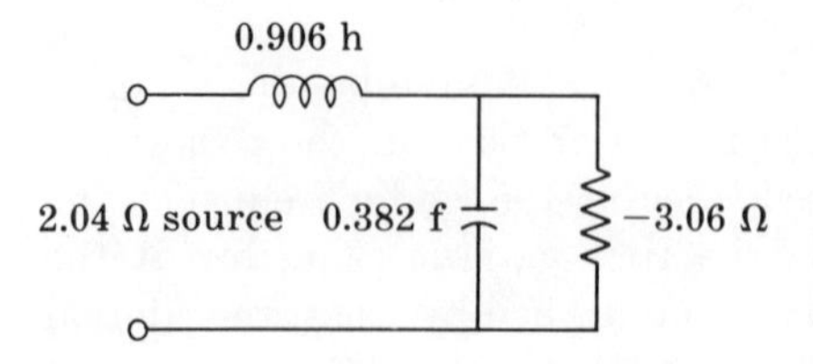

Fig. 6.21 Parametric amplifier prototype.

accommodated by the circuit. The expanded admittance is

$$Y_1 = \frac{0.382 \times 0.04 \times 0.64}{(0.04 \times 0.25)j\omega} + \frac{0.382(0.25 - 0.04)(0.64 - 0.25)j\omega}{0.04\sqrt{0.25}\,[(j\omega)^2 + 0.25]} + \frac{0.382}{0.04}\,j\omega$$

$$Y_1 = \frac{1}{1.02j\omega} + \frac{1.56j\omega}{(j\omega)^2 + 0.25} + 9.55j\omega \tag{6.124}$$

For the impedance that replaces the inductance of the prototype we may as well choose the dual realization. The resulting impedance is

$$Z_2 = 22.7j\omega + \frac{3.71j\omega}{(j\omega)^2 + 0.25} + \frac{1}{0.433j\omega} \tag{6.125}$$

Using (6.124) and (6.125) for the appropriate circuits converts the prototype of Fig. 6.21 to the normalized parametric amplifier of Fig. 6.22. This amplifier can now be converted to an amplifier for a practical frequency range using reasonable components by frequency and magnitude scaling.

To be specific let us suppose the signal frequency is 10^9 rad/sec. This makes the idler at 4×10^9 rad/sec and the pump at 5×10^9 rad/sec. The nominal bandwidth is 10^8 rad/sec since we had a 10 percent signal bandwidth in the low-pass to double-bandpass transformation. Let us again choose a 50-ohm circulator and let the diode have a capacitance swing of ± 10 pf about a nominal capacitance of 40 pf. In the prototype the frequency is scaled so that the pump frequency is 1 rad/sec. Let us first scale frequency by a factor of 5×10^9 and then fix the impedance levels.

Scaling the frequency of Fig. 6.22 gives the values of Fig. 6.23. To bring the circulator impedance up to 50 ohms the impedance level on the left should be raised by a factor of 24.5. To convert the capacitor which

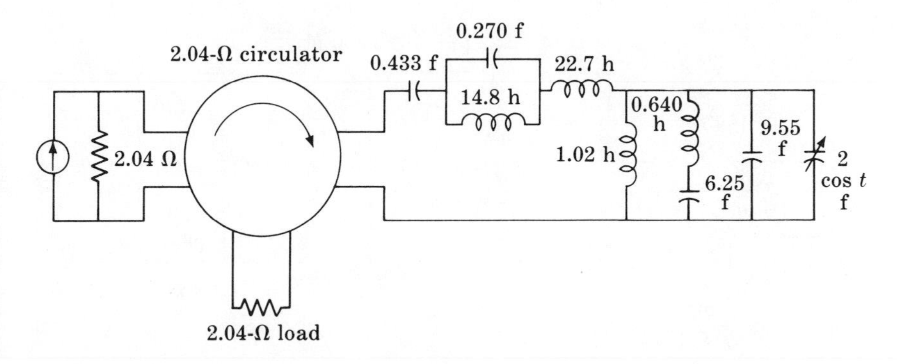

Fig. 6.22 Normalized parametric amplifier.

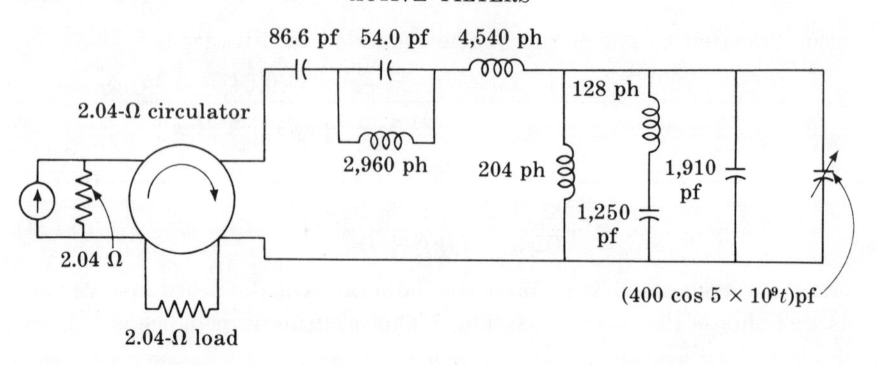

FIG. 6.23 Frequency-scaled parametric amplifier.

swings ±400 pf in Fig. 6.23 to a capacitor that swings ±10 pf the impedance level on the right should be raised by a factor of 40. The ell of inductors can be converted to a transformer to take care of the different levels required at the two ends. The result is shown in Fig. 6.24. Of the 47.7 pf on the right, 40 go with the varactor diode and 6.7 are an added padding capacitor.

The example above worked out quite reasonably with element values that are more or less realistic for UHF circuits. If we had specified a higher gain or a wider bandwidth, or both, would we still have been able to realize the amplifier? The answer is—probably not. The negative-resistance-type parametric amplifiers are subject to gain-bandwidth restrictions that are similar to, although more complicated than, those for conventional amplifying devices with parasitic capacitance for use in cascade stages. Distributed parametric amplifiers that get around this one limitation have been used[14] but the practical problems of building such amplifiers restrict their use.

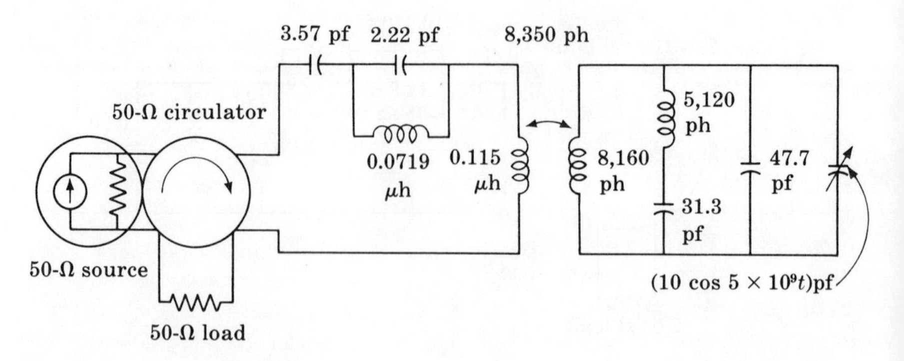

FIG. 6.24 Parametric amplifier.

The quantity that limits the gain-bandwidth product of a negative-resistance parametric amplifier is the pumping ratio; that is, the ratio of the variable part of the capacitance to the fixed part. In computing the fixed capacitance one must include both the diode fixed capacitance and the shunt capacitance in the linear, time-invariant network. Thus for the circuit of Fig. 6.24 the pumping ratio is 10/47.7. For the circuit of Fig. 6.11 the pumping ratio is 2/55. The higher the pumping ratio, the greater the possible gain-bandwidth.

By making certain approximations in addition to those inherent in the analysis above, Kuh and Fukada[15] came up with limitations on gain-bandwidth as a function of pumping ratio and the signal and idler frequencies. These frequencies enter into the problem because they determine how the idler impedance converts to a negative impedance at signal frequency through the formula (6.87). This limit on gain-bandwidth applies to amplifiers designed by the low-pass to double-bandpass technique when the bandwidth is not too large and the pumping ratio is not too high. The formula is

$$\frac{B}{\omega_a} \ln 2G \leq \frac{\pi}{4} \frac{C_1}{C_0} \left(\frac{\omega_i}{\omega_a} \right)^{1/2} \tag{6.126}$$

where B is the bandwidth and G is the overall voltage gain.

Ku[16] has developed an alternative synthesis procedure for negative-resistance-type amplifiers. His method, which is considerably more involved, synthesizes the two bandpass filters directly taking into account the fact that the idler admittance is not real and constant over the whole band. Thus Ku's method is more exact than the one above. The problem is that Ku still requires that the two frequency formulas apply and that the coupling network be lossless. As discussed below, when the bandwidth is broad these assumptions may not be too good. For the moderate-bandwidth cases, the simpler transformation method described above is adequate. In fact, most amplifiers are designed with even less stringent considerations. The reasons for this are discussed in the conclusions after a discussion of noise figure. Ku's gain-bandwidth limitation is

$$G \leq \cosh^2 \left(\frac{\pi \sqrt{\omega_a \omega_i}\, C_1}{4BC_0} \right) \tag{6.127}$$

Both formulas (6.126) and (6.127) show that increasing the pumping ratio increases gain-bandwidth. Since the pumping ratio is invariant through the frequency and magnitude scaling that took us from Fig. 6.22 to Fig. 6.24, one can check at the Fig. 6.22 stage to see if his gain-bandwidth requirements can be met by his varactor. Making the computations to this check point is about as easy as the formulas (6.126) and (6.127).

For large gain-bandwidth products the pumping ratio must be large. In the normalized amplifiers with the variable part of the capacitance set to $2 \cos t$, the pumping ratio is just $2/C_0$. Large pumping ratio means small C_0. The formula (6.87) for the admittance of the variable capacitor is good only when C_0 is fairly large. Otherwise we must use the more general formula (6.86). The approximation that $Y(\omega) \to \infty$ for $\omega > 1$ led to the formula (6.87). The more reasonable approximation $j\omega A(\omega) \approx C_0$ for $\omega > 1$ gives the formula

$$Y_C(\omega_a) = j\omega \left[\frac{-1}{A(\omega_a - 1) - \mu} - \mu \right] \tag{6.128}$$

where
$$A(\omega) = \frac{Y(\omega) + j\omega C_0}{j\omega}$$

$$\mu = \frac{-C_0 + \sqrt{C_0^2 - 4}}{2}$$

In the example above, C_0 is the capacitance in Fig. 6.22. Thus

$$\mu = \frac{-9.55 + \sqrt{91.2 - 4}}{2} = -\frac{0.2}{2} = -0.1 \tag{6.129}$$

But
$$A(\omega_a - 1) = \frac{2.06}{-j0.8} = j2.58 \tag{6.130}$$

Thus for the example above the μ in the denominator of (6.128) is negligible. To check the numerator we examine

$$\begin{aligned} Y_C(\omega_a) &= -\frac{0.8 \times 0.2}{2.06} + j(0.2 \times 0.1) \\ &= -0.078 + j0.02 \end{aligned} \tag{6.131}$$

Since the correction term is in quadrature with the two-frequency admittance, the error in dropping it is small, the error is not negligible. As the pumping ratio increases, the error increases rapidly.

We could carry the discussion of errors further by computing the exact response over the band and seeing what the real bandwidth is. It is most important that the designer knows wherein the approximations lie. Additional computations would not be particularly instructive.

D. Noise Considerations. Considering the problems of designing a bandpass parametric amplifier, as compared to the design of a tunnel-diode amplifier, the parametric devices must have other properties that make them really worthwhile. The property that justifies the extra design difficulties is the good noise figure. The parametric amplifier, unlike the transistor (both bipolar and FET), the tunnel diode, and the vacuum tube, has no quantum electronic carriers that are controlled for the amplification process. There is no source of excess noise in the basic amplifying process. The only source of excess noise is the thermal noise of the resistors of the amplifier. Thus the amplified noise of the idler frequency resistance is the main contribution to this excess noise.

Since the parametric amplifier is a negative-resistance amplifier, one must be careful in computing noise figures. The noise figure is normally defined in terms of available power, and the available power from a negative resistance is infinite. Penfield and Rafuse[17] have made a thorough investigation of noise in parametric amplifiers. For the negative resistance they use the concept of exchangeable gain, which is well defined for negative resistances. For ascertaining fundamental limits on amplifier noise figure, the techniques of Penfield and Rafuse are needed. In the present discussion we shall consider only the circulator configuration of the example above. Then the source impedance and the output impedance of the amplifier are real and positive. In fact the source and load are matched so all gains are available gains. Consequently we need only the well-known noise-figure definitions.

The noise figure for an amplifier with source and load impedances matched to the internal impedances is

$$F = \frac{S/N \text{ out}}{S/N \text{ in}} = \frac{\text{total noise power out}}{\text{thermal source noise} \times G_a} \tag{6.132}$$

where S/N is the signal-to-noise ratio and G_a is the available power gain. For the formula (6.132) the noise at the input is taken as the noise power generated by the source resistance at a standard temperature $T_0 = 290°\text{K}$. To show the noise figure for the negative-resistance parametric amplifier with circulator we shall use the second formula (6.132).

A linear, time-invariant resistor with resistance R has a noise voltage associated with it. The noise voltage spectral density is†

$$\bar{e}_n^2 = KTR \tag{6.133}$$

where K is Boltzmann's constant and T is the absolute temperature of the resistor. Similarly for a conductance G the noise current spectral density is

$$\bar{\imath}_n^2 = KTG \tag{6.134}$$

For a linear, time-invariant passive network, Thévenin's and Norton's theorems can be used for noise calculations. Rather than worry about each resistance, one can use the formulas (6.133) and (6.134) on the real

† Usually the formula has a factor of 4 included. Then the noise voltage is in rms values. Since we use complex amplitudes the voltage is different by a factor throughout. Also a term Δf is often included in the noise formula. Equation (6.133) is a spectral density and must be integrated df to get power.

part of the equivalent internal impedance or admittance provided all resistors are at the same temperature.

To compute the noise figure of our parametric amplifier we use a circuit like that of Fig. 6.10 with the current source a noise source. From this circuit we can compute the noise voltage across the time-variant capacitance. Knowing this voltage we can take it through the circulator to the load for a specific amplifier configuration. Then we have the noise figure for that amplifier.

For the noise computations the first circuit is that of Fig. 6.25. Since the power spectral density of (6.134) is the square of a current, the current source in the figure must be the square root. The admittance $\underline{Y}(\omega)$ includes the average capacitance of the original time-variant capacitor. Thus

$$\underline{Y}(\omega) = Y(\omega) + j\omega C_0 \tag{6.135}$$

The quantity $G(\omega)$ is Re $[\underline{Y}(\omega)]$.

The voltage spectral density at frequency ω_a will be made up of contributions due to the source current at ω_a and at all other frequencies $|\omega_a + k|$ with k taking on all positive and negative integer values. The noise source current at frequency $\omega_k = \omega_a + k$ produces a voltage component at ω_a. This voltage is related to the current by the factor $Z_{-k}(\omega_k)$ from (6.61) and (6.63). Furthermore when $\omega_a < 1$, $Z_{-k}(\omega_k)$ is the coefficient of the negative frequency, $-\omega_a$, term since $\omega_k = k - \omega_a$. Thus when $\omega_a < 1$ the voltage spectral density at frequency ω_a is

$$\overline{v_n(\omega_a)}^2 = KT \Big\{ G(\omega_a)|Z_0(\omega_a)|^2 + \sum_{k=1}^{\infty} [G(\omega_a + k)|Z_{-k}(\omega_a + k)|^2 + G(k - \omega_a)|Z_{-k}(k - \omega_a)|^2] \Big\} \tag{6.136}$$

The squared magnitude of each component of the voltage appears separately in this equation because the thermal noise at different frequencies is uncorrelated. There are no cross products of current amplitudes at different frequencies. If ω_a is between n and $n + 1$ with n an

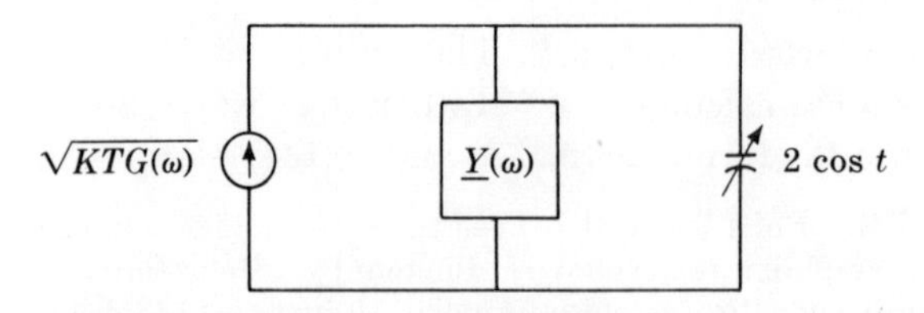

FIG. 6.25 Noise equivalent circuit.

integer, then the voltage spectral density at frequency ω_a is

$$\begin{aligned}\overline{v_n(\omega_a)}^2 = KT \Big[& G(\omega_a)|Z_0(\omega_a)|^2 \\ & + \sum_{k=1}^{\infty} G(\omega_a + k)|Z_{-k}(\omega_a + k)|^2 \\ & + \sum_{k=n+1}^{\infty} G(k - \omega_a)|Z_{-k}(k - \omega_a)|^2 \\ & + \sum_{k=1}^{n} G(\omega_a - k)|Z_k(\omega_a - k)|^2 \Big] \end{aligned} \tag{6.137}$$

At first glance one might be concerned about the convergence of the infinite series in (6.136) and (6.137). A study of the continued fraction expressions for the Z_k shows that each $Z_k(\omega_a + m)$ decays exponentially as the integer m approaches ∞ and that the $Z_k(\omega)$ decays exponentially as $k \to \pm\infty$. The exponential rate is $|\mu|^{|k|}$ in both cases where

$$\mu = \frac{-C_0 + \sqrt{C_0^2 - 4}}{2} \tag{6.138}$$

Since $C_0 > 2$, $|\mu| < 1$.

When the two frequency formulas are valid, the only terms in (6.137) that are significant are $Z_0(\omega_a)$ and $Z_{-1}(1 - \omega_a)$. Then

$$Z_0(\omega_a) = \frac{\underline{Y}^*(1 - \omega_a)}{\underline{Y}(\omega_a)\underline{Y}^*(1 - \omega_a) - \omega_a(1 - \omega_a)} \tag{6.139}$$

$$Z_{-1}(1 - \omega_a) = \frac{j\omega_a}{\underline{Y}^*(\omega_a)\underline{Y}(1 - \omega_a) - \omega_a(1 - \omega_a)} \tag{6.140}$$

With these two expressions in (6.136) the noise spectral density becomes

$$\overline{v_n}^2(\omega_a) = KT \left[\frac{G(\omega_a)|\underline{Y}(1 - \omega_a)|^2}{|\underline{Y}(\omega_a)\underline{Y}^*(1 - \omega_a) - \omega_a(1 - \omega_a)|^2} + \frac{G(1 - \omega_a)\omega_a^2}{|\underline{Y}(\omega_a)\underline{Y}^*(1 - \omega_a) - \omega_a(1 - \omega_a)|^2} \right] \tag{6.141}$$

The denominators of the two terms in (6.141) are identical since the magnitude of the sum of a complex number and a real number is equal to the magnitude of the sum of the conjugate of that complex number and the same real number.

The formula (6.141) gives the noise voltage spectral density across the variable capacitance. To find the noise spectral density at the output we must multiply by the squared magnitude of the transfer function that relates the voltage at the diode to the output voltage. For computation of noise figure in the special case under consideration there is no need to find this transfer function. The available power gain, which goes in the denominator of (6.132), involves the same transfer function. In fact, the first term in (6.141) times this transfer function is exactly the denominator of (6.132). Thus the noise figure† is

$$F = 1 + \frac{G(1 - \omega_a)\omega_a^2}{G(\omega_a)|\underline{Y}(1 - \omega_a)|^2} \tag{6.142}$$

At the center frequency of the amplifier of the previous section,

$$\underline{Y}(1 - \omega_a) = G(1 - \omega_a) = G(\omega_a) = \frac{1}{2.04}$$
$$= 0.491$$

Thus for this amplifier

$$F = 1 + \frac{0.04}{(0.491)^2} = 1 + 0.166 \tag{6.143}$$

This is a noise figure of 0.66 db, an extremely low noise figure.

The formula (6.142) is derived with the pump normalized to 1 ohm. The excess noise term in that formula is proportional to ω_a^2. Thus the excess noise is smaller when the signal frequency is small compared to the pump frequency. When the specific technique used to design the example above is employed, the dependence of the noise figure on frequency and source admittance is more evident than in (6.142). In this case $G(\omega_a) = G(1 - \omega_a) = \underline{Y}(1 - \omega_a)$. Furthermore the formula (6.87) relates this admittance to the frequency. In this special case (6.142) becomes

$$F = 1 + \frac{\omega_a^2}{G^2(\omega_a)} = 1 + \frac{\omega_a}{1.5(1 - \omega_a)} \tag{6.144}$$

For amplifiers designed as above the excess noise is proportional to the ratio of signal frequency to idler frequency. Thus low-noise amplifiers have the pump and idler frequencies as high as feasible for a given signal frequency.

The design procedure of the previous section was based on the assumption that the diode and all filter elements were lossless. The computations that gave the noise-figure formula (6.142) continue this assumption. In an actual diode the spreading resistance may give enough loss, especially at the idler frequency, to make an additional contribution to the noise figure.

A second factor that affects the noise figure is the temperature of the resistors that are sources of thermal noise. In deriving (6.142) we assumed that all resistors were at the same temperature. In practice, by proper design, the resistance at the idler frequency can be cooled.

† This is a single frequency or spot noise figure. It is valid in the situation where a narrowband IF amplifier follows the fairly broad-band, low-noise-figure RF amplifier.

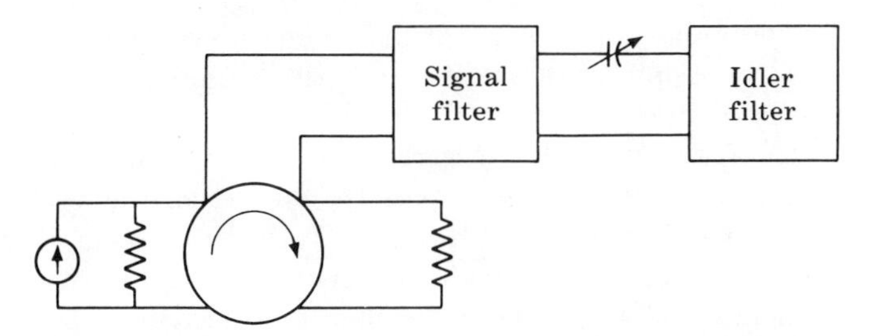

FIG. 6.26 Separate-signal and idler configuration.

Then the noise formula (6.142) becomes

$$F = 1 + \frac{T_i G(1 - \omega_a)\omega_a^2}{T_0 G(\omega_a)|\underline{Y}(1 - \omega_a)|^2} \tag{6.145}$$

where T_i is the temperature of the resistance seen by the diode at the idler frequency and T_0 is the reference temperature 290°. In the circuit of Fig. 6.22, the resistance seen by the diode at both signal and idler frequencies is the source resistance. In the circuit of Fig. 6.11 the resistances seen by the diode at signal and idler frequencies are different resistors. In a circuit such as Fig. 6.11, the resistor for the idler frequency could be cooled.

When the signal frequency is low compared to the pump frequency, the signal and idler frequencies are widely separated. With wide separation of signal and idler frequencies, the basic design of Fig. 6.26 can be used. In this case the signal filter can be designed exactly as the filter for the tunnel-diode amplifier was designed. The design should be such that at the idler frequency the signal filter is virtually a short circuit. The idler filter, which looks like a short circuit at signal frequency, presents the proper impedance at the idler frequency. The spreading resistance and incidental losses in the components are the resistive load of the idler filter. Now if the idler filter and the diode are cooled, the temperature T_i can be made very small. Such cooled parametric amplifiers have almost no excess noise.

6.5 Extensions and Generalizations

In the previous sections of this chapter various analysis and synthesis techniques have been applied to the design of parametric amplifiers wherein the variable parameter was a capacitance. The methods presented can also be used in the design of frequency multipliers, mixers, modulators, and other frequency-converting circuits with variable capacitance, inductance, or resistance. Of course only the reactance elements can have power gain. Some of the more important circuits and the application to them of the techniques discussed above are discussed in this section.

A. Frequency Multipliers. One common method of generating a high-frequency signal is to pass the output of a low-frequency oscillator through a frequency multiplier. There are two good reasons for this method of generating high-frequency signals. Most important is that in many system applications there is a need for signals at several frequencies wherein the precise mathematical relation between these frequencies is important. If all such frequencies are generated from a low-frequency, crystal-controlled, master oscillator, the required relationships can be maintained. Another reason is that the power available from transistor oscillators drops off sharply with frequency as one moves up into the UHF and microwave range. Often one can get more power more conveniently from a transistor oscillator followed by a multiplier than from some other oscillator operating directly at the desired frequency.

The basic analysis method of Sec. 6.1 above is appropriate to frequency multipliers. By studying the method qualitatively, we can develop a few principles of design. A possible circuit for a frequency multiplier with input frequency ω_a and output frequency at $n\omega_a$ is that of Fig. 6.27. With the source and load as shown, the filters must present a high impedance at the designated frequencies ω_a and $k\omega_a$ respectively, and a low impedance at $k\omega_a$ and ω_a respectively.

For the analysis method the block diagram should be set up so that the successive iterates are the voltage across the nonlinearity. Then the first iterate will have a large amplitude. This large voltage will produce comparative large currents in the nonlinear element at the harmonic frequencies of the second iterate and so on. If the nonlinearity is primarily square law, then the second iterate will contain a large current at the second harmonic. If the desired output is at a higher

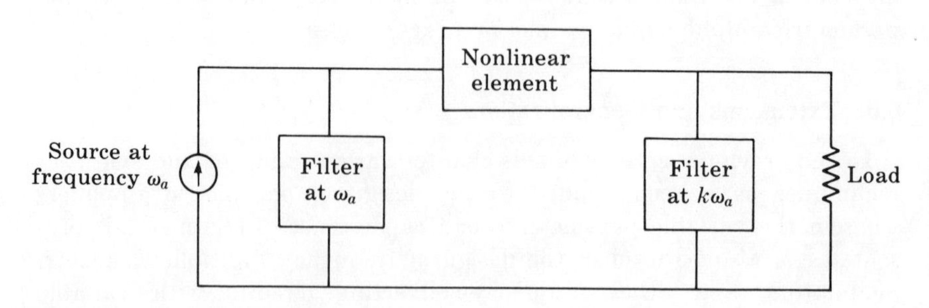

Fig. 6.27 Frequency multiplier.

frequency than the second harmonic, $k > 2$, then there is no filter at the second harmonic in the simple circuit of Fig. 6.27. Consequently the large current in the nonlinearity will not produce a large voltage across the nonlinearity at the second harmonic frequency.

The third approximation to the response of the circuit will contain terms at the third and fourth harmonics. The amplitude of these respective terms increases as the second harmonic voltage increases. If the second harmonic voltage is small, these third and fourth harmonic terms will be small. A continuation of the iteration will not produce large terms at the higher harmonics. Consequently if $k > 2$ in Fig. 6.27, the simple circuit will not be a very good frequency multiplier.

A conclusion of the above heuristic reasoning is that if frequency multiplication is to take place in a device whose nonlinearity is primarily square law and if the output is to be at the third or fourth harmonic, then a filter at the second harmonic should be added to the circuit of Fig. 6.27. For this situation where the filters present high impedances at their resonant frequencies in order to produce a large voltage in response to currents generated in the nonlinearity, the second harmonic filter should be placed in series with the nonlinear element. Such a filter is called an idler filter since it is idle so far as input and output are concerned.

For higher harmonics an extension of the same reasoning shows that more idlers are needed. With idlers at second and third harmonics, the output can be taken at the fifth or sixth. With idlers at the second and fourth harmonics up to the eighth can be obtained, etc. In each design the engineer must compute the trade off between enhancement of the output by adding idlers and losses in the idlers. All high impedance at resonance filters are placed in series as shown in Fig. 6.28*a*. An alternative design utilizes filters that present low impedances at their resonant frequency and high impedances elsewhere. Then each successive approximation is a current, and these currents generate voltages in the nonlinearity. Such a design would have the basic structure of Fig. 6.28*b*.

In the above discussion of multipliers the nonlinear element could be a resistance, an inductance, or a capacitance. The resistance has the disadvantage that it is lossy, except in the case of an ideal diode. Thus there are losses in a nonlinear resistor at each frequency where appreciable currents and voltages exist. In the reactances, the currents and voltages at each frequency are out of phase. Thus there is no loss in the basic nonlinear element. Of course, all devices have parasitic losses.

For any nonlinear element the amplitude of the harmonics generated is larger as the element characteristic is more nonlinear. All filters have parasitic shunt capacitance. When a nonlinear capacitance is used, the parasitic capacitance of the filters adds linear capacitance and thus lessens the percentage of the nonlinear coefficients. When filters at

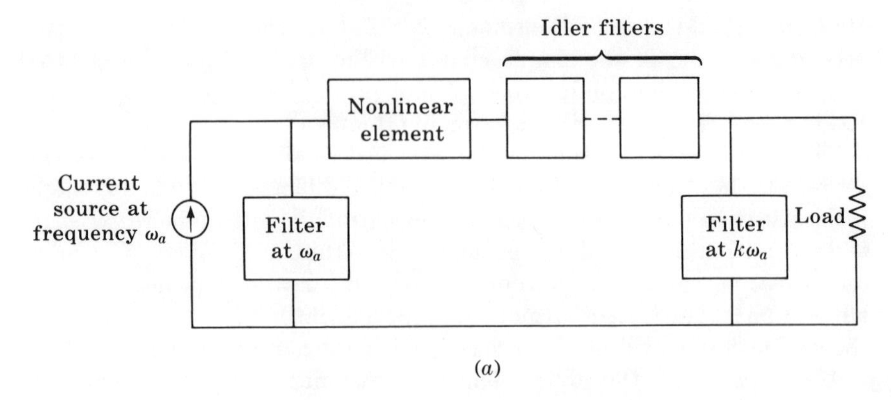

(a)

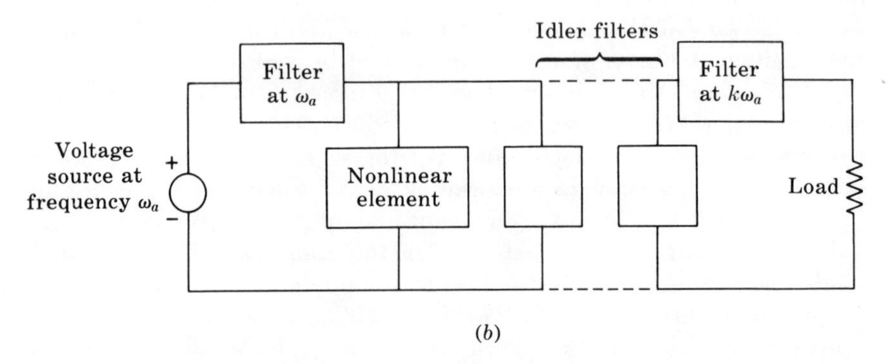

(b)

Fig. 6.28 Multipliers with idlers. (a) High-impedance (shunt-tuned) filters; (b) low-impedance (series-tuned) filters.

several frequencies are needed, then the parasitic capacitance is lessened if a single structure can incorporate more than one resonance. The low-pass to multiple-bandpass transformation, like the low-pass to double-bandpass transformation, described above, is one way to design such structures.

B. Mixers, Modulators, and Other Linear Converters. In the parametric amplifiers of Sec. 6.4 the input and output were both at frequency ω_a. The analysis of Sec. 6.3 and the design methods of Sec. 6.4 both apply if the output frequency is $|\omega_a \pm k\omega_b|$ where k is an integer. The difference equations and continued fractions apply with only minor modification when the sinusoidal time-variant element is a resistance, conductance, inductance, reciprocal inductance, or elastance instead of a capacitance.

The application of mixers and modulators with resistive nonlinear or time-variant elements is well known to anyone familiar with radio communications. Parametric frequency converters are finding wide application because they have the good noise properties of parametric amplifiers and do not require a circulator. The analysis above applies directly to a device with the circuit of Fig. 6.11 if the tank on the right is tuned to frequency $(1 - \omega_a)$, $(1 + \omega_a)$, or any other frequency congruent to ω_a. When the output is taken at a different frequency from the input, there is no need for a circulator to separate input and output.

Parametric amplifiers and frequency converters with more than one time-variant element are characterized by a set of coupled difference equations. Although there is a generalization of the continued fraction solution to these higher-order equations, there are no simple formulas like the ones given above. The same problem applies to the single sinusoidal element where more than one frequency component of the sinusoidal element is significant. Again a difference equation of higher than second order is involved. These more complex parametric signal processors have found little application. Where they have been used, the analysis is based on the a priori assumption that only certain frequencies are significant. Then the difference equations reduce to a set of algebraic equations in the complex amplitudes of the various frequency components of the voltage (or currents) across the various time-variant elements. Design formulas for these circuits are very approximate.

For all parametric signal processors, signal processors where the basic operation depends on the variation of a circuit parameter, the circuit equations are too complex for exact analysis. Exact synthesis is even farther away. Nevertheless, by making certain approximations engineers have evolved good design formulas. For the theorist there are many gaps that must be filled in justifying the approximations and finding the limits of applicability of the design formulas. Because of the usefulness of parametric signal processors, there is hope that a better understanding will produce more applications.

6.6 Acknowledgment

The research reported in this chapter was supported by the National Science Foundation and the Post-Doctoral Program of the Rome Air Development Center (Air Force Systems Command). The specific projects are NSF-GK-2988 and NSF-GK-1235, and Air Force Contract F30602-68-C-0075.

References

1. Hale, J. K.: "Oscillations in Nonlinear Systems," McGraw-Hill Book Company, New York, 1963.
2. Hayashi, C.: "Nonlinear Oscillations in Physical Systems," chap. 7, McGraw-Hill Book Company, New York, 1964.

3. Minorsky, N.: "Nonlinear Oscillations," D. Van Nostrand Company, Inc., Princeton, N.J., 1962.
4. Manley, J. M., and H. E. Rowe: Some General Properties of Nonlinear Elements, Part I, General Energy Relations, *Proc. IRE*, vol. 44, no. 7, pp. 904–913, July, 1956.
5. Penfield, P.: "Frequency Power Formulas," The M.I.T. Press, Cambridge, Mass., 1960.
6. Tricomi, F. G.: "Integral Equations," Interscience Publishers, Inc., New York, 1957.
7. Anderson, D. R., and B. J. Leon: Nonlinear Distortion and Truncation Errors in Frequency Converters and Parametric Amplifiers, *IEEE Trans.*, vol. CT-12, no. 3, pp. 314–321, 1965.
8. Trick, T. N., and D. R. Anderson: Lower Bounds on the Threshold of Subharmonic Oscillations, *Purdue University School of Electrical Engineering TREE*65-16, 1965.
9. Sandberg, I. W.: On the Truncation Techniques in the Approximate Analysis of Periodically Time-varying Nonlinear Networks, *IEEE Trans.*, vol. CT-11, no. 2, pp. 195–201, June, 1964.
10. Adams, J. V., and B. J. Leon: Steady State Analysis of Linear Networks Containing a Single Sinusoidally Varying Capacitor, *IEEE Trans.*, vol. CT-14, no. 3, pp. 313–319, September, 1967.
11. Kuh, E. S., and R. A. Rohrer: "Theory of Linear Active Networks," Holden-Day, Inc., Publisher, San Francisco, 1967.
12. Kuh, E. S., and Y. T. Chan: A General Matching Theory and Its Application to Tunnel Diode Amplifiers, *IEEE Trans.*, vol. CT-13, no. 1, pp. 6–18, March, 1966.
13. Weinberg, L.: "Network Analysis and Synthesis," McGraw-Hill Book Company, New York, 1962.
14. Currie, M. R., and R. W. Gould: Coupled-cavity Traveling Wave Parametric Amplifiers, Part I, Analysis, *Proc. IRE*, vol. 48, no. 12, pp. 1960–1972, December, 1960.
15. Kuh, E. S., and M. Fukada: Optimum Synthesis of Wide Band Parametric Amplifiers and Converters, *IRE Trans.*, vol. CT-8, no. 4, pp. 410–415, December, 1961.
16. Ku, W. H.: A Broadbanding Theory for Varactor Parametric Amplifiers—Parts I and II, *IEEE Trans.*, vol. CT-11, no. 1, pp. 50–58 and pp. 59–72, March, 1964.
17. Penfield, P., and R. P. Rafuse: "Varactor Applications," The M.I.T. Press, Cambridge, Mass., 1962.

CHAPTER 7

PRESENT AND FUTURE TRENDS IN INTEGRATED CIRCUITS

Jan A. Narud
Motorola Semiconductor Products Division
Phoenix, Ariz.

7.1 Introduction

In industry it is a long haul from state of the art to final production. In the particular industry of integrated circuits the long haul becomes a highly complex progressive network, the result of which in annual world sales was undreamed of a mere 10 years ago when the whole business of integrated circuits was indeed an infant.

The following list summarizes the major concepts involved in the various types of integrated circuits and their use in our modern business technology today and their probable use in the future.

1. The step-by-step processing of an integrated circuit
2. Digital integrated circuits
3. Large-scale integrated circuits (LSI)
4. Design automation

These concepts are covered in more detail in the sections which follow.

7.2 Fabrication of the Integrated Circuit

Present integrated circuits, in general, may contain transistors, diodes, resistors, capacitors, and an interconnection pattern between these components. In practice, a large number of these circuits are made simultaneously on a single piece of semiconductor known as a wafer. This wafer is usually about 2 or 3 inches in diameter. The manufacturing process involves epitaxial growth, oxidation, photomasking, impurity diffusion, and metallization. It is the photomasking which provides the necessary selectivity in the diffusion and metallization processes.

The process begins with the starting material or substrate. Today, monocrystalline silicon is used most often because of its relatively low cost and stability over a large temperature range. The substrate provides a heat sink, electrical isolation, and mechanical strength for the circuit. The monocrystalline silicon is obtained from a small, highly perfect silicon crystal which is immersed in molten silicon that is maintained at a temperature just above its melting point. After the large single crystal is grown, it is sawed into slices or wafers. The wafers are then milled to the desired thickness and polished to a mirror finish. A wafer of *p*-type material is chosen. This allows the use of *npn* transistors, which will ultimately result in sufficiently close tolerances on resistors in the circuit.

Next a heavily doped N^+ region is created by the diffusion of donor impurities into the substrate. The purpose of this "buried layer" is to reduce the collector series resistance R_{sc} of the transistors in the integrated circuit. This buried layer bypasses the highly resistive collector regions. Typically, R_{sc} may be decreased from 50 to 5 ohms. A *p* substrate with an N^+ buried layer is shown in Fig. 7.1.

The diffusion cycle includes oxidation, photomasking, and impurity diffusion. First a coating of silicon dioxide is formed on the wafer by heating it in an oxygen-rich atmosphere. The SiO_2 is then removed from those areas into which the *n*-type impurities will be diffused. This is accomplished through the process of contact printing. Here, the wafer is placed on a vacuum chuck, covered with a photosensitive emulsion (photoresist) and rotated at high speed. Surface tension and centrifugal forces resulting from the spinning cause a uniform thin film of the photoresist to form on the surface. The wafer is now illuminated by ultraviolet light through a negative containing the appropriate pattern. Those areas of the photoresist exposed to light undergo a photochemical reaction and polymerize. The unexposed photoresist is then removed with a

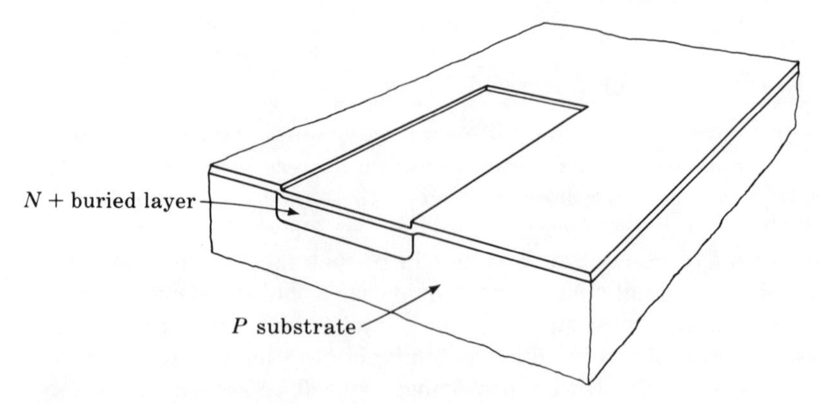

FIG. 7.1 A *p* substrate with an N^+ buried layer.

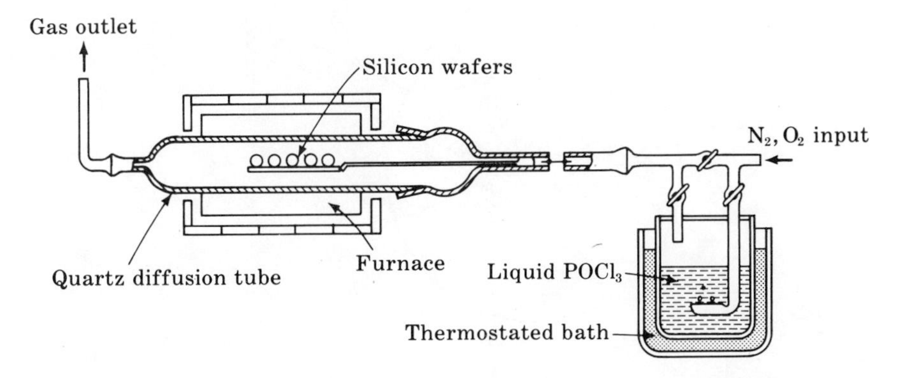

FIG. 7.2 A schematic representation of a typical apparatus for phosphorus diffusion.

solvent leaving a polymer, which is highly resistant to corrosive etches, over the desired areas. An etchant containing hydrofluoric acid is used to remove the unprotected SiO_2, exposing regions where the buried layers are to be created.

An *n*-type impurity is now diffused into the *p*-type substrate. Phosphorus is used as the *n*-type impurity and in later diffusions of *p*-type impurity, boron is employed. These materials are chosen because they do not readily penetrate the SiO_2 mask and have a high solid solubility in silicon. The diffusion is done by heating the wafer in an oven where it is exposed to a gas containing the impurity. The schematic representation of a phosphorus-diffusion system is depicted in Fig. 7.2. It is important that the furnace have a region where temperature is nearly constant so that the depth of diffusion is uniform over the wafer. In diffusion, the density of impurities will vary with distance from the surface. The region closest to the surface has the highest impurity concentration and hence the least surface resistivity. Junctions are formed where the number of *n*-type impurities equals the number of *p* impurities. When the diffusion is completed, the polymerized photoresist and silicon dioxide are removed from the wafer.

An epitaxial film of *n*-type silicon, which will form the collector region of the transistors, is then grown on the wafer. Epitaxial growth is a means of arranging atoms in a single-crystal fashion upon a single-crystal substrate so that the lattice structure of the resulting layer is an exact extension of the substrate crystal structure. Figure 7.3 represents a system used for the growth of silicon epitaxial films. This system includes a reaction chamber consisting of a long cylindrical quartz tube encircled by an RF induction coil. The single-crystal silicon wafers are placed upon a quartz sleeve in a rectangular graphite rod called a boat and put in the so-called reaction chamber. The boat is heated by RF induction

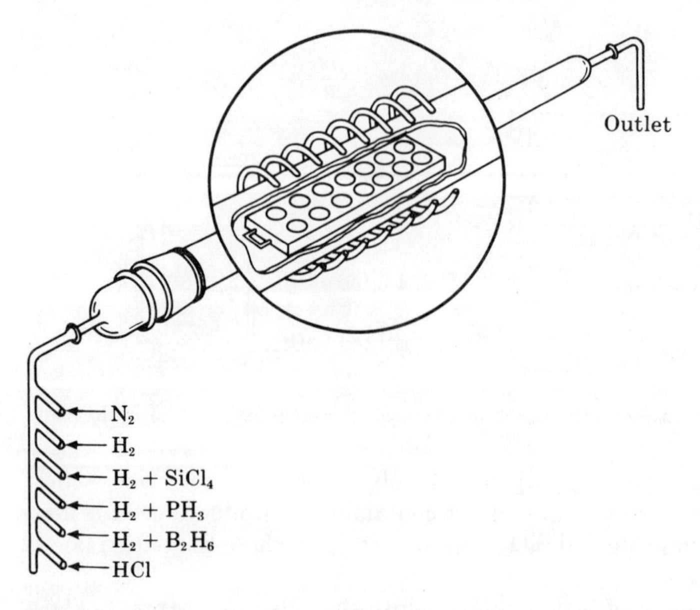

FIG. 7.3 A system for the growth of silicon epitaxial films.

in the presence of gases needed to produce epitaxial films of the desired impurity type and concentration. Unlike diffusion, epitaxial growth results in impurity distributions which do not vary with distance from the surface. Epitaxial films can be made very uniform with typical resistivity variations of 0.1–0.3 ohm-cm. However, the reason for using epitaxial growth to form the collectors is to avoid impurity saturation in the wafer. The wafer with this epitaxial layer is shown in Fig. 7.4.

Several more diffusion cycles are now done. The process is the same

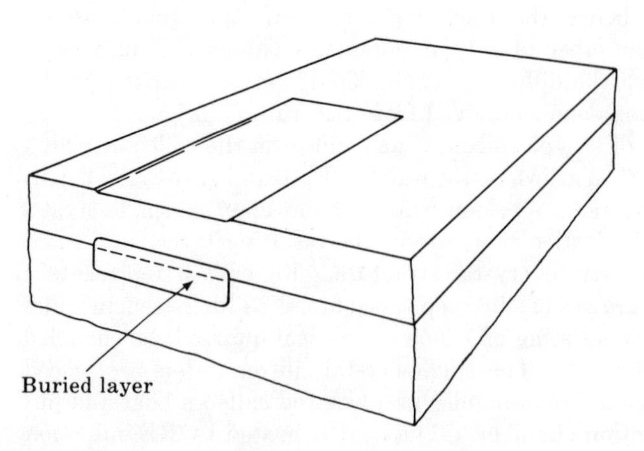

FIG. 7.4 A silicon wafer with an epitaxial layer.

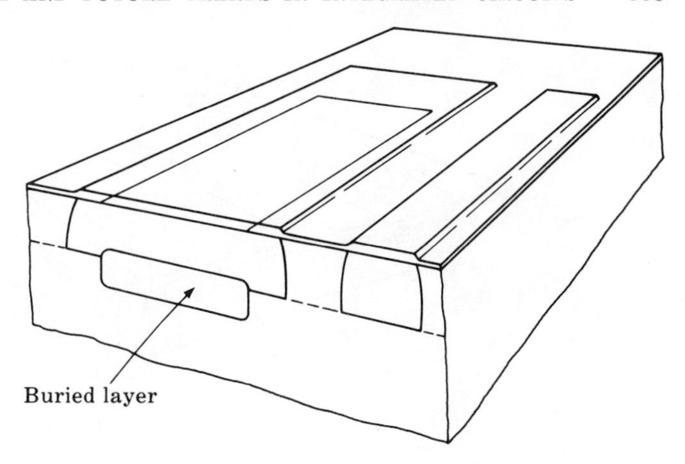

FIG. 7.5 Channel diffusion.

as described in the creation of the buried layer except that the SiO_2 is no longer removed at the end of the cycle. In the first of these operations, an acceptor impurity is diffused into the n-type epitaxial layer all the way to the p substrate, as seen in Fig. 7.5. These p-type channels provide isolation between the different components by separating them with back-to-back diodes.

In the second of these diffusions, p impurity is introduced into the n-type collector regions to form the base of the transistors and any resistor in the circuit. Resistivities in the range of 50 to 600 ohms per square are usually obtained. Figure 7.6 shows the wafer after the second diffusion process.

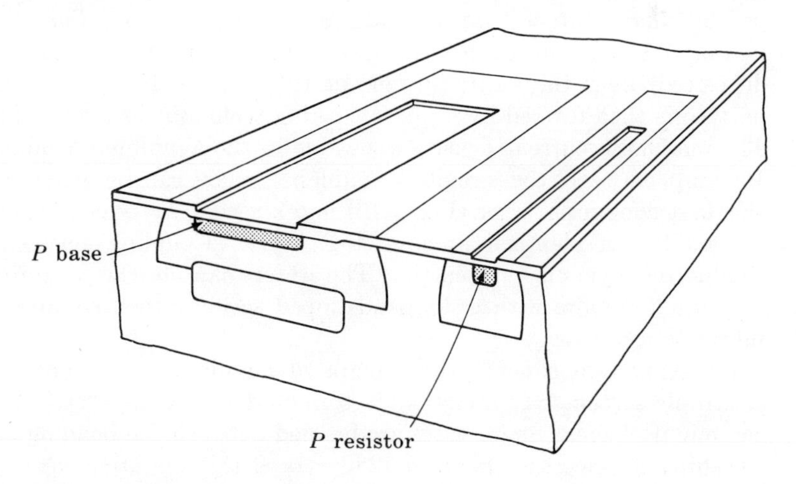

FIG. 7.6 Base diffusion of the silicon wafer.

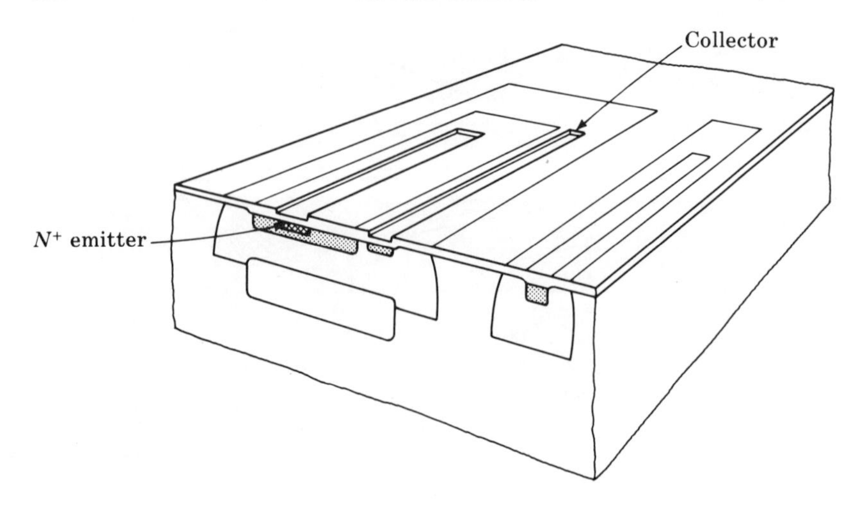

FIG. 7.7 Complete junction formation for a monolithic circuit.

An n-type impurity is now diffused into the base regions to form the transistor emitters. In this step windows are also opened in the collectors to form a high concentration of n impurity near the surface. This is necessary because the aluminum used for contacting and interconnecting is a p impurity soluble in silicon. Thus the N^+ region prevents rectifying contacts. Now the junction formation of the monolithic circuit is completed as illustrated in Fig. 7.7.

The next step is to interconnect the various components on the integrated circuit. A fourth set of windows is etched in the SiO_2 layer at points where contact is to be made to the components. This is shown in Fig. 7.8. A thin even coating of aluminum is evaporated (vacuum deposited) over the entire surface of the wafer. The interconnection pattern is then formed by a photomasking technique as seen in Fig. 7.9. The maximum current density allowable in the aluminum conductors is 10^5 amp/cm^2. If the circuit is nonplanar, glass can be evaporated on this first conductor layer (Fig. 7.10) and a second layer of metallization formed by an identical process (Fig. 7.11). Coupling between these conductive layers is very small. The wafers can now be separated into individual circuits with a diamond-tipped scribing tool, completing the fabrication process.

It remains to connect the integrated circuit as a part of a larger electronic system. One method is to mount the IC in a TO-5 can or a ceramic package. Small wires are bonded between the bonding pads on the chip and the pins. Figure 7.12 illustrates this mounting and indicates

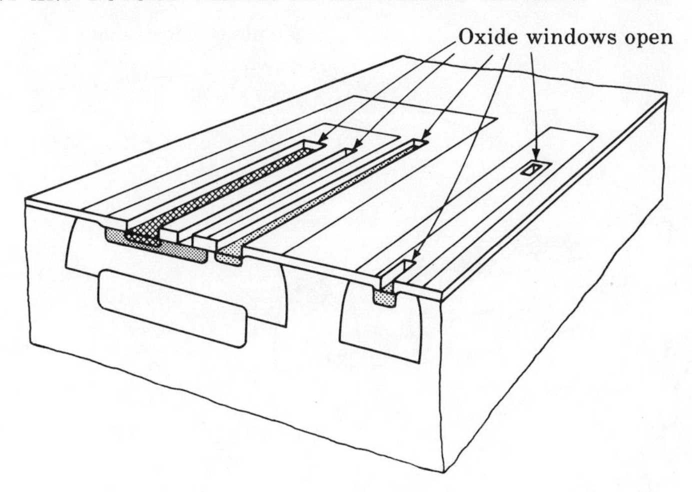

FIG. 7.8 Etching of windows for contacts.

the parasitics which result. Another technique is flip-chip mounting, shown in Fig. 7.13. Here, windows are etched in the SiO_2 along the edge of the chip where bonds are to be made. An identical pattern of raised aluminum contacts is evaporated on the chip. Each of the contacts on the substrate has a lead which can be used for external

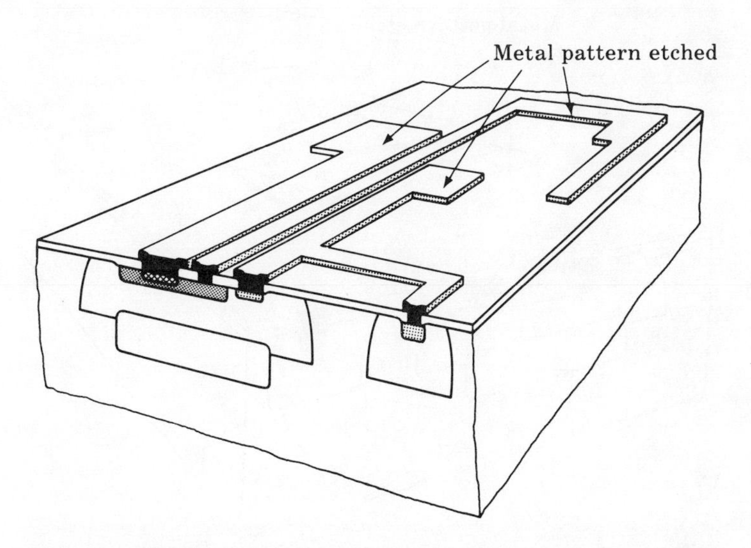

FIG. 7.9 The interconnection pattern formed by a photomasking technique.

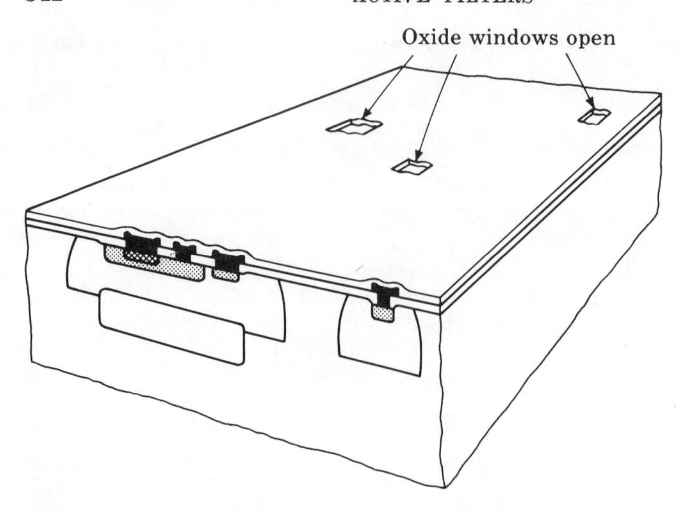

FIG. 7.10 Glass layer used for nonplanar circuit connections.

interconnections. The semiconductor chip is then placed upside down on the glass or ceramic substrate and joined to the aluminum contacts by ultrasonic bonding. The final method is the beam lead technique. Again, windows are etched in the SiO_2 at the desired points of contact. Layers of tantalum, palladium, and gold are then evaporated on the top

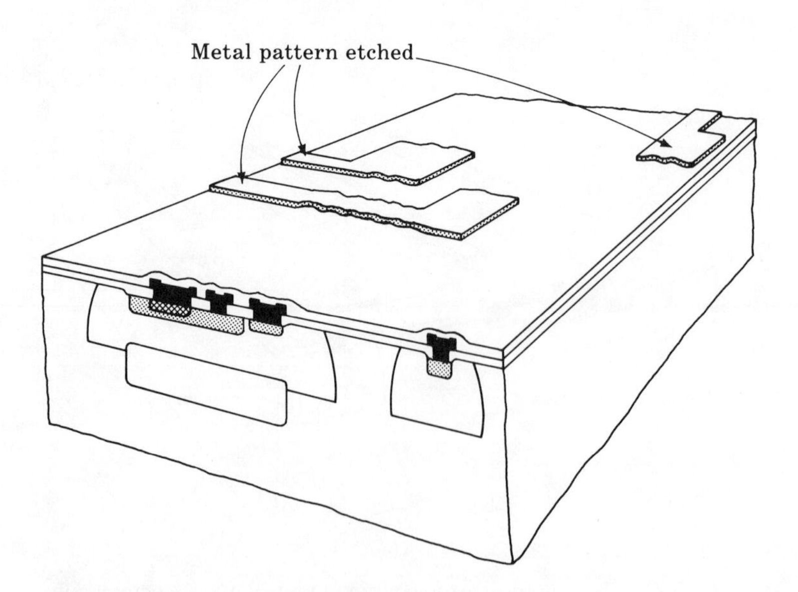

FIG. 7.11 The second interconnection pattern.

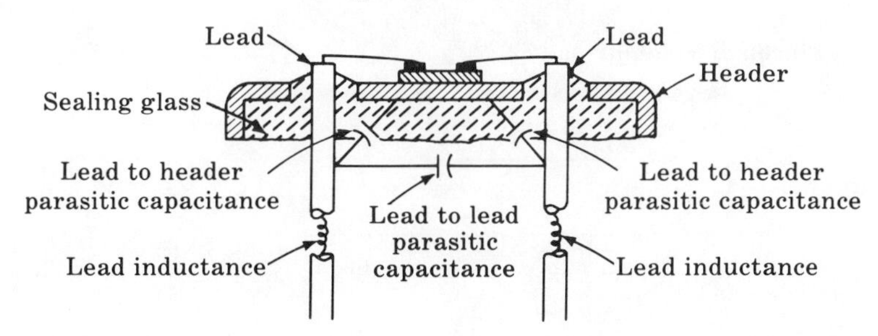

FIG. 7.12 Schematic representation of the mounting of an integrated-circuit die.

of the entire chip (Fig. 7.14) and the desired pattern is etched. Next the chip is turned over and all the excess semiconductor material is etched away, leaving only the beam interconnections and the individual components (Fig. 7.15). Interconnections can be made between a number of different chips in this manner, making the beam lead technique attractive for hybrid large-scale integration.

Thus far, only monolithic integrated circuits have been discussed. A monolithic circuit is one which is fabricated in a single crystal of semiconductor. Unfortunately, the passive components made in this manner are limited in their accuracy and range of values. Typical (p-type) resistors have tolerances of 15 percent and are restricted to values between 50 ohms and 20 kilohms. The lower limit is set by the resolution attainable in the photomasking processes. Larger resistors take more area and are more liable to faults. This will result in a lower yield. Monolithic capacitors are formed from back-biased junctions. They are leaky and have low breakdown voltages. Usual capacitance per unit areas are 0.1 to 0.3 pf/mil^2 with tolerances of 30 percent. Thus

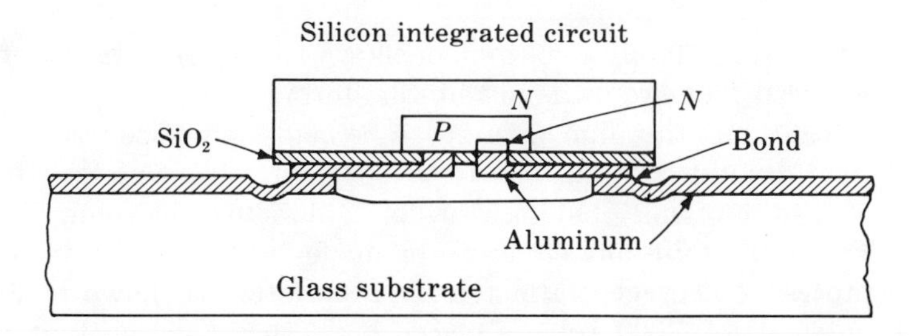

FIG. 7.13 Flip-chip mounting of an integrated-circuit die.

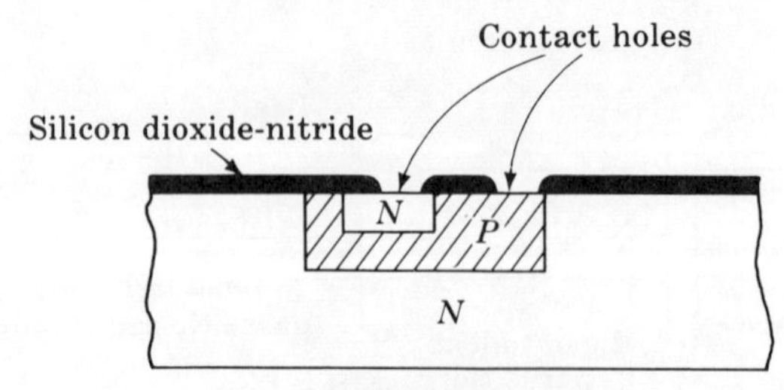

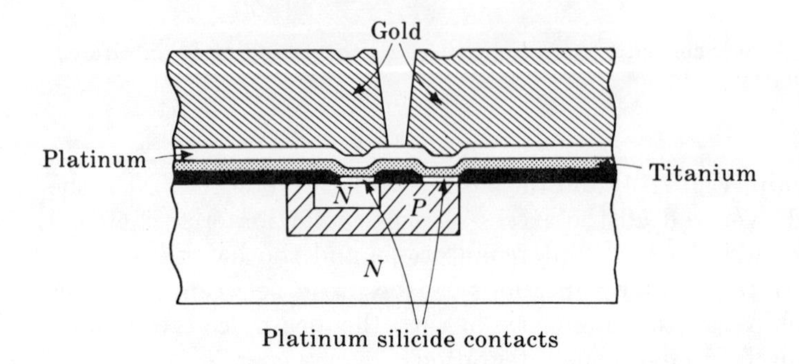

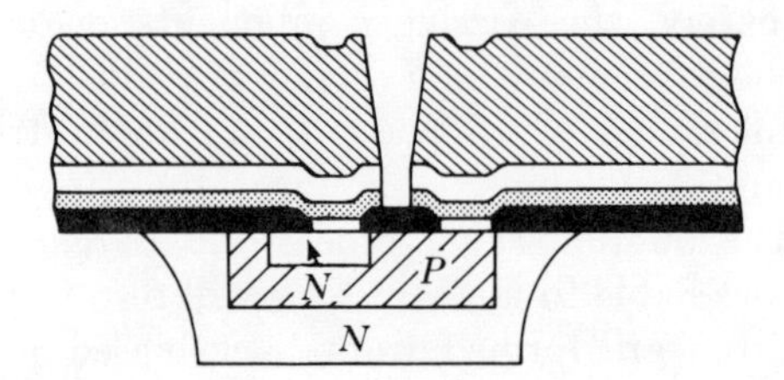

FIG. 7.14 The beam lead process.

when accurate passive components with a large range of values are required they are made in thin-film form.

Tantalum thin-film circuits can be made to realize resistors and capacitors accurate within 0.01 to 0.1 percent. Capacitances per unit area of 3 to 4 pf/mil^2 and resistors of 1 ohm to 1 megohm are attainable. Tantalum thin-film circuits are made by an anodizing or sputtering process. The schematic for such a system is shown in Fig. 7.16. A potential of 5,000 volts is placed between a block of tantalum and the substrate which are in an atmosphere of argon. This potential is

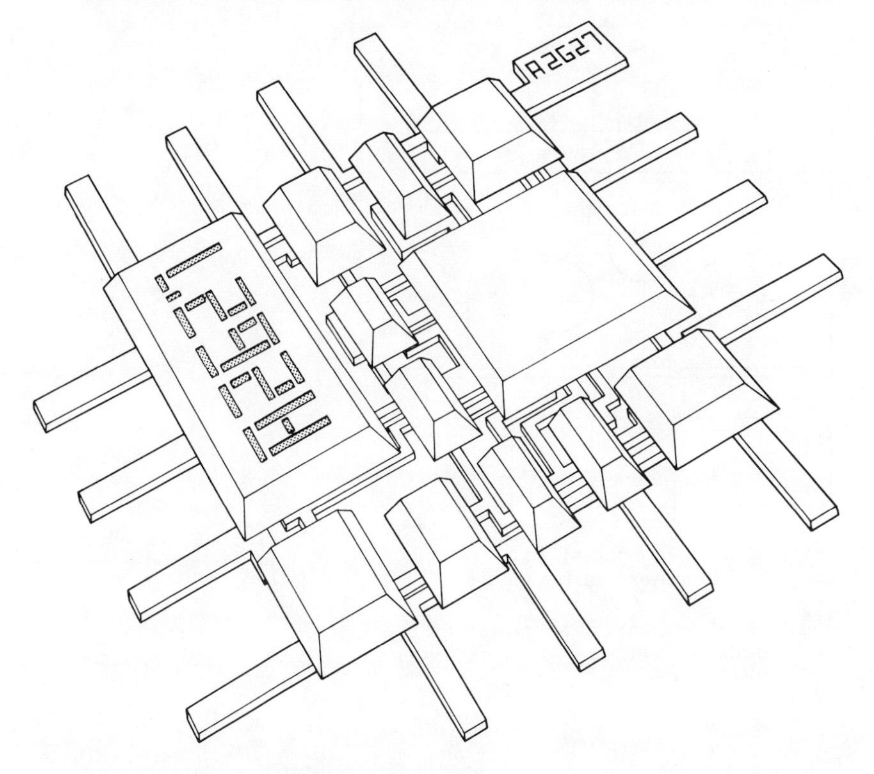

Fig. 7.15 Beam interconnections between individual components of a circuit.

sufficient to ionize the argon. The Ar^+ ions are accelerated toward the cathode, bombarding it with sufficient energy to free Ta^- ions. The tantalum ions move to the substrate where the passive components are to be formed. Since the process is "cold" (done at room temperature) it can be interrupted to monitor the resistance values. It is this feature which permits components with such strict tolerances. A tantalum thin-film circuit is shown in Fig. 7.17.

Due to the high packing densities realizable with integrated circuits, they are widely used in satellite systems. Here they are subjected to an environment with high radiation. In the presence of radiation, hole-electron pairs are formed in semiconductors causing the junctions to become leaky. This is extremely critical with the collector-substrate junction which has the highest nuclear cross section. For this reason, SiO_2 isolation is employed. Such an integrated circuit is shown in Fig. 7.18.

In this section, the techniques used in the fabrication of integrated

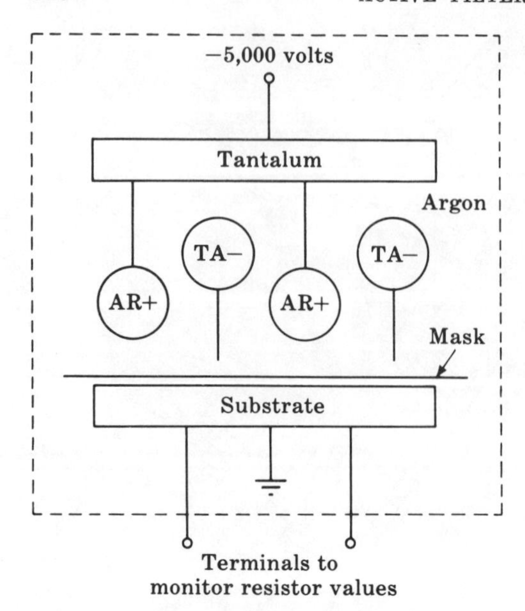

FIG. 7.16 A schematic of the process for making thin-film circuits.

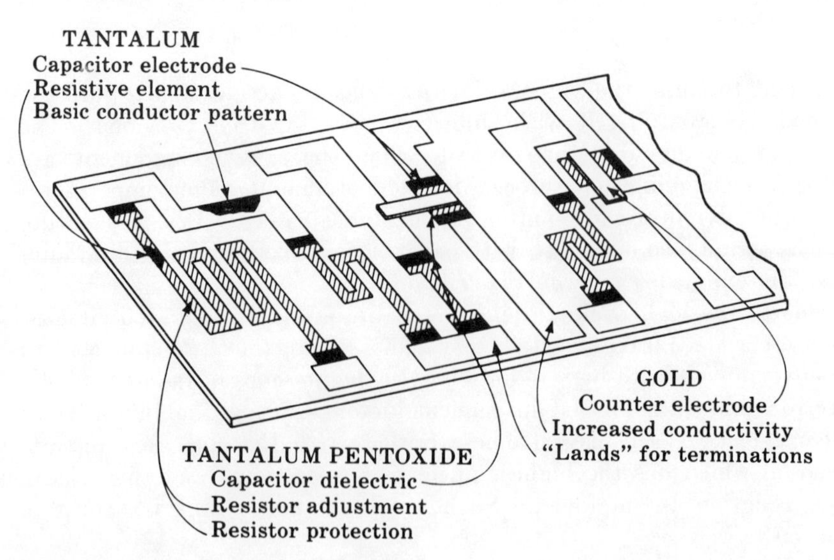

FIG. 7.17 A tantalum thin-film circuit.

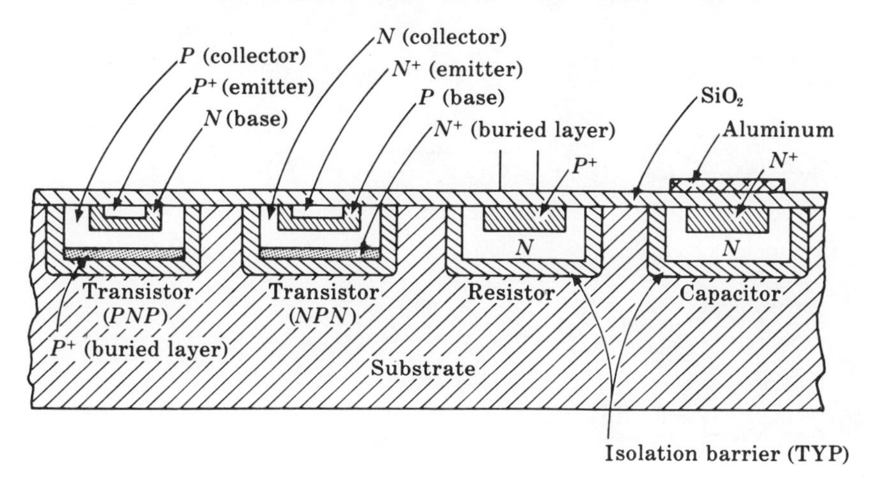

FIG. 7.18 An integrated circuit with SiO_2 isolation.

circuits have been reviewed. The presentation of this broad and sophisticated topic has been brief and necessarily general. For more detailed information, the interested reader is referred to the literature.

7.3 Digital Integrated Circuits

We shall now discuss very briefly four basic logic circuits which have found wide usage in industry, namely the resistor-transistor logic circuit† (RTL), the diode-transistor logic circuit (DTL), the emitter-coupled logic circuit (ECL), and the transistor-transistor logic circuit (T^2L). A comparison of the four types will then follow.

Consider first the RTL circuit shown in Fig. 7.19. The inputs V_1, V_2, . . . , V_n are obtained from the output $V_{0,i}$ of a previous RTL stage. Likewise, the output V_0 is fed into input transistors of succeeding stages. Using positive logic, that is, representing a high voltage by a logical 1 and a low voltage by a logical 0, one readily finds that the circuit performs an NOR operation. This follows since any input voltage V_i exceeding the base-to-emitter turn-on voltage (approximately 0.7 to 0.8 volt) causes the transistor Q_i to saturate, hence causing the output voltage $V_0 \approx V_{sc}$ (typically 0.1 to 0.3 volt) = logical 0. On the other hand, if all input voltages are below the turn-on voltage, that is, each input is at a voltage level corresponding to a logical 0, then negligible amount of collector

† The RTL is really a special case of the direct-coupled transistor logic circuit (DCTL) whereby base resistors R_b are added to the input transistors to alleviate "current-hogging" effects.

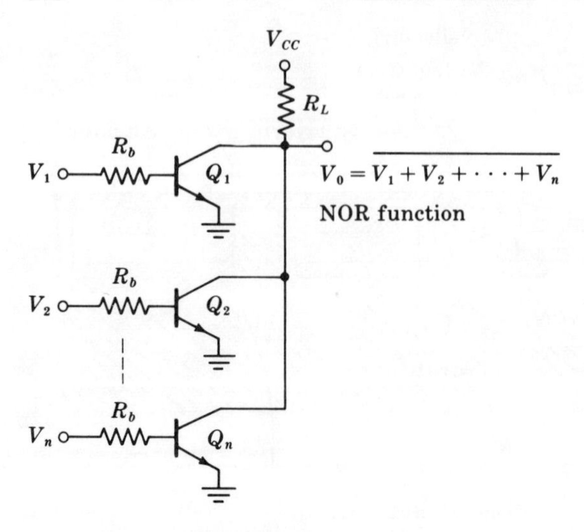

FIG. 7.19 An RTL circuit.

current is drawn through $Q_1, Q_2, \ldots, Q_n$ and $V_0 \approx (V_{cc}R_b + nV_{be}R_L)/(R_b + nR_L)$ logical 1, where V_{be} is the base-emitter voltage and n is the fan out. If the base resistors R_b are eliminated, then the circuit is called a direct-coupled transistor logic (DCTL) circuit. However, the overall operation of the logic circuit then becomes dependent on the transistor characteristics, seriously affecting the fan out (the number of stages to which V_0 can be connected). The addition of R_b helps alleviate this problem, but with a degradation of the transient response. Figure 7.20 shows a typical DCTL stage and its characteristics.

A typical DTL circuit is shown in Fig. 7.21. Basically, the input diodes and resistor R_b act as an AND gate which is then followed by a transistor inverter, thus overall giving an NAND operation. Assume that a forward-biased diode has a voltage V_D and that the collector voltage of the transistor when in saturation is $V_{sc} < V_D$. Now assume any one of the input voltages $V_i = V_{sc}$ = logical 0. Kirchhoff's voltage law then yields $V_{BE} = V_{sc} - V_D < 0$, where V_{BE} is the transistor base-to-emitter voltage. Thus, Q_1 is cutoff and $V_0 = V_{cc}$ = logical 1. If, however, all input voltages V_i are high (logical 1s) then the input diodes are reverse biased and the transistor is operating in the saturation region. The input diodes are reverse biased since the potential at point A is larger than $3V_D$ (assuming $V_{BE} \approx V_D$) and the voltage level of a logical 1 (the input voltage) is greater than $3V_D$. Also the transistor is guaranteed to be in the ON state since the base supply voltage is much larger than $3V_D$.

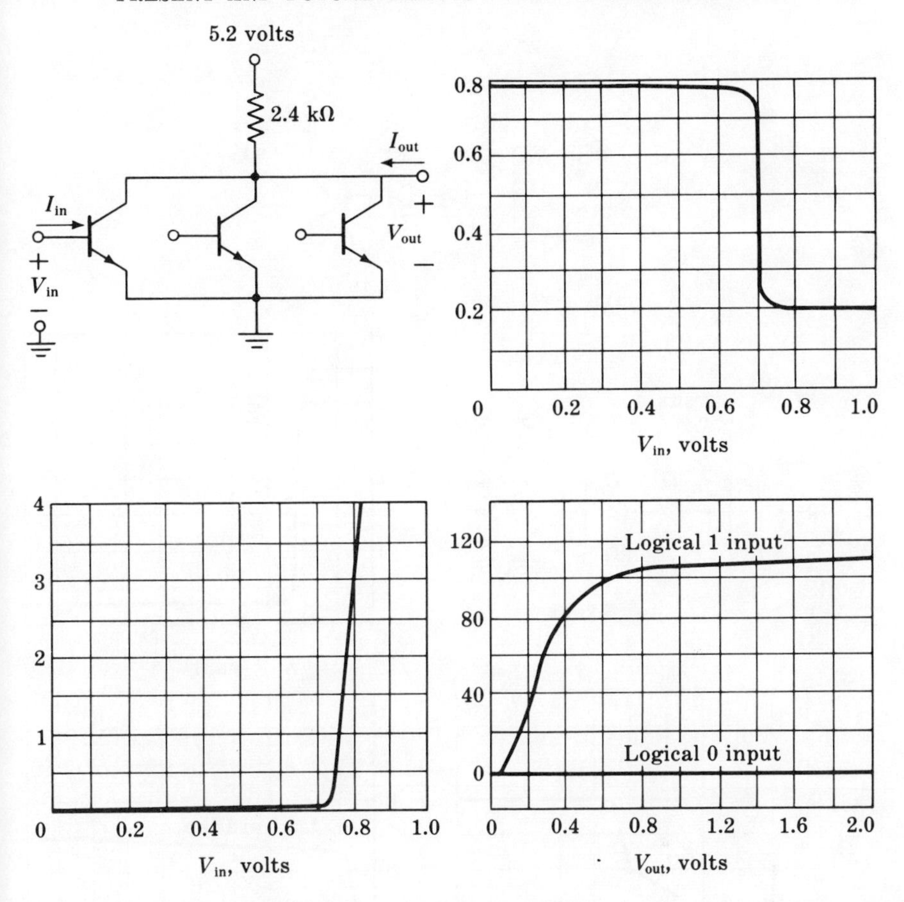

FIG. 7.20 A typical DCTL stage and its characteristics.

Diodes D_1 and D_2 (called offset diodes) are used to provide an offset voltage so that a good noise margin is obtained. These diodes are also useful in speeding up the operation of switching the transistor from saturation to cutoff.

Let us now consider the ECL circuit shown in Fig. 7.22. The internal transistors of the circuit act essentially as a difference amplifier. If we now define the voltage levels of a logical 1 and a logical 0 to be substantially more and less positive than V_{BB}, we see that the emitter current I_0 flows entirely through Q_{BB} if all input voltages V_i are logical 0s. Then the collector voltages V_{c1} and V_{c2} are equal to V_{cc} and $V_{cc} - \alpha_0 I_0 R_c$ respectively. If, however, any one of the input voltages V_i changes to the logical 1 state, then the emitter current I_0 will flow only through

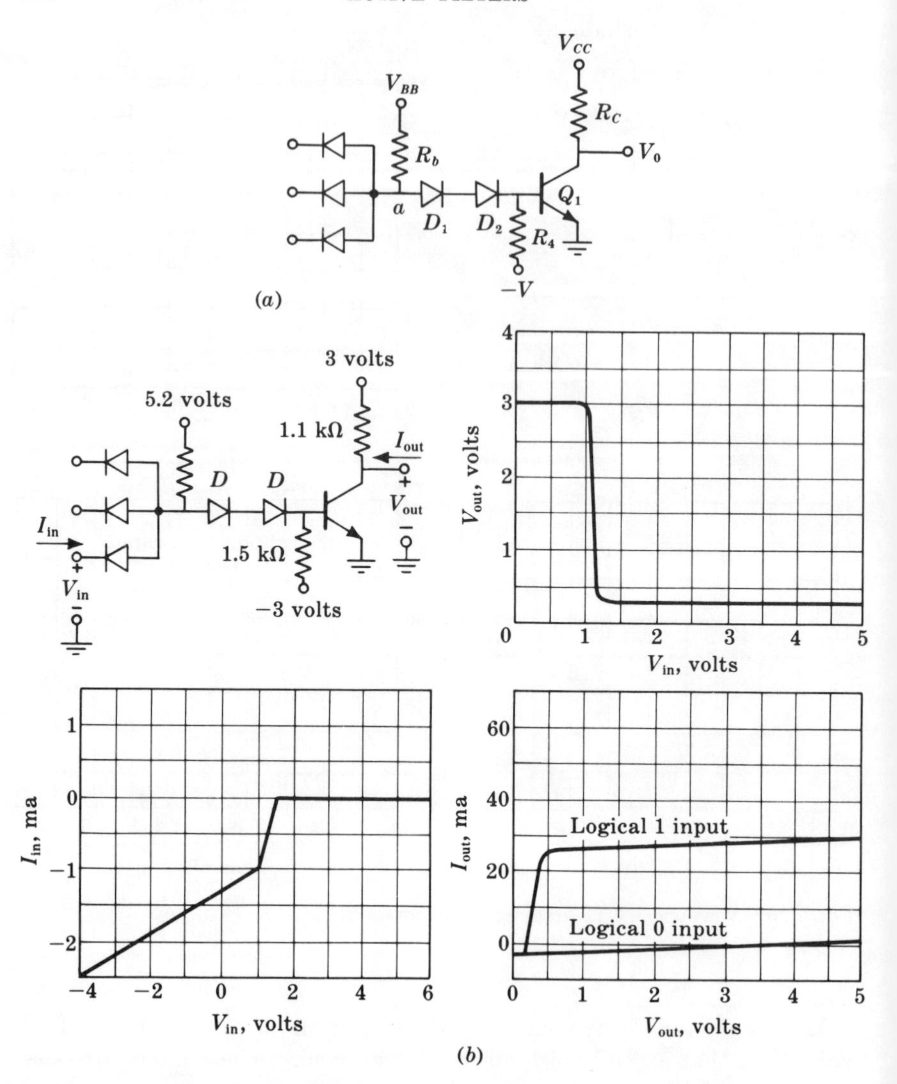

FIG. 7.21 (*a*) A DTL stage; (*b*) typical values and characteristic curves.

Q_i instead of Q_{BB}. Thus $V_{c1} = V_{cc}$ and $\alpha_0 I_0 R_c$ and $V_{c2} = V_{cc}$. Therefore, we see that V_{c1} and V_{c2} assume opposite states (0 and 1) and that we may choose R_c such that $I_0 R_c$ equals the desired logic swing. To translate the voltage levels down by the proper amount and also to obtain a larger fan-out capability of a low-impedance driving source, the collector voltages V_{c1} and V_{c2} are fed into emitter-follower stages. The

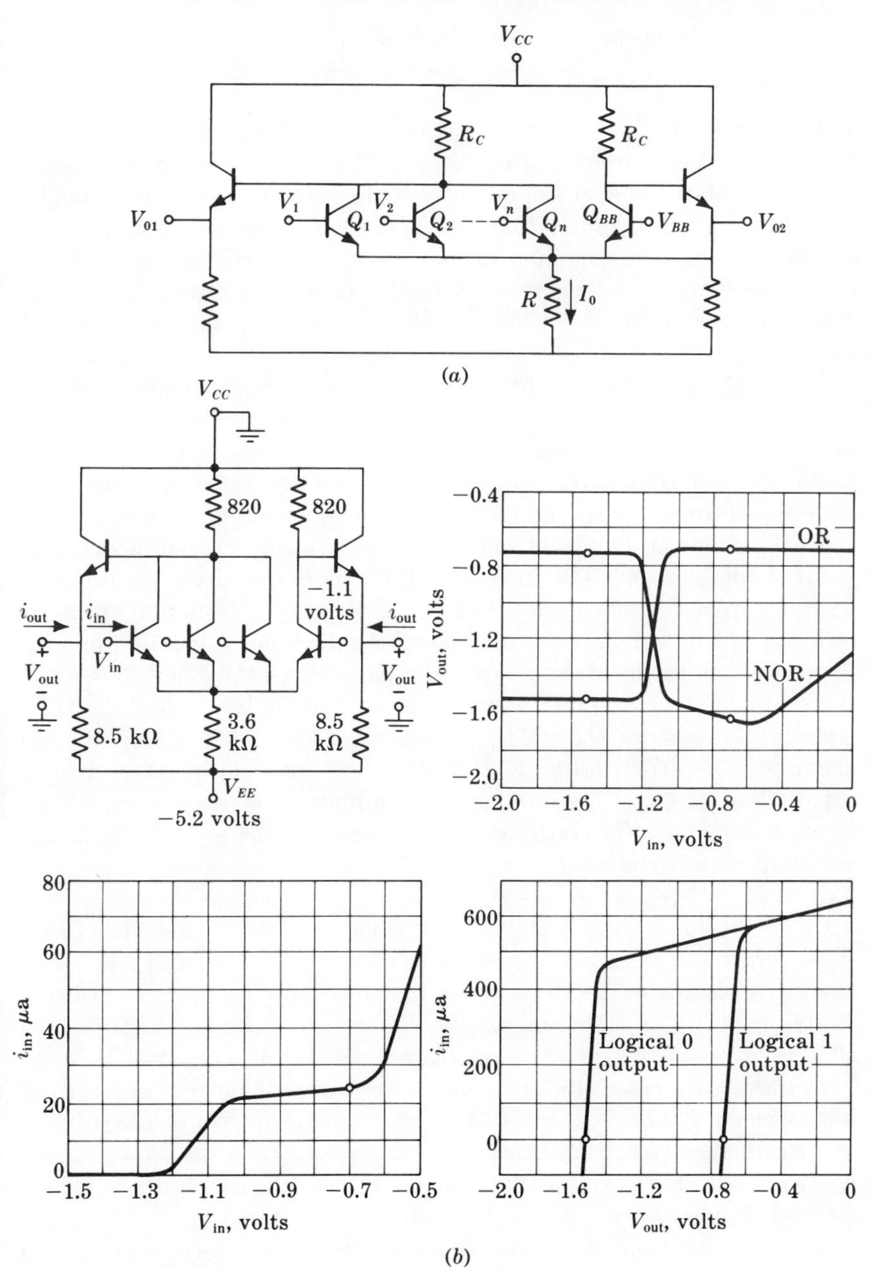

FIG. 7.22 (a) An ECL stage; (b) typical values and characteristic curves.

emitter-follower output voltages V_{01} and V_{02} are given by

$$V_{01} = \overline{V_1 + V_2 + \cdots + V_n}(\text{NOR})$$

and $V_{02} = V_1 + V_2 + \cdots + V_n(\text{OR})$. One of the important advantages of the ECL circuit is that none of the transistors ever enters the saturation state, and therefore ultra-high-speed switching is possible. Also one can readily show that the logic swing is primarily determined by the ratio R_c/R rather than the actual values of these resistors. Since resistance values may vary substantially in an integrated circuit, but ratios remain constant within excellent tolerances, this is a distinct advantage.

The final logic circuit to be discussed is the transistor-transistor logic (T^2L) circuit shown in Fig. 7.23. Actually, this circuit can be looked upon as a DTL circuit modified to eliminate as much parasitic capacitance as possible. For example, the function of the input diodes of the DTL is now performed by the multi-emitter-base junction of transistor Q_1 of the T^2L. This arrangement minimizes the parasitic capacitances of the input diode cluster. The base-collector and base-emitter junctions of Q_1 and Q_2 play the role of the offset diodes of DTL. [Note that the base-collector junction of Q_1 is forward biased when an input voltage is grounded (logical 0) since steady-state reverse base current cannot flow from Q_2.] Finally, Q_5 is an inverter stage and is thus similar to Q_1 of the DTL circuit. Transistors Q_3 and Q_4 (Darlington pair) are added to: (1) alleviate a "current-hogging" problem, (2) provide greater driving capabilities for larger fan outs, and (3) provide a low output impedance when Q_5 is cut off. The latter item is an important aspect in lowering the switching time (the output voltage charges through a low-impedance path).

In equilibrium, Q_3 and Q_4 are ON when Q_5 is OFF and vice versa. However, in the transition region, both Q_4 and Q_5 will be ON and hence a large current i_{ps} will be drawn from the power supply, the current being limited only by the 150-ohm resistor. However, when Q_4 starts to turn off, the process is regenerative, causing the collector current of Q_5 to go from a large value to zero in a very short time. This accounts for the large spike in i_{ps} shown in Fig. 7.23. The rapid change in i_{ps} necessitates the use of large capacitors (discrete 0.1-μf typically) on the power-supply bus to bypass the corresponding noise voltage which appears. This is a major disadvantage of the T^2L circuit.

A brief comparison of the four types of logic circuits is now made. To begin, one notes that RTL is quite inexpensive and indeed is the cheapest circuit on the market. Its use is restricted to low-speed applications in low-noise environments. Also, its fan out is more limited. The DTL circuit, on the other hand, provides good noise protection and is particu-

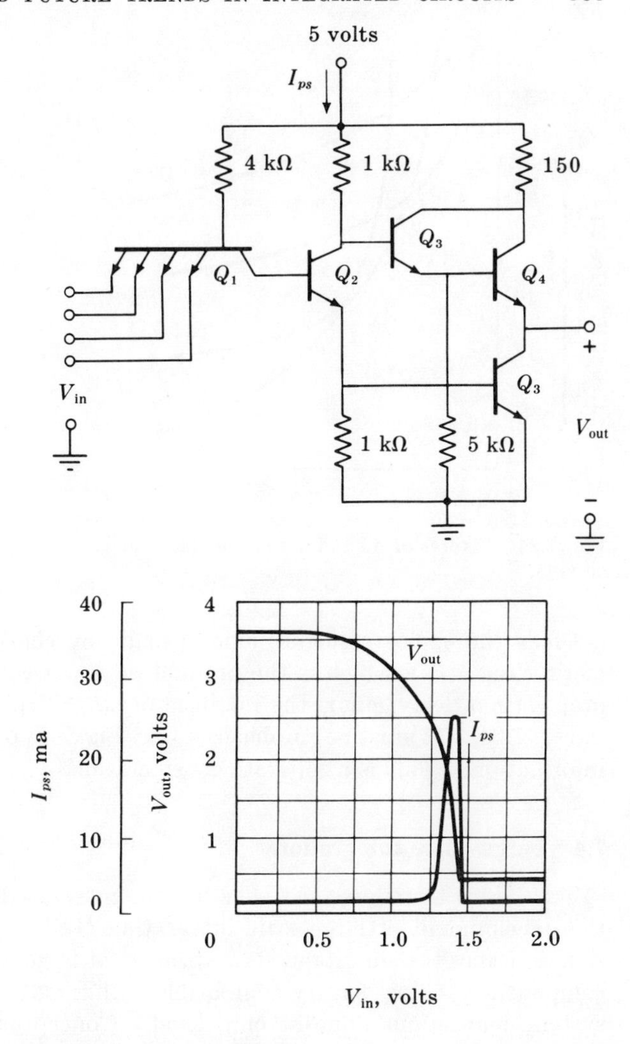

FIG. 7.23 A T²L stage.

larly useful in peripheral equipment. It is not a high-speed gate. ECL is used primarily in high-speed applications due to its nonsaturating mode of operation. Also it has good fan out due to the emitter-follower stages and it provides both the output function and its complement. Finally, the T²L is faster than the DTL and also has a higher driving capability. It is not as fast as ECL, but does consume less power. But the T²L has a noise problem due to the spike in i_{ps} and often requires a greater system hardware "debugging" effort.

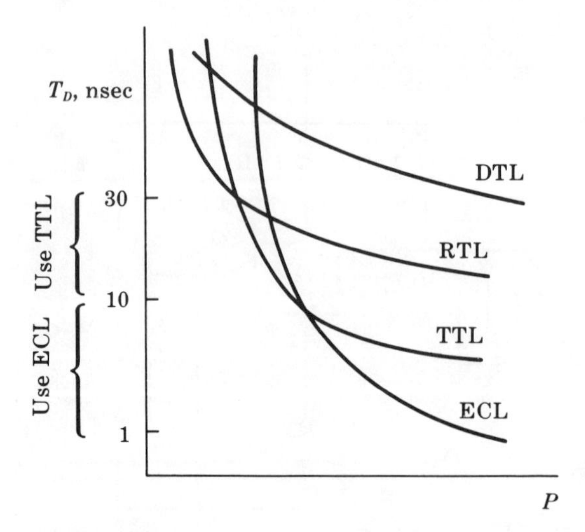

FIG. 7.24 Figures of merit for the four basic logic circuits.

Often the choice of logic mode is made by considering the figure of merit $P(\tau_d - \tau_{di})$, which is the product of power consumption times the propagation delay minus the intrinsic delay. Typical curves are shown in Fig. 7.24. It must be emphasized that Fig. 7.24 does not contain noise information. This is a separate consideration.

7.4 Large-scale Integration

The desire to reduce costs has led the integrated-circuits industry to the development of large-scale integration (LSI). Large-scale integration describes the simultaneous realization of large-area circuit chips and component packing density compatible with a maximum number of subsystem connections done at chip level. Conceptually, LSI is not new but rather is an extension of the monolithic integrated circuit.

Figures 7.25 to 7.28 serve to place LSI in one particular historical perspective. Figure 7.25 shows the progress which has been made in reducing the average area of integrated components. As shown in Fig. 7.26, the stacking factor S_F (the ratio of total component area to chip area) has improved during the same period. Figure 7.27 shows the increase in the number of circuits per chip.

It is interesting to compare the average die area shown in Fig. 7.25 with Fig. 7.27. The increase in number of circuits per chip from 1962 to 1964 is seen to be caused by enlarging the die area rather than reducing component size. The problems of attempting to hold tight tolerances

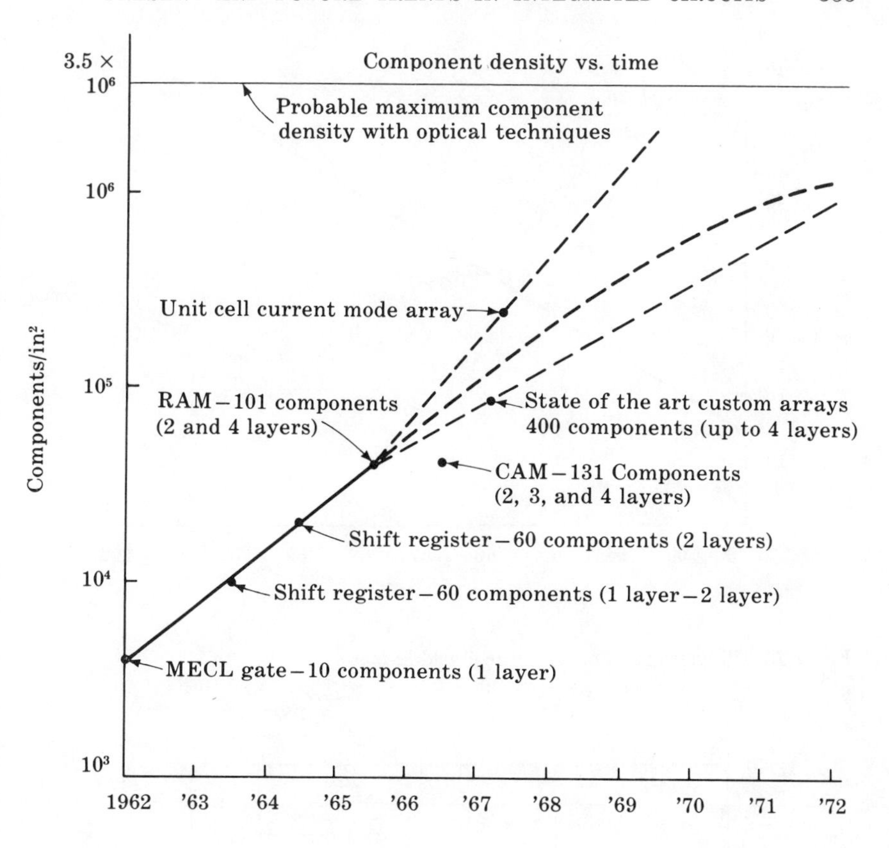

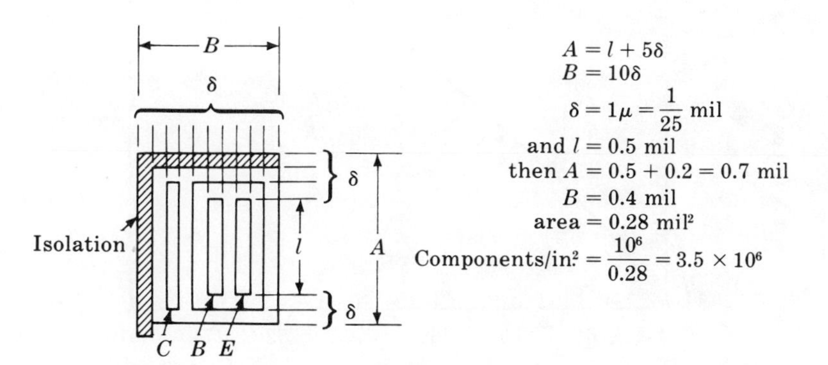

$$A = l + 5\delta$$
$$B = 10\delta$$
$$\delta = 1\mu = \frac{1}{25}\text{ mil}$$
$$\text{and } l = 0.5\text{ mil}$$
$$\text{then } A = 0.5 + 0.2 = 0.7\text{ mil}$$
$$B = 0.4\text{ mil}$$
$$\text{area} = 0.28\text{ mil}^2$$
$$\text{Components/in}^2 = \frac{10^6}{0.28} = 3.5 \times 10^6$$

Fig. 7.25 Progress in reducing the average area of integrated components.

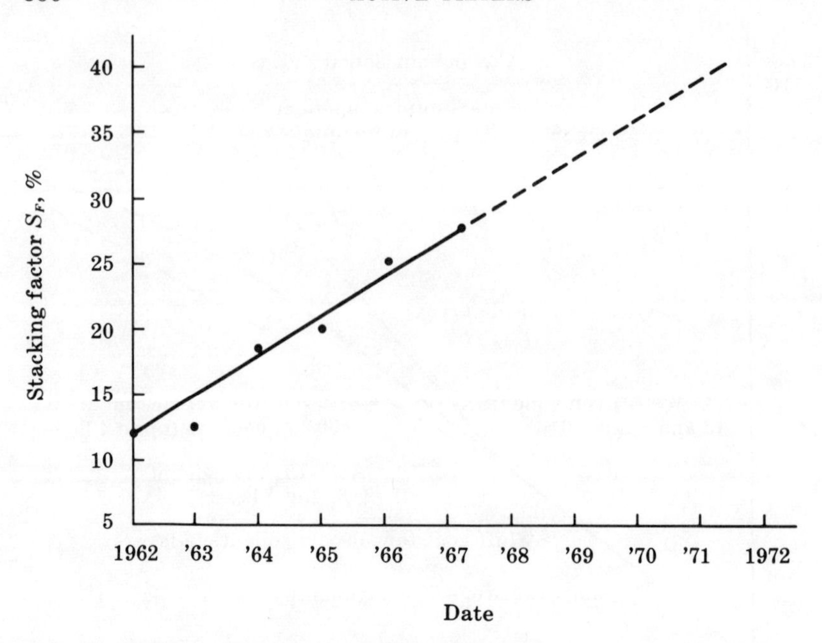

Fig. 7.26 Progress in improving the stacking factor.

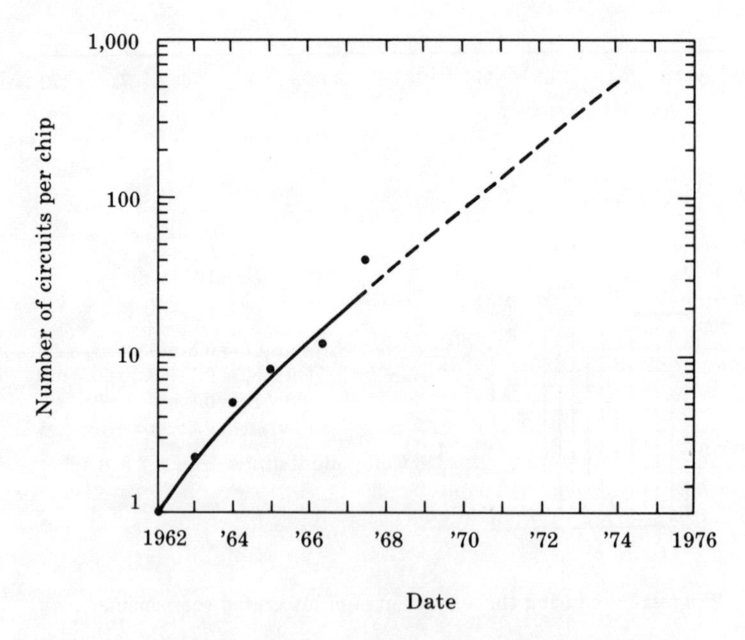

Fig. 7.27 Progress in increasing the number of circuits per chip.

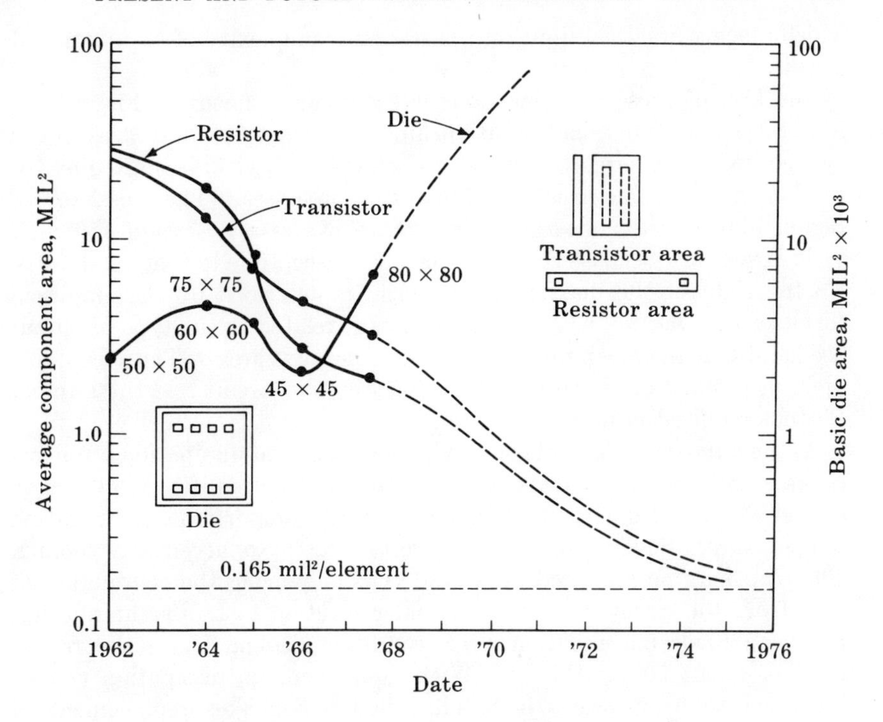

FIG. 7.28 Comparative progress in average component area and basic die area.

over large areas caused the halt of this trend in 1964 and attention was focused on reducing component size. From 1964 to 1966 significant progress was made in this field and so, although die area decreased, the number of circuits per chip actually increased. Today, the number of circuits per chip is still increasing due to continued advances in reduction of component area and use of multilayer interconnections.

What does the future hold? Most likely optical limitations will eventually set 1.0 μ as the smallest mask dimension. This would imply transistor areas of about 0.165 mil². Achievement of component densities above 3×10^5/in.² will be difficult in view of yield, complexity of interconnections, cost, ohmic drops, power dissipation requirements, etc. Stacking factors of 0.50 are envisioned. That is, half the chip area will be used for isolation and interconnection of the components.

LSI promises a way of producing more reliable integrated circuits at lower cost. Conceivably a single chip could contain not only components and circuits, but whole systems and subsystems. However, it is important to realize that in view of cost and reliability the IC having the most components is not necessarily the optimum. There are several factors which place a practical limit on the size and complexity of an integrated circuit.

The first of these considerations is economic in nature. Figure 7.29 illustrates the functional relationship between yield and area for a typical integrated circuit. For areas below 5 mil^2, yield is limited by the resolution achievable in the photomasking processes. The crosshatched region indicates the most practical range of areas. Beyond 150 mil^2, yield is very low because the circuit is more susceptible to random defects. To the right of the maximum, the yield is described by the empirical formula $Y = k_1e^{-k_2A}$, where k_1 is a constant related to the type of circuit used and k_2 is the mean number of defects per unit area. The acceptable yield in production is strongly influenced by the circuit and the number of steps required in its manufacture.

As the number of elements on a chip increases, heat dissipation problems become more severe. It is undesirable for junction temperatures to exceed 90°C. Also, circuit lifetime is inversely proportional to operating temperature. Several techniques are available to accelerate cooling. The chip may be mounted on a stud thus enhancing the conduction of heat from the circuit. So far, customer reaction to this technique has not been favorable and moreover, condition cooling has a theoretical limit of about 500 watts/in^2. With liquid cooling, dissipation can be increased to 6,450 watts/in^2. This, however, makes replacement of components very inconvenient. A more promising approach is HLSI or hybrid LSI. The general idea is to attach a number of chips to a ceramic base which provides heat conduction and interconnections between the

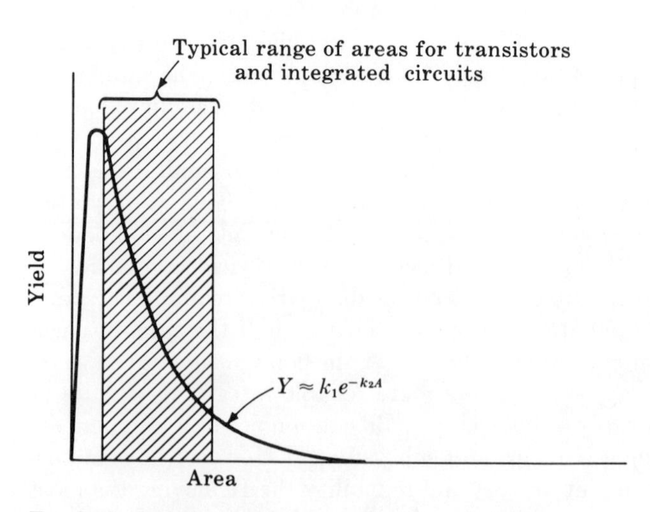

FIG. 7.29 Relationship between die size and yield.

chips. However, thermal mismatch problems discourage the use of large chips. Here again, there is a constraint on the dimensions on a chip and the number of elements it may contain.

Interconnections constitute one of the most serious problems encountered in LSI. Generally, several layers of interconnections are required on the chip. Fabrication of each successive layer gets progressively more difficult since the surface to which the metallization is applied becomes more irregular. At present, integrated circuits in production are limited to two layers of metallized interconnections. Because of this, the semiconductor material is used for constructing crossunders. This technique introduces a parasitic series resistance and a parasitic capacitance to ground. The series resistance causes additional power dissipation and degrades noise immunity, thereby limiting the number of elements on the chip.

The design of high-speed logic circuits also favors the use of smaller chips and HLSI. It has been found that the mode of propagation in the metallized connections on the silicon chip is predominantly a dispersive *RC* type. The intrapropagation delays so introduced can become the most significant limitation on circuit speed. Those interconnections on the ceramic substrate support the TEM mode, which has no appreciable delay.

It is seen that such factors as chip area occupied by metallized connections and connection delay are important in the design of an integrated circuit. In order to be able to estimate these quantities, studies have been made on the distribution of connection lengths. There are actually two types of connections, internal circuit intraconnections and circuit-to-circuit interconnections. Each type has its own distribution function. The results are shown in Fig. 7.30. Note that the distance indicated in this graph is actual length normalized to the size of the chip diagonal. The overall probability distribution is approximated by two Poisson distributions, the first maximum caused by the intraconnections and the second by the interconnections. The probability of very short intraconnections is small due to the minimum spacing that must be maintained between components. Also, on a well-laid-out chip the likelihood that two components are intraconnected is a decreasing function of the distance between them, thereby reducing the probability of long intraconnections. Similar arguments can be advanced to explain the behavior around the second maximum caused by the circuit-to-circuit connections.

If LSI is to become entirely practical and commercially viable, there is need of a process to quickly and economically fill small orders of specialized circuits. Since mask generation is the most expensive and time-consuming step, the irregular-array technique seems promising. Here, an unmetallized chip (standard-cell or macro) containing a mixture of

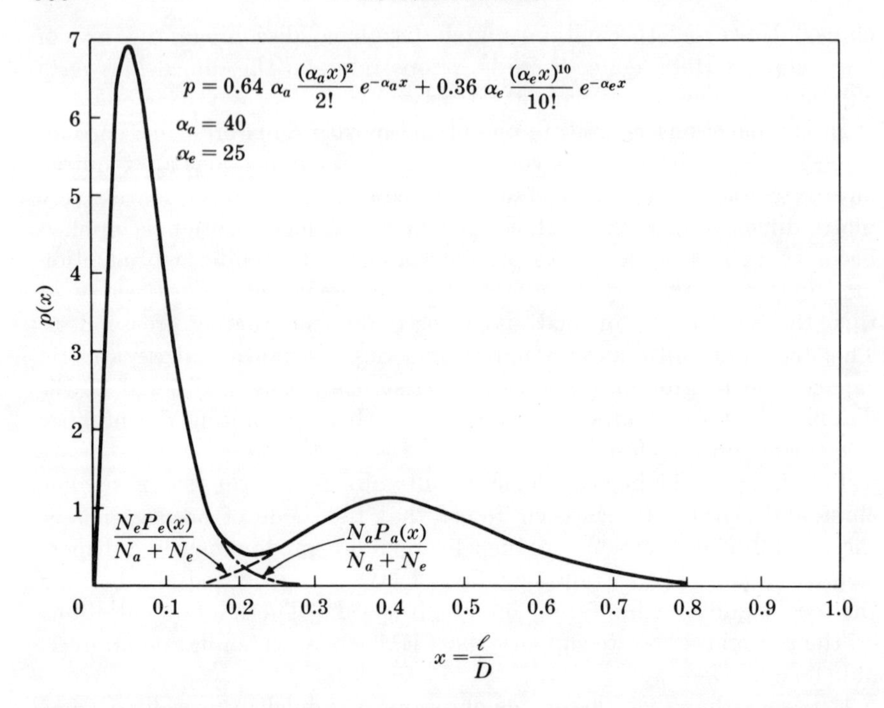

FIG. 7.30 Distribution functions for integrated-circuit interconnection lengths.

gates and flip-flops is used as the basic element. Thus for small quantities of specialized circuits only the interconnection mask need be changed. Due to the redundant components, however, it is doubtful that this method will find general application.

The present growth of computer and communication systems has stimulated the need for larger and more complex integrated circuits. It has been shown that chip size and stacking factor are a compromise with yield, power dissipation, and propagation delay. These limitations, however, reflect present technological capabilities. What, if any, will be the ultimate limitations? To seek answers to this question, scientists and engineers are examining atomic and physiological systems. Particular attention has been focused on the neuron, nature's basic logical building block. The neuron is more sophisticated than an electronic logic circuit. Besides its logic properties, it has other characteristics such as the increase of its output frequency with input amplitude. Figure 7.31 contains a comparison of neuron and electronic logic circuits. For a well-written, engineering-oriented book on this interesting topic, "The Machinery of the Brain" by D. E. Wooldridge (McGraw-Hill, 1963) is recommended.

	Microelectronic circuits	*Neurons*	*Relative merits of neuron and microcircuit characteristics*
Propagation delay	10^{-9} sec	2×10^{-4} sec	2×10^{5} times slower
Power dissipation	10^{-2} watt	10^{-9} watt	10^{7} times less power dissipation
Power delay product	2×10^{-11} joule	2×10^{-13} joule	10^{-2} times smaller
Linear size	$\sim 10^{-3}$ m	$\sim 10^{-5}$ m	10^{-2} times smaller
Failure rate	$\frac{4 \times 10^{-9}}{\text{hr}}$	$\frac{60 \times 10^{-10}}{\text{hr}}$	Same approximately
Packing densities	$\frac{10^{-6}}{\text{m}^3}$	$\frac{10^{13}}{\text{m}^3}$	10^{7} times larger
Number of atoms	10^{19}	2×10^{13}	2×10^{-6} less atoms
Information-processing mode	Parallel serial	Highly parallel and redundant	

FIG. 7.31 A comparison of neuron and electronic logic circuits.

7.5 Design Automation

The complexity of LSI has motivated interest in design automation. The increasing demand for custom rather than standard circuits gives additional impetus to this investigation. Automated design can deal with cases that are too complex for hand computation and it yields more accurate predictions of response while consuming far less time.

In the manufacturing process, it is important to provide an automatic testing procedure which yields the highest level of confidence with the least number of tests. For example, consider a combinational circuit with 40 inputs. This circuit has 2^{40} input combinations and hence could require 2^{40} tests. If this circuit is tested for 10 different values of bias voltages, assuming an optimistic 10^{-6} sev/test, it would take approximately 100 days for exhaustive testing. Obviously, exhaustive testing is not practical. However for a smaller number of tests, what is the optimum selection to yield the highest confidence level? This important information is generated as a by-product of the computer design procedure.

Automated design can be divided into two areas, circuit design and automatic mask generation. The circuit design is based on Maxwell's equations, the diffusion equation, current-carrier relations, and the law of junctions. Available algorithms for solving these partial differential equations are unreliable and consume a great deal of computer time. Therefore modeling is done in terms of lumped elements, simplifying the analysis. Numerical solution of these equations is accomplished by predictor-corrector methods. Also, computer time is minimized by replacing certain involved functional relationships by average or extremal values as shown in Fig. 7.32.

In any practical circuit realization there will be deviations from design center element values. This contingency is accounted for by including a Monte Carlo worst-case analysis in the design program. Typically, 1,000 runs are used in each Monte Carlo analysis. Figure 7.33 shows the range of possible responses corresponding to a given set of element tolerances. In this particular example, it is important that the logical

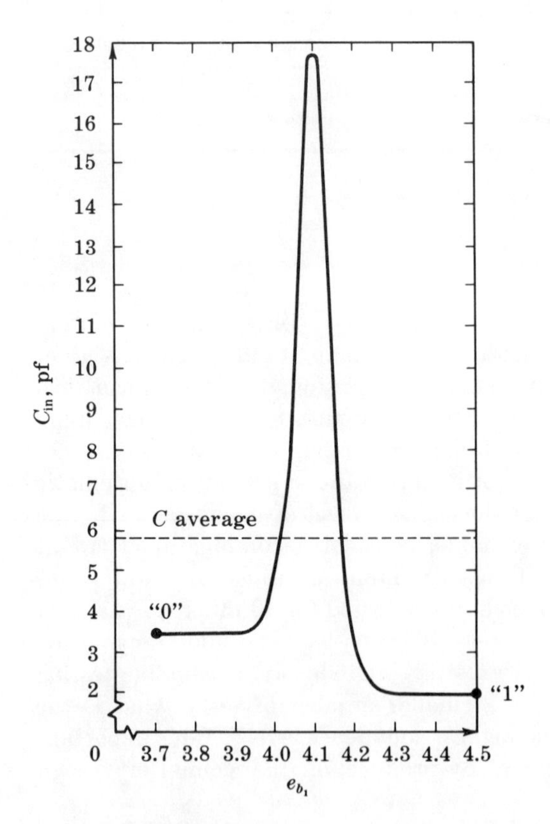

FIG. 7.32 Input capacitance vs. input voltage.

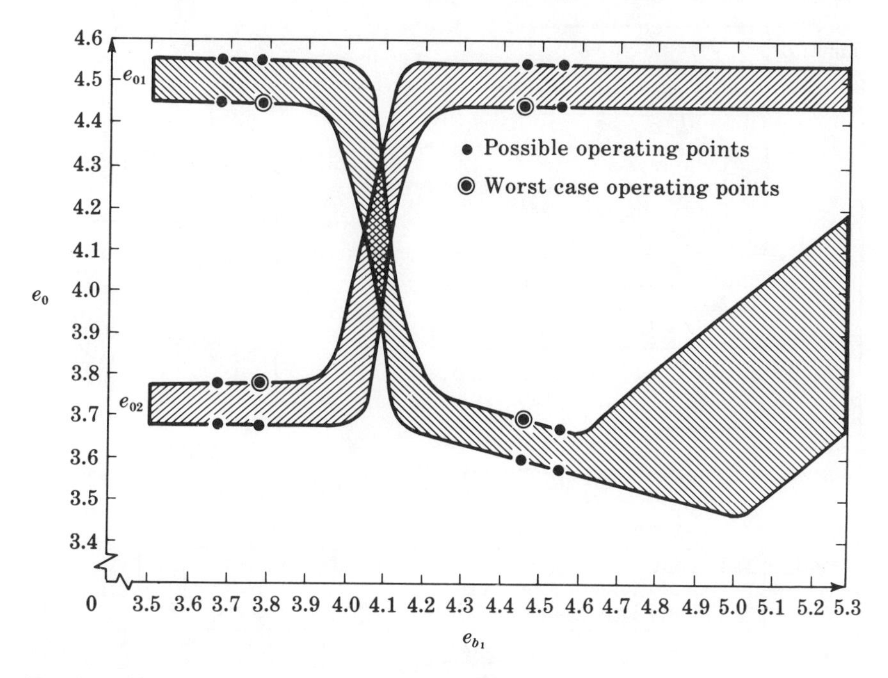

FIG. 7.33 Tolerance variation of transfer characteristics.

1 or 0 levels do not fall in the transition region. SCEPTRE has proved to be the most convenient of the available network analysis programs since it provides a great deal of flexibility in the specification of functional relationships. Figure 7.34 depicts a typical circuit-design process.

Mask generation is one of the most time-consuming steps in integrated-circuit design. In automated design, the computer is used to control a Gerber drafting machine which generates the masks. The machine makes the negatives by first outlining the appropriate areas with a narrow light beam, inlining twice with successively wider beams, and then gross painting to fill in the area. In addition to the masks, an outline composite is always drawn to check alignment. Man-machine interaction during the design process can occur by means of a CRT display and light pencil. The time saving in automatic mask generation is substantial. One complicated problem was given both to the computer and a group of draftsmen. The computer took two days, while five draftsmen required three to four weeks. Moreover, the man-made masks contained numerous errors.

Referring to Fig. 7.34, the design of an integrated circuit usually commences with either a conception of a new circuit or a set of custom specifications for a particular configuration. In on-line systems the

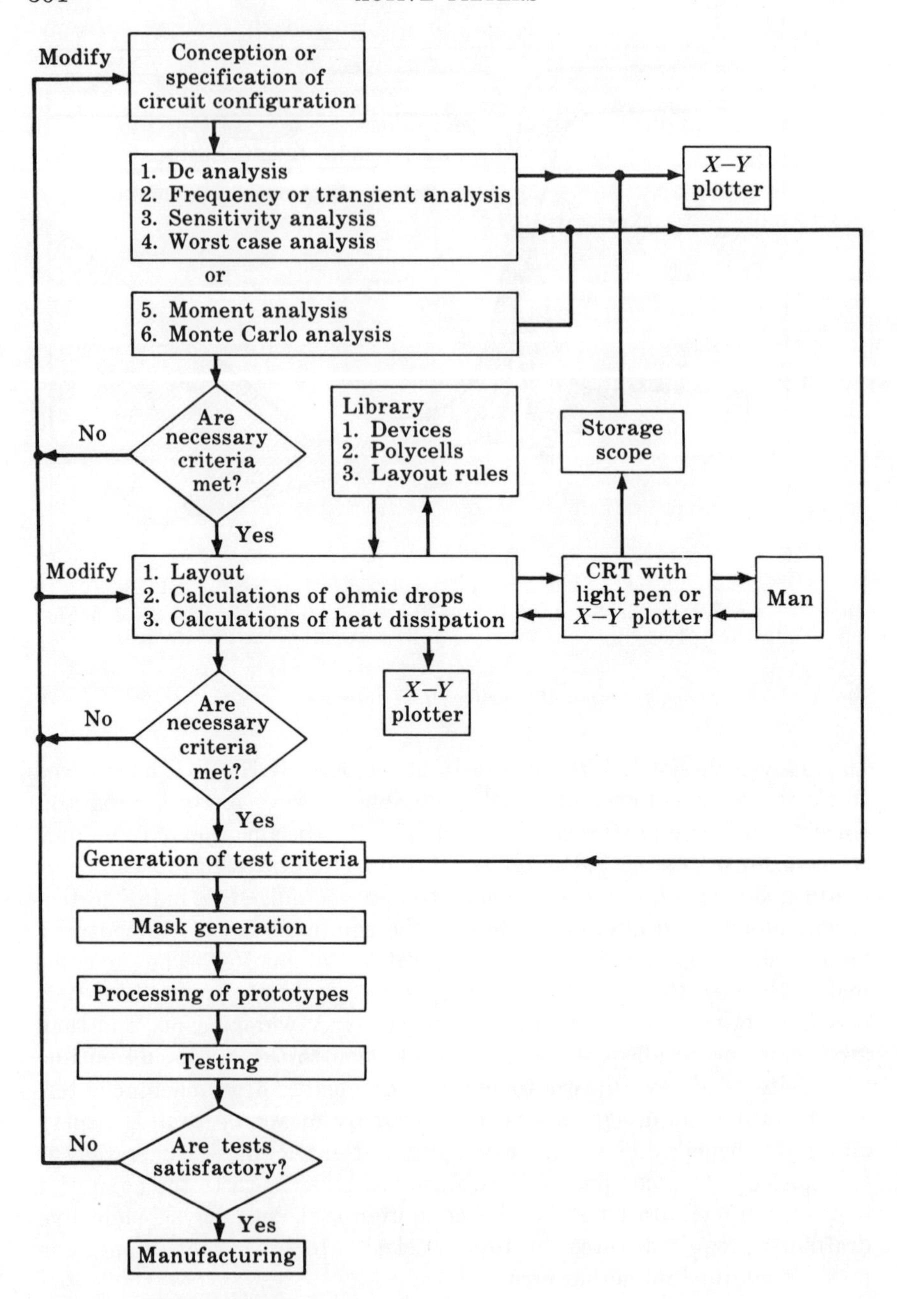

FIG. 7.34 A typical circuit-design process.

appropriate model of a particular circuit topology together with its specifications is transmitted to the computer via some terminal such as teletype or some other kind of keyboard. Depending upon what type of circuit is involved, the design automation system has in its memory routines for performing dc analysis, frequency or transient analysis, sensitivity analysis, worst-case analysis, moment analysis, and/or Monte Carlo analysis. Which of these analysis routines is utilized depends, as mentioned before, on the type of circuit and the reliability specifications required. Although there are a number of analysis algorithms, there does not exist at the present any system that is capable of performing these analyses for practical integrated circuits required in computers and linear applications.

The main problem areas are:

1. Iterations that require the inversion of large matrices.
2. The error introduced by the numerical analysis used.
3. Integration routine.
4. Numerical instability introduced by the fact that one is converting a differential equation into a set of discrete different equations.
5. The simultaneous large and small time-constant problem.
6. Worst-case analysis frequently produces a result that indicates the circuit does not meet the specifications.
7. Monte Carlo analysis frequently requires a lot of computation time when the circuit is of medium or large complexity.

For the dc-, frequency-, or transient-analysis case there is a compromise between the complexity of the model used and the accuracy of the result. A too-complex model results in long analysis time as well as being costly. Moreover, in circuits with small and large time constants the integration interval has to be made small enough for a numerical process to convert, thus compounding the problem despite the use of corrector and predictor routines. The presence of nonlinearities of various kinds makes matters even worse. In particular cases where nonlinearities exhibit themselves rather abruptly, as in flip-flops, various threshold circuits, etc., the use of a corrector-predictor method that varies the size of the integration step is an absolute necessity and complicates things still further. So far SCEPTRE seems to be the analysis routine that comes closest to meeting the requirements for dc and transient analysis, while modified ECAP is most applicable for analysis in the frequency domain.

So far as worst-case, moment, and Monte Carlo analysis are concerned the problem quite often has the effect that one does not know the distribution of the various parameters involved accurately enough. Of the three, however, moment analysis seems to be the best compromise at the present where a symmetric or skew gaussian distribution is assumed for the components.

As suggested in the referenced paper,† macroanalysis, where the macro is limited to gates or flip-flops, seems to offer a solution to these problems. In this method a complex structure is not analyzed on the micro level taking into account the properties of each element, but rather an approximate model is generated (linear or nonlinear) based upon which the whole complex structure is evaluated. For example, a structure consisting of 25 gates was analyzed with good accuracy and with a computation time on a GE 635 computer of only about 20 seconds. However, the reader should be cautioned that much work remains to be done in order that such algorithms may be generally used. Another solution to the problem is the use of hardware-type integration. However, it is still too early to evaluate the feasibility as well as the relative cost of this approach.

Before leaving the subject of analysis, we should touch on optimization. Here, techniques used in control theory, such as steepest-descent method, etc., have been applied to the optimization of integrated circuits. However, problems such as local minima and maxima create a big bottleneck and a lot of work remains to be done in order that optimum parameter values, as well as the configurations (synthesis), can be realized by computer-aided design.

Returning to Fig. 7.34, after the first part of the analysis is complete a decision is made as to whether the necessary criteria have been met. If the answer is no, modifications are made and the previous cycles are repeated. On the other hand, if the answer is yes, one proceeds to lay out the masks in preparation for the construction of the circuit. Essentially, this consists of a library of component devices, polycells, and layout rules together with the actual layout calculations of ohmic drops and heat dissipation across the chip. Again, a man-machine interrelationship has to take place and several attempts have to be made before a satisfactory layout is generated, either around the local loop of generating a layout or modifying both the circuit topology and the layout. Also, at this point test criteria, mask generation, processing of prototypes, and testing follow. If the tests turn out to be unsatisfactory, modification of the previous steps is repeated. On the other hand, if the tests meet the necessary specifications the product goes either to pilot-line production or to manufacturing.

In spite of the amount of homework that remains to be done in order to make design automation an economic practicality, it represents a powerful tool for the design engineer even today. As a matter of fact,

† J. A. Narud and C. S. Meyer, Characterization of Integrated Logic Circuit, *Proc. IEEE*, vol. 52, no. 12, pp. 1551–1564, December, 1964.

it has accelerated the turnaround time for the design engineer by an order of magnitude and relieved him of much of the drudgery involved in computations, layout, testing, and diagnosis. Any way one looks at it, design automation is here to stay and much work will go into software, hardware, and peripheral equipment during the next few years. It is a fast-moving field attracting much attention, with new approaches being generated almost on a daily basis. It is, therefore, a foregone conclusion that design automation will play an important role in the design of integrated circuits. In any case, one thing is certain: LSI or MSI and custom circuits with fast turnaround will not be a reality without an interactive design automation procedure, preferably with a 5-second maximum turnaround time.

INDEX